LIPID PHARMACOLOGY

Volume II

MEDICINAL CHEMISTRY
A Series of Monographs

EDITED BY

GEORGE DESTEVENS

CIBA Pharmaceutical Company, A Division of CIBA Corporation
Summit, New Jersey

Volume 1. GEORGE DESTEVENS. Diuretics: Chemistry and Pharmacology. 1963

Volume 2. RODOLFO PAOLETTI (ED.). Lipid Pharmacology. Volume I. 1964. RODOLFO PAOLETTI AND CHARLES J. GLUECK (EDS.). Volume II. 1976

Volume 3. E. J. ARIENS (ED.). Molecular Pharmacology: The Mode of Action of Biologically Active Compounds. (In two volumes.) 1964

Volume 4. MAXWELL GORDON (ED.). Psychopharmacological Agents. Volume I. 1964. Volume II. 1967. Volume III. 1974. Volume IV. 1976

Volume 5. GEORGE DESTEVENS (ED.). Analgetics. 1965

Volume 6. ROLAND H. THORP AND LEONARD B. COBBIN. Cardiac Stimulant Substances. 1967

Volume 7. EMIL SCHLITTLER (ED.). Antihypertensive Agents. 1967

Volume 8. U. S. VON EULER AND RUNE ELIASSON. Prostaglandins. 1967

Volume 9. G. D. CAMPBELL (ED.). Oral Hypoglycaemic Agents: Pharmacology and Therapeutics. 1969

Volume 10. LEMONT B. KIER. Molecular Orbital Theory in Drug Research. 1971

Volume 11. E. J. ARIENS (ED.). Drug Design. Volumes I and II. 1971. Volume III. 1972. Volume IV. 1973. Volumes V and VI. 1975. Volume VII. in preparation

Volume 12. PAUL E. THOMPSON AND LESLIE M. WERBEL. Antimalarial Agents: Chemistry and Pharmacology. 1972

Volume 13. ROBERT A. SCHERRER AND MICHAEL W. WHITEHOUSE (EDS.). Antiinflammatory Agents: Chemistry and Pharmacology. (In two volumes.) 1974

In Preparation

LEMONT B. KIER AND LOWELL H. HALL. Molecular Connectivity in Chemistry and Drug Research

LIPID PHARMACOLOGY

Volume II

EDITED BY

RODOLFO PAOLETTI

INSTITUTE OF PHARMACOLOGY
UNIVERSITY OF MILAN
MILAN, ITALY

AND

CHARLES J. GLUECK

DEPARTMENT OF INTERNAL MEDICINE
UNIVERSITY OF CINCINNATI MEDICAL CENTER
CINCINNATI, OHIO

1976

ACADEMIC PRESS · New York · San Francisco · London

A Subsidiary of Harcourt Brace Jovanovich, Publishers

ACADEMIC PRESS, INC.
111 Fifth Avenue, New York, New York 10003

United Kingdom Edition published by
ACADEMIC PRESS, INC. (LONDON) LTD.
24/28 Oval Road. London NW1

Library of Congress Cataloging in Publication Data

Paoletti, Rodolfo, ed.
    Lipid pharmacology.

    (Medicinal chemistry; a series of monographs,
v. 2.   )
    Vol. 2 edited by R. Paoletti and C. J. Glueck.
    Includes bibliographies and index.
    1.   Lipid metabolism.   2.   Drug metabolism.
I.   Glueck, Charles J.   II.   Title.   III.   Series.
QP751.P33   1964   612'.397   63-21396
ISBN 0–12–544952–6

CONTENTS

Chapter 1
CLINICAL EVALUATION OF ATHEROSCLEROSIS
· *G. Schettler and H. Mörl*

Chapter 2
ESSENTIAL FATTY ACIDS—RECENT DEVELOPMENTS
· *Roslyn B. Alfin-Slater and Lilla Aftergood*

v

Chapter 3

METHODS FOR EVALUATION OF HYPOLIPIDEMIC DRUGS IN MAN:
MECHANISMS OF THEIR ACTION • *Tatu A. Miettinen*

Chapter 4

DIETARY AND DRUG REGULATION OF CHOLESTEROL METABOLISM IN
MAN • *Scott M. Grundy*

Chapter 5

DIET AND DRUGS IN OBESITY CONTROL
• *Daniel L. Azarnoff and Don W. Shoeman*

Chapter 6

ETHANOL AND LIPID METABOLISM • *C. S. Lieber*

Chapter 7

CHOLESTYRAMINE AND ION-EXCHANGE RESINS
• *H. Richard Casdorph*

Chapter 8

TREATMENT OF HYPERLIPOPROTEINEMIA IN CHILDREN
                    • C. J. Glueck, F. W. Fallat, and R. C. Tsang

Chapter 9

FEMORAL ANGIOGRAPHY TO EVALUATE HYPERLIPIDEMIA THERAPY
                                    • David H. Blankenhorn

CONTRIBUTORS

Numbers in parentheses indicate the pages on which the authors' contributions begin.

LILLA AFTERGOOD (43), *School of Public Health, University of California, Los Angeles, California*

ROSLYN B. ALFIN-SLATER (43), *School of Public Health, University of California, Los Angeles, California*

DANIEL L. AZARNOFF (161), *Departments of Medicine and Pharmacology, Clinical Pharmacology-Toxicology Center, University of Kansas Medical Center, Kansas City, Kansas*

DAVID H. BLANKENHORN (277), *Department of Medicine, Cardiology Section, University of Southern California School of Medicine, Los Angeles, California*

H. RICHARD CASDORPH (221), *Lipid Research Foundation, Long Beach, California*

F. W. FALLAT (257), *Department of Internal Medicine, Lipoprotein Research Section, University of Cincinnati Medical Center, Cincinnati, Ohio*

C. J. GLUECK (257), *Department of Internal Medicine, University of Cincinnati Medical Center, Cincinnati, Ohio*

SCOTT M. GRUNDY (127), *Department of Medicine, Division of Metabolism, University of California School of Medicine, and The Veterans Administration Hospital, San Diego, California*

C. S. LIEBER (183), *Laboratory of the Section of Liver Disease and Nutrition, Bronx Veterans Administration Hospital, Bronx, New York, and Department of Medicine, Mount Sinai School of Medicine of the City University of New York, New York, New York*

TATU A. MIETTINEN (83), *Second Department of Medicine, University of Helsinki, Helsinki, Finland*

H. MÖRL (1), *Oberarzt, Medizinische Universitäts-Klinik, Heidelberg, Germany*

G. SCHETTLER (1), *Oberarzt, Medizinische Universitäts-Klinik, Heidelberg, Germany*

xi

DON W. SHOEMAN (161), *Department of Pharmacology, Clinical Pharmacology-Toxicology Center, University of Kansas Medical Center, Kansas City, Kansas*

R. C. TSANG (257), *Department of Pediatrics, Fels Division of Pediatric Research and Newborn Division, University of Cincinnati Medical Center, Cincinnati, Ohio*

PREFACE

More than fifteen years after the publication of "Lipid Pharmacology," this volume with the same title but with vastly different content appears.

The rapid development of our basic information on the nature, fate, and significance of the different lipoprotein fractions, the understanding of the essential fatty acids as precursors of hormonelike agents such as prostaglandins and thromboxanes, the availability of new selective agents acting not only against the formation of endogenous cholesterol but also as a controlling agent on the absorption of cholesterol and the metabolic rate of lipoproteins represent just a few of the advances in the area of lipid pharmacology during this comparatively short period of time.

The field is now well beyond the stage of speculation and animal experimentation, and accordingly more space is devoted to the clinical use of hypolipidemic drugs and to the quantitative methods now available for a precise determination of lipid metabolism in human subjects.

We sincerely hope that this new publication will update the informations of many clinicians and biologists in this important practical area. We are particularly grateful to the distinguished authors for their ability to present in a clear and authoritative way many complex areas of basic and clinical research and to the staff of Academic Press for their usual care in publishing the volume.

Rodolfo Paoletti
Charles J. Glueck

CONTENTS OF VOLUME I

Chapter 1

CLINICAL EVALUATION
OF ATHEROSCLEROSIS

G. Schettler and H. Mörl

Medizinische Universitäts-Klinik
Heidelberg, Germany

I. General Introductory Remarks

Atherosclerosis is manifested clinically only with the appearance of sequelae, i.e., disturbances of blood supply in the dependent organs. The arterial reorganization can have various consequences; these are responsible for the mixed clinical picture of atherosclerosis. Besides a gradual asymptomatic reorganization of vessels with age, there are changes in the vessels which are pathological and which begin at an early age, often with deleterious consequences. Consequently, at present atherosclerosis is of sociomedical importance rather than just a problem affecting the individual. Functional and morphological adaptation, temporal and topographical relationships, and the metabolic activity of the organ concerned play a role in the hemodynamic effect of stenosis or occlusion of a vessel. The most important factor is the availability and adaptability of the collateral circulation, although the rate of vessel narrowing is crucial.

According to the WHO classification, atherosclerosis is "a variable combination of changes in the arterial intima consisting of the focal accumulation of lipids, complex carbohydrates, blood and blood products, fibrous tissue and calcium deposits associated with changes in the media" [Classification of Atherosclerotic Lesions, Report of Study Group on Definition of Terms (1958)].

The arteries have already been altered considerably for years prior to the first manifestation of circulatory impairment. Time of manifestation and clinical picture are variable and depend upon the location and extent of the occlusion as well as the functional capacity of the collateral circulation. There are often discrepancies between the pathological-anatomical material and clinical data (Hild, 1966; Widmer, 1963; Widmer et al., 1967). Many anatomical vessel occlusions remain clinically silent, as shown by angiographic studies. For example, two-thirds of 64 men and 11 women with arterial occlusive disease of the extremities from a pool of 6400 workmen were free of symptoms in the Basle study project of Widmer (1963). It has been stated (Widmer, 1963) that peripheral vascular disease and coronary arterial disease have about the same frequency and occur in about the same age. I do not think this conclusion is justified. Hild (1966), however, found a statistical correlation between peripheral vascular occlusive disease and coronary arterial disease in Heidelberg. Arterial occlusive disease in patients with myocardial infarction was demonstrable by arteriography only in approximately 10% of all cases. Only one-half of these patients had symptoms. No definite correlations exist between arteriosclerosis of the aorta and coronary arteries.

Neither in patients with aortic aneurysm nor in patients with aortic occlusive disease is there a significant increase of coronary arterial occlusive disease (Groom et al., 1964; Strong and McGill, 1963). One cannot draw conclusions from aortic calcifications as to an increased incidence of coronary artery sclerosis (Chapman et al., 1960). Also there are no definite relationships between the incidence of peripheral artery, coronary artery, and aortic occlusive disease and cerebral vascular occlusive disease. Even arteriographically demonstrable occlusions of cerebral arteries may remain clinically silent. The occurrence of arteriosclerotic occlusive disease in extracranial or intracranial arteries (which is rather high beyond the fourth decade) does not always correlate with clinical manifestations (Drake and Drake, 1960; Bernsmeier and Gottstein, 1965). Clinical symptoms in the region of the cerebral vessels are mainly dependent upon in the region of the aorta and the peripheral arteries. The renal and coronary arteries, however, are possible exceptions. Arteriosclerotic occlusive changes in the renal arteries are rarely isolated findings, although via liberation of renin and angiotensin they may lead to hypertension with secondary sequelae. If hyptertension is associated with arteriosclerotic renal arterial disease, the incidence of symptomatic arteriosclerotic disease in other vascular regions is high. Clinical studies (Wollenweber et al., 1968) revealed a positive correlation between the extent of arteriosclerotic renal arterial disease and the frequency of symptomatic coronary, cerebral vascular, and peripheral arterial disease. These episodes are frequently multiple and reduce the life expectancy.

The general development of an arterial stenosis initially corresponds only to a restriction of the reserve blood supply, i.e., the maximum capacity to increase blood flow. The early stage of an arterial occlusion disease is manifested only in periods of elevated oxygen requirement, e.g., there is intermittent disturbance of blood flow in exercise. These periods appear clinically as intermittent lameness, angina abdominalis, angina pectoris in exercise and as transient syncope. If the occlusion process progresses and even normal resting metabolism in the tissue can no longer be maintained, chronic ischemia and finally necrosis results. Basically three forms of disturbed blood supply can be distinguished: ischemia, lesion, and necrosis. The former two processes are reversible; the latter constitutes an irreversible state.

Moreover, the variable pathokinetics of atherosclerosis is of clinical significance. There are sporadic as well as chronically progressive, stationary, and fulminating forms. Atherosclerosis does not have an uniform course; rather, it has a remittent character, tending to scarring and relapses.

II. Aortic Atherosclerosis

At an advanced age, a diffuse atherosclerosis of the aorta is general, the most predisposed site being the distal abdominal aorta. Ulcerations, thromboses, and aneurysms are most pronounced here. The aorta is elongated and dilated, its X-ray shadow density is increased, and it frequently contains calcium deposits. This Mönckeberg (1903) form of atherosclerosis usually does not stenose.

As the rigidity of the wall increases, the pressure reservoir of the formerly elastic aorta loses its capacity to absorb a considerable part of the stroke volume of the heart elastically and to transform it into a continuous stream of blood. This results in practically the whole stroke volume of the heart flowing through the rigid aorta during systole, thus leading to elevation of systolic blood pressure with fall of diastolic pressure. With reduced air chamber function efficiency, the heart must do more work since a larger volume of blood must be moved.

A. Atherosclerosis of the Ascending Aorta

This part of the aorta is weakly involved, and mesoaortitis luica and cystic idiopathic medionecrosis of the aorta predominate (Gsell, 1928, Erdheim, 1929). Atherosclerosis-induced aneurysms of the Valsalva sinus aortae usually do not have clinical symptoms; only after perforation into the right auricle or the right ventricle is cardiac insufficiency with continuous murmur abruptly produced.

More pronounced dilatation of the ascending aorta can lead to stretching of the aortic valve with relative aortic valve insufficiency.

A dissecting aneurysm in the ascending aorta due to atherosclerosis is rare. Dissecting aneurysms and spontaneous ruptures of the aorta generally arise after a medionecrosis. In 12 of 41 cases of dissecting aneurysms, of which the pathology and anatomy were investigated, we found only one caused by atherosclerosis (Mörl, 1965a). Complaints of very severe thoracic pains are usual in unilateral or bilateral rupture of the aorta, frequently giving rise to a false diagnosis of myocardial infarction. There are spontaneous ruptures of the aorta in about a one-quarter of patients with aortic coarctation—usually atherosclerosis is of hypertensive origin (Reifenstein et al., 1947; Gross, 1953; Mörl, 1965).

B. Atherosclerosis of the Aortic Arch

Next to coronary sclerosis, occlusion processes in the branches arising from the aortic arch (brachiocephalic trunk, carotid artery, vertebral artery, and subclavian artery) are the most important potentially fatal manifestation of atherosclerosis. About one-third of all states of insufficient cerebral blood supply result from stenoses or obliterations of the extracranial afferent arteries, stenoses near the point of origin from the aortic arch being relatively frequent. This is referred to as an aortic-arch syndrome, also known as Takayasu syndrome (1908), Martorell-Fabré syndrome (1944), or pulseless disease. The Takayasu syndrome is restricted to obliteration due to inflammation, which is chiefly found in younger women. In contrast to the assumption that this aortitis would occur mainly in Southeast Asia, studies (chiefly angiographic in nature) have shown that this disease also appears more frequently in Europe than was assumed (Deutsch, 1974).

Attention must be paid to congenital malformations of the aortic arch, although at an advanced age atherosclerosis is the most frequent etiological factor (Irvine et al., 1963 among other authors). Possible etiological factors are syphilis, tuberculosis, aneurysms, traumas, embolisms, and extravasal compressions (Bernsmeier and Held, 1970). A complete and an incomplete aortic-arch syndrome are distinguished.

The clinical picture of supraaortic arterial occlusions corresponds to the loss of function in three main supply areas, the carotid, subclavian, and vertebral-basilar. The main symptom of a blood flow insufficiency of carotid type is contralateral motor and sensory unilateral disorders, especially of the upper extremities. Very much more rarely, there are simultaneous ipsilateral monocular disturbances of vision ("veil phenomenon"), aphasias, headaches, deterioration of mental agility, orthostatic syncopes, and intermittent cramps. The appearance of unilateral cramplike pain in the jaw musculature during eating and trophic disturbances of one-half of the face indicate insufficiency of the external carotid artery. The vertebral-basilar insufficiency syndrome is characterized by ipsilateral disturbances of cranial nerves, contralateral hemiparesis, atactic symptoms and dissociated sensory disturbances as well as disorders of vision, attacks of vertigo, and inner ear hardness of hearing.

Stenosis or obliteration of the subclavian artery at the point of origin from the aortic arch can result in a subclavian steal syndrome[1] significant not in terms of the relatively rare brachial claudication but of cerebrovascular insufficiency which may be associated with transient loss of consciousness and apoplexylike manifestations. Besides the absence of pulse in one arm and the consequent difference in blood pressure, there are occasional ophthalmological changes resulting from the hypoxemia caused by low blood pressure. The ipsilateral vertebral artery takes over the blood supply of the arm due to the fall in pressure behind the ob-

---

[1] First described by Reinich et al. (1961).

struction to flow; the blood flow in this vertebral artery is reversed. During muscular work of the arm the pressure gradient between the circulus arteriosus cerebri and the postocclusive section of the subclavian artery is increased by the fall in peripheral resistance to such an extent that blood may be removed from the basilar circulation (Simon *et al.*, 1962; Sproul, 1963; Siekert *et al.*, 1964; Toole, 1964; Marshall and Mantini, 1965). If such an occlusion process is suspected, the most important diagnostic measures are palpation and auscultation of the carotid and subclavian arteries.

The Doppler ultrasound probe has recently provided a means of testing the passability of the vertebral artery. However, final clarification requires an upper aorto arteriography. This is always the method of choice since there are often combined occlusions and stenoses of vessels.

The numerous extra- and intracranial collateral connections frequently prevent functional effects, even of total removal of individual afferent arteries. It is known that up to about the twentieth year of life a ligature of the carotid artery can be carried out without loss of cerebral function. An insufficient blood supply to the brain is hence to be expected only with multiple stenoses of occlusions or with severe atherosclerotic changes in the aortic arch and in the vessels leaving it. However, cases of complete aortic-arch syndrome are known in which the blood supply to the brain was only maintained via a vertebral artery but in which severe neurological deficits were not demonstrable (Porstmann, 1971). On the other hand, moderate stenoses often result in a grave disturbance of cerebral blood supply if there is an abrupt fall in the systemic arterial blood pressure. This danger should be considered especially in treatment of hypertension as well as in states of shock and disturbances of cardiac rhythm.

C. Atherosclerosis of the Large Afferent Extracerebral Arteries (Disorders of Cerebral Blood Supply)

Ule and Kolkmann (1972) justified the special position accorded to cerebral arteriosclerosis by the vital importance of the brain, on the one hand, and by the relative frequency of arteriosclerotic changes in the cerebral arteries, on the other. The latter is to be viewed as a special characteristic of this organ. Ule and Kolkmann also attached great importance to the behavior of sclerosis of the cerebral arteries, which deviates from that of arteriosclerosis; its correlation with extracerebral arteriosclerosis is not subject to any definite rules.

Atherosclerotic changes located in the internal carotid and vertebral arteries are almost twice as frequent as in the arteries at the base of the brain. The most susceptible sites in the internal carotid artery are the carotid sinus with the bifurcation spur at the branching point of the common carotid artery and the carotid siphon, the four bends of the artery as it passes through the base of the skull, the carotid channel of the temporal petrosa portion of bone, and in the sinus cavernosus. According to Mathur and co-workers (1963), sclerosis develops first in the basilar artery, then in the arteriae cerebri mediae, and only later and to a smaller extent in the arteriae cerebri anteriores et posteriores.

Although constituting only 2% of the total body weight, the brain requires 20% of the heart minute volume to cover its oxygen and metabolic requirements. Metabolically, it is therefore a particularly active organ. For example, the heart itself requires only 5% of the heart minute volume.

In order to cope with these differentiated metabolic processes, the brain has a particularly stable, delicately reacting blood flow dominated by autoregulation. This autoregulation of cerebral circulation consists of the interplay of two reciprocal components: the perfusion pressure and the resistance of the cerebral vessels. The brain is known to be supplied by the large four extracerebral vessels. These form the circulus arteriosus of Willis, which can be demonstrated to enable a collateral supply between the flow area of the carotid and vertebral arteries. This circle acts as a fine-reacting equilibrating valve for hemodynamic pressure. The functioning of the abundantly available and preformed collaterals depends on the following factors: (1) the width of the collateral connections; (2) the elasticity; (3) the pressure relationships in the collateral circulation; (4) the blood pressure and thus the general circulatory relationships; and (5) the rate of arterial occlusion.

The blood required per 100 gm brain tissue is 55–60 cm³/minute, with a critical limit at 30 to 35 cm³/minute. In persons with healthy vessels there is a significant reduction of brain blood flow only if the pressure falls to about 70 mm Hg. In patients with cerebral sclerosis, the critical pressure threshold is reached even at 110 to 120 mm Hg. Ischemic damage, even of the cerebral region, and the resulting clinical symptoms do not necessarily indicate irreversible tissue damage. An acute diminution of oxygen and energy supply results initially in a reversible functional disturbance of the ganglion cells provided that the minimal turnover for maintenance is met.

Infarcts today occupy the third place in the statistics of causes of death. The most frequent is apoplexy (62%) elicited vascularly or hemodynamically. Two facts are of crucial practical importance.

1. In most patients it is possible to recognize these extracranial vessel diseases without special technical aids (typical anamnesis: previous

ischemic attacks with maintained consciousness, weak or absent pulse, auscultation with high-frequency stenotic murmur, measurement of blood pressure on both upper arms, apparatus search tests with ophthalmodynamography, and supraorbital sonography).

2. About 75% of these extracranial vessel blockades are accessible at present to surgical correction (Morris and DeBakey, 1966; Vollmar, 1974).

The causes of deficient cerebral blood supply are luminal narrowing or occlusions of cerebral arteries. These can develop *in situ,* as in atherosclerosis or in vasculitis, can arise from displacement of material from proximal vessel sections or from the heart, as in emboli, or they can result from pressure, tension or formation of kinks in arteries. Nonobstructive atheromatous changes in walls, especially at the carotid fork, have been accorded more attention in recent years after the assumption gained ground that some patients get their ischemic deficits from platelet or fibrin emboli. These originate at an atheromatous deposit on which thrombi can develop at any time (Whisnant *et al.,* 1961; Hollenhorst, 1962; Wylie and Ehrenfeld, 1970). However, atherosclerosis is the most frequent cause of deficient cerebral blood supply. Finally, deficient blood flow can also arise from disturbed distribution of blood, as in the steal mechanisms which depend on altered pressure gradients in arteries or arterioles. The latter are caused by stenoses or occlusions or by a local disorder of vessel tone regulation, a paralysis of autoregulation. The kind and distribution of the cerebral defects permit conclusions as to the localization and extent of the ischemic area, but only very limited conclusions as to the site and extent of the arterial displacement (Gänshirt, 1972).

Consideration of defective cerebral blood flow in terms of its clinical consequences requires distinction of (1) asymptomatic stenosis; (2) asymptomatic occlusion; (3) ischemic attacks; (4) amaurosis fugax; (5) progressive incipient apoplexy; and (6) complete brain infarct.

Asymptomatic carotid or vertebral stenosis is only discovered if a vessel murmur is heard during a physical examination and arteriography carried out as a result. It has become apparent that an asymptomatic stenosis does not constitute a preinsult situation and thus does not indicate the need for medication or surgical intervention. However, it must be ascertained that the other brain arteries are free from obliterative processes. The upper body aorto arteriography which is always referred to serves this purpose; both carotid and both vertebral circulations as well as the subclavian arteries and the brachiocephalic trunk are shown. However, continued monitoring of this finding is recommended.

Ischemic attacks inevitably lead to a complete cerebral infarct. Patients with ischemic attacks should consequently be subjected immediately to cerebral angiography. If there is unilateral carotid stenosis contralateral to the neurological symptoms of the attacks and ipsilateral to the amaurosis fugax, surgery should be recommended.

If a bilateral carotid stenosis is found in patients with lateralized ischemic attacks, the stenosis on the side responsible for the symptoms should be operated on, even if it is the one with less narrowing of the lumen. In a bilateral carotid stenosis in which the symptoms are not lateralized, indicating an insufficiency in the vertebrobasilar area, either one or both the stenoses should be removed in order to improve the collateral circulation via the circulus arteriosus cerebri.

Removal of the stenosis is also advisable in patients with ischemic attacks, unilateral carotid occlusion, and carotid stenosis, irrespective of whether the symptoms are related to the side with the occlusion or stenosis or whether they have anything at all to do with the carotid circulation. Cases with carotid stenosis, vertebral stenosis, and intermittent insufficiency in the area supplied by the vertebral artery can be freed of symptoms merely by removal of the carotid stenosis. Interventions in the vertebral artery are more rarely indicated.

Progressive incipient apoplexy is a brain infarct which develops slowly or within 1 to 2 days; the increase and extension of deficits reflect the growth of the infarct in the brain. Up to now, vascular surgeons have considered that such cases should be thrombendarteriectomized within 6 hours after the symptoms have set in. Statistics show that the mortality is 50% (DeBakey, 1962; Heberer et al., 1966), so an active approach is not advisable. The conservative treatment measures available include immediate heparinization, infusion of NaCl-free plasma expander for hemodilution, reduction of the blood viscosity and the fibrinogen level, and dispersal of the erythrocytes aggregated in the region of the static blood in order to improve the cerebral microcirculation. Heparinization is the more risky method and not clearly superior to therapy with plasma expanders. A progressive brain insult should be subjected as soon as possible to carotid angiography on the opposite side in order to confirm the diagnosis. Progressive brain insult is quite commonly confused with an acute subdural hematoma or a bleeding glioma.

In complete brain infarct (discussed by Dorndorf and Gänshirt, 1972), all surgical measures are ruled out if the tissue degradation which suddenly appears in the ischemic area has gone to completion. Anticoagulation is also initially precluded. The first concern is to normalize blood pressure. There is often hypertension, which must be cautiously reduced to systolic values of 160 to 170 mm Hg. Reserpine is suitable for this purpose. If the blood pressure is too low, it must be raised carefully.

Use of drugs to reduce blood pressure in hypertension does not worsen cerebral blood flow but improves it; the hypertension leads to subintimal and perivascular hemorrhages at arterioles and capillaries and to brain edema, turning a white into a red infarct. Treatment of cardiac insufficiency or disturbances of heart rhythm is preeminent among the conservative measures. Infusions with hyperosmolar solutions (40% sorbitol or, in severe cases, 20% mannitol) are used to combat cerebral edema.

D. Atherosclerosis of the Thoracic Aorta

Atherosclerotic processes are usually least advanced in this section of the aorta. Thrombotic deposits are a possible source of emboli. If aneurysms due to syphilis are (or were) more frequent in this region, there are also aneurysms due to atherosclerosis. The leading symptom is pain resulting from compression of neighboring organs, such as the spine, trachea or esophagus. According to Joyce et al. (1964) a good one-third of these aneurysm-carriers die from rupture and more than one-half from concomitant cardiovascular diseases. X ray of the thorax provides pertinent information. Since the prognosis of atherosclerotic aneurysm of the thoracic aorta is relatively good and the risk of surgery on the thoracic aorta is generally greater than that on the abdominal aorta, surgery is undertaken with greater reluctance in the former than in the latter condition (Heberer and co-workers, 1966).

E. Atherosclerosis of the Abdominal Aorta
   (Angina Abdominalis, Stenosis of Renal
   Arteries, Aneurysms)

1. ABDOMINAL ANGINA

Favoring of the vessel origins due to formation of vortices gives atherosclerosis a particular importance in certain diseases of the abdominal region. Despite wide distribution, extensive disturbances of blood flow in abdominal organs are rare because the usually slow development of the stenosis allows a fully functional collateral circulation to be formed.

Single stenoses of visceral arteries and total occlusions of the inferior mesenteric artery are clinically almost silent. An acute occlusion syndrome gives rise to the clinical picture of an acute abdomen. The triple combination of sudden abdominal pain, shock and intestinal cramps, often with bleeding, point to acute insufficiency of visceral blood flow.

Visceral angiography should be undertaken immediately for rapid and precise localization (Wenz, 1972) and is an obligatory precondition for surgical intervention (Morris *et al.*, 1962).

Isolated obliterations of the celiac trunk generally cause only slight complaints. Obliteration processes in the superior mesenteric artery and the celiac trunk can lead to a disease picture characterized by the triple combination of symptoms "intermittent belly pain—malabsorption syndrome—vascular murmur." The lowered blood supply to the whole small intestine and the colon up to the flexura lienalis is manifested clinically in more than one-half of the cases by intermittent attacks of postprandial pain. The author who first described the condition (Ortner, 1902), therefore, spoke of dyspragia intermittens angiosclerotica intestinales, though the term "abdominal angina" is in more current usage today. Other designations are "intestinal angina" (Mikkelsen, 1957), "mesenteric arterial insufficiency" (Derrick and Logan, 1958), and "syndrome d'ischèmie intestinale paroxystique" (paroxysmal intestinal ischemia syndrome; Leymarios, 1960). The similar picture arising through reflex vasoconstriction of the splanchnic vessels in cardiac insufficiency should be differentiated from true abdominal angina. Today a selective angiography of the large unequal visceral arteries is necessary finally to confirm the clinical diagnosis, which is often an exclusion diagnosis. It is important that the Riolan anastomosis (collaterals formed from the middle and left colic arteries) be included. Demonstration of this connection makes the often lacking clinical symptomatology understandable. Classically, the pains begin 15–30 minutes after ingestion of food and last from 1 to several hours. The intensity of these pains depends on the amount and composition of the food and ranges from a feeling of pressure or fullness in the upper abdomen to coliclike pains over the whole abdomen. The loss of weight (which can lead to a severe cachexia) is often not due to an absorption disorder alone but also to abstinence from food because of fear of the pain which follows a meal. Fecal passage of blood indicates ischemic damage to the intestinal mucosa (which can heal leaving circular cicatricial stenoses). In sudden thrombotic or embolic occlusion of an unequal artery ("mesenteric infarct") manifesting the clinical picture of acute abdomen, it is absolutely necessary to intervene surgically as soon as possible.

The "mesenteric steal syndrome" is of importance in differential diagnosis. It is due to occlusion of the distal aorta which can be bypassed by anastomoses between the mesenteric arteries and the internal iliac artery. Violent intermittent body pains caused by withdrawal of blood from the visceral arteries can appear during walking.

2. STENOSES OF THE RENAL ARTERIES

Stenoses and occlusions in the renal arteries are frequently of athero-sclerotic origin and rarely also of embolic nature (Gore and Collins, 1960; Eliot and Edwards, 1964; Kassirer, 1969). Aseptic renal infarction or nutritive atrophy of a kidney can result, and, in addition, renovascular hypertension, according to the fundamental investigations of Hartwich (1930) and Goldblatt *et al.* (1934).

Two phases in the genesis of renovascular hypertension can be dis-tinguished. In the first phase, the high pressure is caused solely by the kidney behind the stenosis, while in the second phase a state arises in which the hypertension is an independent condition due to autoregula-tory processes and secondary atherosclerotic changes in the nonstenosed kidney. In the second phase, the hypertension can no longer be treated by correction of the primary lesions alone.

The proportion of renovascular hypertension among all forms of secondary hypertension is 2–4%. Assuming that 80% of all hypertensives have no discernible cause for their high blood pressure (i.e., are "essen-tial hypertensives"), there remain 20% with secondary hypertensions; i.e., of the total number of hypertensives, only 0.4–0.8% have renovascular hypertension. The number is nevertheless sufficiently large that such cases are found in every practice and it is worth anticipating them since successful surgical therapy is possible if an early diagnosis is made (Stefanini *et al.*, 1964; Morris *et al.*, 1966). Of course, it should be pointed out here that not every renal arterial stenosis causes high pressure, i.e., has a pathophysiological effect. Therefore, only the patients with renovas-cular hypertension who are diagnosed in time can be cured by operation. If high blood pressure is detected, especially with markedly raised diastolic values, one should look for stenosis of a renal artery. Experience has shown that auscultation of the abdomen is not a very certain method, and that renovasography is the best technique.

In all hypertensives less than 50 years old an iv pyelogram with an early urogram is necessary. However, since false positive and also false negative results are found in 10% of the cases, an angiography should be carried out in every case under 40 years old. Kidney renin should be determined separately on each side if a stenosis is present. If the quotient of the renin activity in the stenosed side and the renin activity on the nonstenosed side is more than 1.5, then a cure or improvement of the hypertension can be expected from a vessel reconstruction in 90% of the cases (Kaplan, 1971).

In patients over 40 years old and in stenoses which are unsuitable for surgery because of their site, conservative therapy today stands in the foreground. Renovascular hypertension responds just as well as essential (primary) hypertension.

3. ANEURYSMS

The abdominal aorta is the most frequent site of atherosclerotic aneurysms. It has been considered up to now that every aneurysm of the abdominal aorta which is diagnosed must be resectioned since rupture after diagnosis gives a survival time of only about 1 year. According to Heberer and co-workers (1966), every penetrating and ruptured aneurysm of the abdominal aorta should be operated on since the patient will certainly be lost without surgical intervention. The indication for surgery is only a matter for discussion if the aneurysm is asymptomatic. In view of the changed etiology, one must be clear that an operation only improves the life expectancy if it prevents rupture of the aneurysm. Since an average of about 30% of atherosclerotic aneurysms of the abdominal aorta rupture, the average life expectancy without operation will differ only by this percentage from patients who are operated. According to the studies of DeBakey and co-workers (1964) the operative mortality is determined essentially by the age of the patient and the state of the myocardium; the life expectancy without operation crucially depends on the size of the aneurysm. An age greater than 60 years and damage to the heart muscle signify a considerably raised surgical risk. Aneurysms of the abdominal aorta are clinically manifested in pressure on the adjoining organs resulting in abdominal pain. If the adipose layer covering the abdomen is thin, they can be readily palpated as an expressive-pulsating tumor which is frequently painful if pressed. A pulse-synchronous murmur can be auscultated over them; this is caused by the turbulence in the aneurysm. Definitive demonstration is possible by aortography (Ryan et al., 1964). The prognosis is substantially improved by surgeons operating manually, the rate of primary mortality then being kept within justifiable limits (DeBakey and Cooley, 1953; Dennis et al., 1965; DeBakey, 1974). The aneurysms are often coated with thrombotic masses so they do not completely fill with contrast medium and can consequently give rise to incorrect interpretations. In more than 80% of cases, the diagnosis can be confirmed by simple X rays of the lumbar vertebral column at two levels, since aneurysms are almost always calcium-containing and, therefore, give good contours.

A total occlusion of the aortic fork or of the distal abdominal aorta (usually by a thrombosis) is designated Leriche syndrome and results in disturbed blood supply to both legs in a considerable number of cases.

The clinical picture of the condition is dealt with in the appropriate section below.

III. Atherosclerosis of the Coronary Arteries (Coronary Heart Disease)

In the highly industrialized countries cardiovascular diseases have attained the first position in the statistics of morbidity and mortality (50–70%). In western European males, 60–70% of all cardiovascular deaths are due to coronary heart disease (women, 20–25%), in the United States 70% (women, 30–55%); the percentages increase with age. Coronary heart disease accounts for 30% of the total mortality in men aged from 45 to 75 years in western Europe (women, 15–20%) and 40% in the United States (women, 25%). At present almost 100,000 people die of myocardial infarct every year in the Federal Republic of Germany.

Among diseases resulting from arteriosclerosis, coronary heart disease is the most important. In the United States there are about one million heart attacks per year, about 700,000 of which are fatal (400,000 sudden and unexpected). About 175,000 involve people under 65 years old. In 1967, coronary heart diseases caused direct costs of 2.1 billion dollars in the United States; the indirect costs were 13.5 billion dollars. According to J. Stammler it must be assumed that today the total costs amount from 25 to 30 billion dollars per year, constituting an enormous burden on the entire economy. Since 1972 the corresponding figures for West Germany lie between 500,000 and 600,00 illnesses per year with about 125,000 deaths. The number of sudden unexpected deaths from heart attack can only be estimated, though it would be about 50,000 per annum. The percentages are similar in most industrial countries.

At present, the North American male has a 20% probability of suffering a coronary attack before he is 50 years old. According to J. Stammler (1974) and O. Paul (1974), 20% of the affected persons die within 3 hours after the onset of symptoms. An additional 10% die within the first week. Our own statistical results are reproduced in Fig. 1. The prognosis of a person who has suffered a heart infarct depends on various factors—"risk factors" and "risk constellations." The WHO characterized the situation in the following statement:

> Coronary heart disease has reached enormous proportions, striking more and more at younger subjects. It will result in coming years in the greatest epidemic mankind has faced unless we are able to reverse the trend by concentrated research into its cause prevention. The Board expressed its wish that countries most affected by cardiovascular diseases increase their efforts both to set up efficient services for control and carry out more extensive research programs.

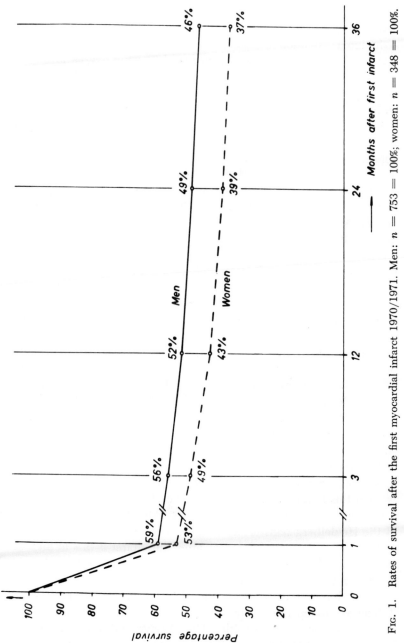

Fig. 1.   Rates of survival after the first myocardial infarct 1970/1971. Men: $n = 753 = 100\%$; women: $n = 348 = 100\%$.

In most cases of acute myocardial infarction, sudden deaths, or other clinical expressions of CHD (angina pectoris, congestive heart failure, arrhythmia, conduction defect, etc.), the underlying disease process is severe atherosclerosis. This can be found to some degree in most adults, male and female, but disability and death are almost always the consequence of atherosclerosis sufficient to result in marked narrowing or occlusion of the arterial lumen.

Particularly important is the anticipation of coronary sclerosis in younger men observed during the last 3 years and the more frequent appearance of "juvenile myocardial infarction" before the 40th year of life (Davia et al., 1974). However, the majority of infarcts continue to occur in the older age groups over 60, partly because many experience their atherosclerotic complications due to the high life expectancy in developed countries. The sex difference in attack by coronary sclerosis is quite unequivocal up to about the 50th year of life. In menopausal women the extent of atherosclerosis increases significantly and reaches the male level with the 70th year of life, though sex-specific predominant localizations and anatomical differences remain. However, men are clearly susceptible to infarct 10 years earlier than women. Most heart infarcts doubtless arise as a result of coronary sclerosis. There is much discussion as to whether coronary thrombosis is the cause or consequence of myocardial infarction. The incidence of coronary thrombi fluctuates between 54 and 96.5% as shown in Table I (according to Chandler et al., 1974). Most cases of sudden unexpected death also result from coronary heart disease (see Paul, 1974). There is also worldwide agreement here. The data which we collected in 1970–1974 in Heidelberg are representative and show the high risks to which the heart-disease patient is subject.

The survival rate is particularly unfavorable in the first 4 weeks after infarction. Of the men, 59% are still alive after 4 weeks and while in women this figure is 53%. This difference is very probably an age effect. The women are, on average, 10 years older than the men in the area of the Heidelberg register at the time of the first infarct.

The rate of survival for 1 year after the infarct is 52% in men and 43% in women. The two curves are almost parallel in the next 2 years. The survival rate falls by 3 to 4% in this phase.

The differences in course in the different age classes is shown in Fig. 2. The age-independence of survival during the third and fourth years after infarction is remarkable. The survival rates in this observation interval did not show any statistically significant differences when there was high blood pressure, diabetes mellitus, or high cigarette consumption at the time of infarction. Of the infarct cases, 48% died in the first hours after the acute event (Fig. 3). If it is further considered that 17%

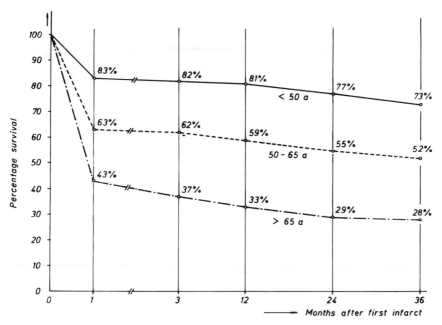

FIG. 2.   Rates of survival after the first myocardial infarct 1970/1971. Men: more than 50 years old, $n = 145$; 50–65 years old, $n = 314$; more than 65 years, $n = 294$.

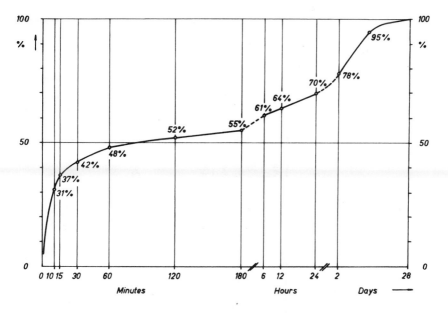

FIG. 3.   Time interval between myocardial infarct and death within the first 28 days. Men and women: $n = 465 = 100\%$.

TABLE I

CORONARY THROMBOSIS IN REGIONAL TRANSMURAL INFARCTION[a]

Summary report	Infarcted hearts with coronary thrombi	Age limit of infarcts (week)	Spatial relations	Underlying arterial lesions	Temporal relations	Conclusions
Chapman	257/282 (91%)	4[b]	Occlusion of extramural artery usually subtends infarct and is separated from it by uninvolved segment of artery	Thrombus forms on eroded intimal surface	Not reported	Thrombus precedes and causes infarct
Erhardt	7/7[c]	2[d]	Infarcts occur in region supported by occluding artery	Thrombi frequently associated with atherosclerotic stenosis and ulceration	Thrombi incorporate [^{125}I]fibrinogen after onset of clinical manifestations of infarction	Coronary thrombus may be secondary event
Roberts	40/74 (54%)	6[d,e]	Site of myocardial necrosis corresponds properly with artery containing thrombus	All thrombi over old stenosing atherosclerotic plaques	Not reported	Coronary thrombus follows rather than precipitates acute myocardial infarction

Schwartz	18/21 (86%)	2[b]	Thrombi often in two arterial segments. Most thrombi proximal and anatomically subtend infarct	Infarcts related to degree of arterial stenosis	Not reported	Anatomic dependence of infarcts on coronary thrombi indicates causal relation between thrombi and infarcts
Sinapius	164/170 (96.5%)	4[d]	Constant local relation between thrombosed artery and supported area infarcted	Most thrombi occur over tears in atheromata	Many thrombi have multiple layers of episodic growth. Oldest portion develops before or with onset of symptoms	Acute infarction caused by complete thrombotic occlusion
Spain	45/50 (90%)	2[d]	Infarcts associated with regional coronary arterial thrombi	Thrombi form at site of well established atherosclerosis	Both infarcts and thrombi have multiple and variable ages that can lead to misinterpretation of temporal relations	Infarcts are preceded by recent and regional coronary thrombi

[a] From Chandler et al. (1974).
[b] Age estimated by histological criteria.
[c] Cases selected only for study of radioactive thrombi.
[d] Age estimated from time of onset of clinical manifestations of infarction.
[e] Only 6 cases had infarcts from 3 to 6 weeks old.

of our infarct patients were already in shock and 13% already dead before a doctor could intervene, it becomes clear how important it is that optimal emergency arrangements in the prehospital phase be organized.

A. Prehospital Phase

It is of primary importance that the person affected by an infarct and his relatives recognize the danger as soon as possible and quickly decide to call a doctor. During the cardiac infarct registration study in the Heidelberg area we found the following result (Fig. 4): about 40% decide to call the doctor in the first 15 minutes, 55% by 30 minutes, and 65% by 60 minutes. The decision took more than 12 hours in 10% of the cases.

Sociological analysis showed that the decision time in reinfarct cases was not shorter than in patients experiencing an infarct for the first time. It is, moreover, worth mentioning that the decision time is much longer in infarcts which occur during the night than those occurring during the day.

How long must the patient wait for the doctor once he has been called?

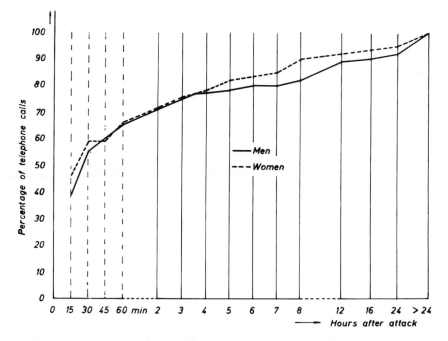

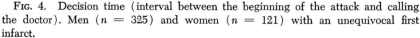

FIG. 4.   Decision time (interval between the beginning of the attack and calling the doctor). Men ($n = 325$) and women ($n = 121$) with an unequivocal first infarct.

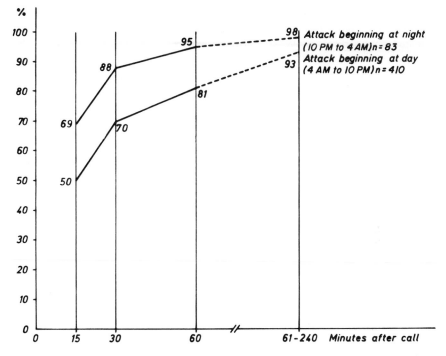

FIG. 5.   Delay period (time from telephoning doctor to the first examination) in the day and at night for men with myocardial infarction.

The waiting period was up to 5 minutes in 50% of the cases studied (Fig. 5). The physician arrived within 30 minutes in 70% of cases. If the doctor was informed after 22 hours, the waiting period was markedly shorter than during the day. At night 69% of the patients waited 15 minutes at most for the physician. It is important that emergency cars should be used to reach the patient quickly, especially during the day since the peak incidence of cardiac infarction is between 8 and 10 AM.

B. Risk Factors

As in the United States, hypercholesterolemia, high consumption of cigarettes, and hypertension are first-order risk factors in West Germany. They are today attributed with a causal relation to cardiac infarction. The second, less significant, category includes the following criteria indicating a raised infarct risk: diabetes mellitus, gout, adiposity, lack of exercise, and stress. Hypertension, diabetes mellitus, adiposity, and cigarette consumption are readily demonstrable and thus of practical significance for effective prevention on a wide scale. In our study only

3% of a total of 735 men with a first infarct were free from these four factors. The "multicentric cardiac infarct study" (Schettler, 1972) carried out in four hospitals in West Germany showed that 4% of 804 men with first infarct were free of these factors.

Early recognition of these four factors is particularly important. As shown by Fig. 6, they are especially prominently represented in male patients with a cardiac infarct before the 40th year of life. The control group is typical for patients of a general practice, but not for the general population. In discriminant analysis, smoking, hypercholesterolemia and hyperuricacidemia contributed most to the separation of these two groups. In factorial analysis, smoking constitutes an independent factor with respect to the other parameters. A high level of cigarette smoking correlates significantly with the average age at the first infarction (Fig. 7). Nonsmokers and ex-smokers had an average age of 63 years at the time of the first infarction; heavy smokers were, on the average, 10 years younger.

In addition, we examined the effect of smoking in patients whose body weight, blood pressure, cholesterol, and blood sugar were normal. There was also an age difference between nonsmokers and heavy smokers with respect to the first infarct.

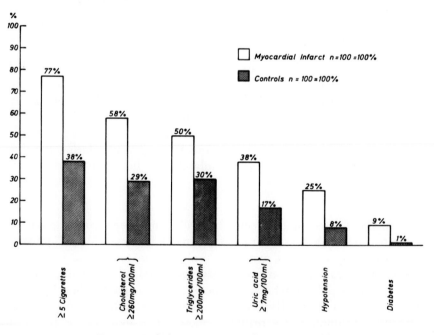

FIG. 6.   Risk factor in men under 40 years old.

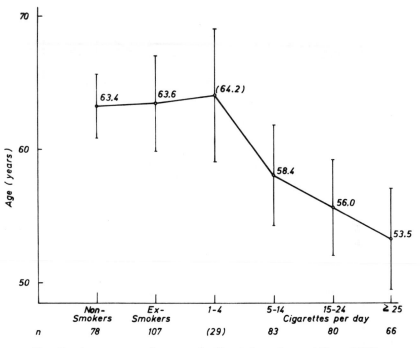

FIG. 7.   Average age of men at the first infarct ($n = 443 = 100\%$).

Lack of exercise is a controversial risk factor, but is especially difficult to determine. We graded the professions of our infarct and control patients according to the degree of physical exercise and independent of the working situation, using a system which we had developed (Kreiger, 1969). It was consistently shown that the infarct patients exercised less than control groups of the same age (Nüssel *et al.*, 1968). Figure 8 shows an example from the series of investigations which we carried out in cooperation with colleagues from the University of Montpellier. In each case, the physical exercise in working life was determined retrospectively with respect to the main activity over 5 years. The groups are correlated with age. The infarct patients in Montpellier, Heidelberg, and Rastatt were less physically active than the control patients.

C. Psychological Constellations

On the basis of the psychosomatic studies of Zukel *et al.* (1969) and Christian and Hahn (1972) and using the Rasch (1961) model for logistic measurement, Hehl (1974) (a member of our work group) developed a personality questionaire. This system of 25 personality scales has already

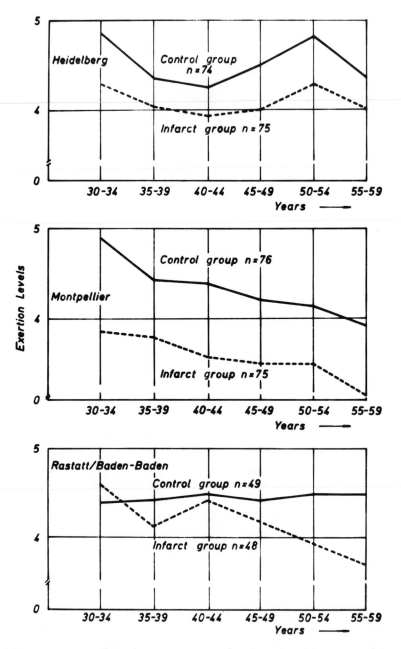

Fig. 8.   Average physical exertion connected with work (in 5-year periods).

been used in 141 patients with cardiac infarct and 163 subjects with an un-
diseased cardiovascular system. Hehl brought the personality traits which
were found particularly frequently in the infarct patients into a dynamic
relationship. Several current projects of our work group are concerned with
the hypothetical model of the psychodynamics of myocardial infarct patients which Hehl developed as a working hypothesis. According to the
hypothetical model, it was to be expected that patients would have gone to
the doctor particularly rarely before their first infarct. This assumption
was confirmed by our inquiry; a relatively high proportion (25%) of the
patients under 55 years old said that they had never been medically
examined before the first cardiac infarct.

Hehl also related to his psychological parameters the risk criteria
angina pectoris, hypertension, diabetes mellitus, overweight, and smoking
in 264 patients of a general practitioner. He found that the more
"endogenous" risk criteria angina pectoris, hypertension, and diabetes
II). These factors correlated with the risk factors smoking or overweight
or the personality characteristics. The more "exogenous" risk criteria,
smoking and overweight, on the other hand, formed a common factor
with the psychological characteristics "overadaptation," "drive to play
activity," and "neglect of health." Smoking and overweight are accordingly explained by these three motives.

In a further analytical step the individual consistency of the personality
traits was tested, and the "neglect of health" was found particularly
inconsistent and thus readily influenced. This constitutes a promising
possibility for health education intervention.

TABLE II

FACTOR ANALYSIS: CRITERIA WITH HIGH WEIGHTING FACTORS II AND V

Criteria	Factor	
	V	II
Age	0.55	0.00
Angina pectoris	0.60	—0.10
Diabetes	0.57	0.10
Hypertension	0.64	0.10
Smoking	0.03	0.69
Overweight	0.26	0.54
Drive to play activity	0.18	0.48
Overadaptation	—0.04	0.45
Neglect of health	0.16	0.50

D. Pathophysiological Bases

Coronary sclerosis was formerly only open to pathological anatomical determination; the clinician was either unable to detect it at all or only through indirect manifestations. The clinician is now able, with the aid of coronary angiography, to diagnose a large number of cases morphologically.

Coronary insufficiency, which can be classified as "relative" and "absolute" signifies the result of a disequilibrium between demand and supply of blood and oxygen. It is a pathophysiological concept which is difficult to verify with the usual methods of clinical examination. According to Doerr (1972), coronary insufficiency is the collective term, while the effects on the heart muscle (transmural infarct, the "simple," i.e., minor damage to the inner layer and miliary necrosis of fibers) are subordinate phenomena, with the same pathogenesis but different morphology.

Diagnosis of coronary heart disease is difficult because coronary sclerosis is masked by the coronary reserve, which thus eludes timely diagnosis. Early diagnosis of coronary sclerosis has, hence, remained a diagnostic illusion. An attack of angina pectoris only occurs when a deficit arises between the requirement and supply of oxygen. However, this already indicates a manifest critical coronary sclerosis with permanent danger of infarction. Nevertheless, the presence of coronary sclerosis or its extent does not reflect the restriction of the coronary reserve or the incidence of attacks of pain. Coronary blood flow is a functional quantity which does not depend on the state of the main vessels alone, but also on the functional efficiency of the collaterals and anastomoses. Therefore, the blood supply of the heart may still be adequate despite substantial coronary stenosis. The second decisive functional quantity is the oxygen requirement. The interrelationship between the two quantities can be illustrated by forming the quotient of oxygen supply and oxygen demand, making the balance problem clearer. The factors which regulate the O_2 supply in detail are (1) coronary flow; (2) O_2 content of the arterial blood; and (3) diffusion path. Among these factors, coronary flow is by far the most important. It is determined by (1) vessel sclerosis; (2) diminution of the coronary perfusion pressure; (3) shortening of the coronary-effective phase of the heart (diastole); (4) increase in the blood viscosity; (5) elevation of the left ventricular end-diastolic pressure.

The oxygen consumption of the heart muscle is determined by (1) wall tension; (2) contractility; (3) external work of the heart; (4) size of the heart; and (5) basal metabolism. Of the quantities named, the coronary flow, wall tension, and contractility are the most important from a therapeutic viewpoint. However, as a rule, coronary flow cannot be varied decisively because of coronary sclerosis and the resulting poor vessel adaptability.

E. Definition and Clinical Symptoms

This disproportion between blood supply and demand is clinically manifested as angina pectoris, and the syndrome is stenocardia.

Angina pectoris gravis is a particularly severe, long-lasting form of true angina pectoris occurring at short intervals. With increase in the frequency and intensity of attacks it is termed infarct-threatening or pre-infarct syndrome. A severe attack of angina pectoris lasting a long time is also called "status anginosus."

Further special forms of true angina pectoris are angina decubitus, which appears in the lying position, and Prinzmetal angina, appearing without discernible cause and accompanied by an elevation of ST interval as in a fresh infarct.

The character of the pain is stated as a feeling of pressure, burning, feeling of cutting and stabbing behind the sternum or strangling the whole chest. Shortness of breath is often associated with it. The patient almost never senses the pain as being highly localized. Exact localization of the point of pain with the finger is atypical for true angina pectoris.

Statements on the radiation of the pain contribute further to recognition. Substernal pain was reported by 50% of the patients, while pain in both arms was reported by only 12% of those questioned. Radiations to the upper abdomen, the throat, and the lower jaw area are less frequent.

Typical retrosternal pain with radiation into the left arm is the familiar classical picture of severe angina; if the attack lasts more than 20 minutes and nitroglycerine is ineffective, then myocardial infarct is indicated. Our own studies and those of other workers (Mörl, 1964; Anschütz, 1973) found these typical symptoms in only a good 50% of all those affected with myocardial infarct, however. Twenty-five percent of infarcts are associated with atypical symptoms and another 25 can be entirely silent. The same observations were made by Rosenman et al. (1967) in the western collaborative group and Kannel (1973) in the Framingham studies; one case in every four infarcts was asymptomatic or had such asymptomatic symptoms that it was not recognized as an infarct. Schettler and Wollenweber (1969) state that comparative anamnestic autopsy studies have shown that of 100 moderately severe to severe coronary sclerotics only 25% complained of angina. In most cases, the event was totally unexpected. This is also confirmed by the most recent investigations from the Heidelberg infarct study of Schettler and Nüssel (1974) in which no mention of an infarct is found in 25% of the cases.

It is also important for the physician who is carrying out the treatment to know that the origin and localization of pain, i.e., the focus of disease and pain sensation, are not necessarily topographically identical (Anschütz, 1968). The radiation must by no means always be continuous. The pain of a myocardial infarct may be expressed under certain conditions under the left shoulder, in the left arm, in the left elbow joint, or in the left wrist. Dyspnea can also be the most prominent sign of left ventricular insufficiency. Further circulatory reactions such as fall in blood pressure, decrease in amplitude of blood pressure, a sudden fall in venous pressure, and a feeling of weakness with or without signs of abdominal excitation can also be observed.

"Disguised" myocardial infarcts on which other diseases are superimposed must be mentioned. This designation covers infarcts which are so masked by symptoms of other diseases that a myocardial infarct is either diagnosed by chance or not at all. This is the case in manifest cardiac insufficiency, apoplectic conditions, abdominal diseases, pulmonary embolism, disorders of peripheral blood flow which appear suddenly, postoperative states, and severe cachexia resulting from malignant diseases.

Since angina pectoris is a symptom which may have diverse causes further differential diagnostic measures with respect to diseases of the nervous system, the movement apparatus, the diaphragm, the digestive tract, and the aorta must be employed.

A precise and detailed case history is of primary importance in the diagnosis of angina pectoris even today. The points stated above must be given particular attention in talking with the patient. The information obtained in clinical investigation of the patient is disappointing. There are no conclusive specific findings. There are signs of heart enlargement, and left ventricular insufficiency only in advanced cases.

Coronary angiography, which is the most certain current procedure for demonstrating coronary sclerosis, obviously cannot be used in an acute myocardial infarct. In a postinfarct state with continuing angina pectoris symptoms and in a "threatening infarct" it is desirable to determine the exact morphological situation in the coronary arteries by means of coronary angiography with a view toward a possible operation. This has increasing importance today because surgery is being more frequently used to treat angina pectoris or "impending infarct." Of course, strict assessment of indications is still as indispensable as ever in order to keep the rate of complications as low as possible (Abrams and Adams, 1969a,b). It is thus not possible to use this technique on a wide scale for the early detection of coronary heart disease (Wilhelmsen, 1974).

IV. Atherosclerosis of the Extremities (Occlusive Arterial Diseases)

A. General Introductory Remarks

Disturbances of peripheral blood flow have substantial sociomedical importance (next to coronary heart disease). The morbidity is estimated as 1% of the population. The incidence of premature invalidism is about the same as that due to diabetes mellitus. The Basle study (Widmer, 1963; Widmer *et al.*, 1972) has shown that arterial occlusion in the limbs and coronary heart disease have about the same incidence. The age of manifestation is also similar (sixth decade). The high mortality in patients with chronic arterial occlusion in the limbs is traceable to the concomitant coronary heart disease (Widmer and Duchosal, 1968). Only 5% of those with atherosclerotic occlusion of the lower extremities are women (Hess, 1959; Preuss and co-workers, 1969).

The almost exclusive involvement of the lower extremities is remarkable. The hydrostatic pressure is thought to be the main factor in this. It is interesting in this connection that in a few bird species the thoracic aorta and the brachiocephalic trunk show more marked changes, clearly because of the higher demands made on the upper extremities in locomotion (Meyer, 1968).

Ratschow (1965) differentiates the obliterating angiopathies of the upper limbs into shoulder, upper arm, and peripheral types. According to Hasse (1974), 10% of all occlusive arterial diseases belong to these categories; this 10% consists of 27% shoulder, 2.4% upper arm, and 70.6% peripheral type.

In the lower extremities three main groups of angioorganopathies are distinguished:

1. *The pelvic type*, which accounts for about one-third of all blood flow disorders (Hasse, 1974). This category includes distal occlusion of the aorta below the renal arteries (Leriche syndrome, 1923). Kappert (1972) states an incidence of about 25% for this location.

2. *The upper thigh type*, which affects the superficial femoral artery, the deep femoral artery, and the popliteal artery [in about 40% of all cases according to Kappert (1972) and about 50% of all cases according to Hasse (1974)]. Isolated occlusion of the deep femoral artery is rarely encountered (4–5%).

3. *The peripheral type*, comprising chronic arterial occlusions of the lower leg, the foot, and the extremities. According to Kappert (1972) and Hasse (1974) it occurs in about one-fifth of cases. This figure doubtless

refers to isolated peripheral occlusions. Angiographic studies revealed, in addition, the combined occlusion types, which with 38 and 52% come closer to the actual relationships. However, it must be emphasized that combined obliterations have a much higher incidence than previously assumed. This was shown by angiographic data (Münster *et al.*, 1966; Jungblut and Günther, 1968).

All stages may have in common the symptoms of "feeling cold" and disturbances of sensitivity. The stages III and IV as well as stage IIb (which is characterized by a walking distance less than 100 meters) should be tackled surgically if possible (Humphries *et al.*, 1963).

In patients with symptoms of intermittent claudication, gangrene develops within 10 years in 15% of cases according to Kappert (1972). Basically, however, deaths following reconstructive surgery to remove chronic arterial occlusions must be considered in perspective. These interventions (with the exception of ascending thrombosis of the aorta) neither save life nor serve directly to prolong life. The success rates in the aortoiliac region are 80–90% immediately postsurgery, 70–80% after 3 years, and about 70% after 5 years, according to Heberer *et al.* (1966). In the femoral-popliteal section, the occlusions and their consequences can be removed for an interim period in 70 to 80%. The longer-term results are not as good; two-thirds survive for 3 years and up to one-half survive after 5 years. Combined occlusions in the aortoiliac and femoral popliteal section are particularly unfavorable. However, it should be borne in mind that there are also cases which do not progress over years and decades.

In the course of a chronic arterial occlusion disease, symptoms of acute occlusion can appear at any stage as a result of complete obliteration. These symptoms are characterized by the six "P's": pain, paleness, paresthesia, paralysis, prostration, and pulselessness. If there are no general contraindications, an acute occlusion should be handled surgically.

According to Fontaine, four degrees of occlusive disease are distinguished:

I   Asymptomatic stage which is only occasionally detected during a medical examination carried out for some other reason

II   Intermittent claudication stage, i.e., a characteristic exercise pain after walking a certain distance

III   Pain during night rest. The resting blood flow in the lying position is no longer sufficient, giving rise to intolerable pains and sleeplessness for weeks on end. An improvement is sometimes effecting by hanging the legs over the side of the bed, which raises the hydrostatic pressure above the critical tissue pressure

IV   Complicated by necrobioses, i.e., moist or dry gangrene

B. Atherosclerosis of the Upper Extremities

The stenoses of the great arm arteries near their origin have already been dealt with under Section II,B. Ten percent of occlusive arterial diseases involve the upper extremities. According to Rau (1970) occlusion processes in the subclavian aorta are two to three times more frequent on the left side than on the right. They are usually proximal to the branching point of the vertebral artery. Discomfort in the arm only occurs with severe muscular work, particularly with the raised arm. Signs indicating a relative vertebral-basilar insufficiency through the cerebral tapping syndrome, such as vertigo and disturbances of vision, are more grave.

A subclavian occlusion distal to the branching point of the vertebral artery does not have such good chances of compensation, since the development of collaterals is less favorable anatomically and functionally. This is expressed by disturbances of muscular nutrition during muscular work (cold feeling and muscular atrophies).

Atherosclerotic changes in the axillary artery are rare. More frequent here are localized thrombi or emboli with corresponding acute symptoms. Atherosclerotic occlusions of the brachial artery are also rare.

All arterial changes peripheral to the elbow are classified as arm-type peripheral occlusion. Occlusions of the upper limbs are encountered with increasing frequency in the proximal to distal direction. The ulnar artery is occluded twice as frequently as the radial artery at the wrist level, irrespective of side. Obliteration of the digital artery is almost always associated with such occlusions.

Independent of their localization, all solitary occlusions frequently have asymptomatic stages or stages with very minor symptoms because of the good collateralization. If the arm is tired by work, the shoulder is probably affected. Feeling cold, pallor or bluish skin, and deafness indicate that the occlusion is more probably localized peripherally; these symptoms may also be accompanied by inflammations and necroses of nail beds.

Besides anamnestic information, data from vessel auscultation, vessel palpation, blood pressure measurements, hand closing with the Allen test, oscillography, and Doppler ultrasound probe can be informative. Early diagnosis is possible by means of Doppler ultrasound probe pressure measurements, in addition to plethysmographic methods (Strand-ness, 1974).

C. Obliterative Atherosclerosis of the Aortoiliac Vessels

A total occlusion of the distal aorta below the renal arteries (which usually has a thrombotic basis) is termed Leriche syndrome. It is char-acterized by a cramplike pain of the pelvic musculature frequently radiat-ing into the upper thigh. Like bilateral occlusions of the great pelvic arteries, this type of occlusion results in impotentia coeundi with main-tained libido. Hypoxemic resting pain does not belong to the picture of circumscribed aortic thrombosis, but indicates additional obstacles in the subsequent vessels. Apart from the intermittent claudication pains in the lower leg, which typically appears at the same time, abdominal pains can appear during walking with obliteration of one pelvic axis. Analo-gously to the subclavian steal syndrome, this is designated the mesenteric steal syndrome or aortoiliac steal syndrome. This collateral circulation removes the blood from the inferior mesenteric artery via the superior rectal artery—inferior rectal artery—internal iliac artery in favor of the leg musculature.

Stenoses of the pelvic arteries cause exercise-dependent pains in the hips, in the upper thigh musculature and even more distally, giving rise to pulse-synchronous rough stenosis sounds which are further conducted to the femoral arteries. The ischemic syndrome is determined by the rate and extent of the occlusion process.

D. Atherosclerosis Obliterans of the Femoro-Popliteal Vessels

Typical intermittent claudication of the lower leg musculature usually results from occlusion of the superficial femoral artery, which is the most frequent isolated arterial occlusion. Subjective statements on the walking distance generally do not stand up to objective testing, so the walking ability should be determined by a "monitored walking distance." The patient walks 75 meters on the level (with a stopwatch) for 1 minute and registers S 1, start of pain; S 2, walking inhibited by pain; S 3, stopping due to pain. This is also a semiquantitative method, since subjective parameters enter into the result.

The decisive factor is the development of collaterals to compensate the occlusion. An incomplete stenosis of the femoral artery (up to 75% of the lumen) does not result in a change of the blood flow or diversion through collaterals even with light work. Under resting conditions the blood flow due to the poststenotic fall in pressure is more or less unchanged even when the lumen is narrowed by 95%. This is because of self-regulation by the distressed periphery (which is in a state of metabolic acidosis) via nervous and humoral regulatory mechanisms to dilate the supplying arterioles. In severe forms of disturbed blood flow, the dilatation of the arterioles can be so extreme that the "arteriolar elasticity" is lost. If the blood flow is hardly adequate under resting conditions, it collapses abruptly upon exertion.

This can be shown very well by venous occlusion plethysmography with the reacting hyperemia (Bartusch and co-workers, 1970). The resting blood flow does not distinguish persons with healthy vessels from those with diseased vessels. However, the values of reactive hyperemia after shutting off an artery for 3 minutes show a typical behavior which permits conclusions as to the reserve blood flow. A second quantitative procedure is ^{133}Xe tissue clearance.

E. Obliterative Atherosclerosis of the Peripheral Vessels

A lower leg type affecting one or more arteries is differentiated from a peripheroacral type with occlusion of the dorsal artery of the foot. Diabetic microangiopathy should be distinguished from these. These peripheral types are mostly inaccessible to surgical intervention, and are associated with diabetes mellitus to a greater extent than would be accounted for by coincidence. A cold feeling in the feet and toes is clinically prominent, and pains in the soles of the feet during walking are reported less frequently. There is a distinct tendency to early ulceration in the acral region.

Ratschow's (1936) test is important for the diagnosis of disorders of peripheral blood supply. The patient lies on his back, raises his legs straight, and performs circular movements or bending and stretching of the ankle. The patient can support his upper legs with the hands. The soles of the feet and the toes should be observed during the test. Even with moderate restriction of the arterial blood supply, there is whitening of the sole of the foot and individual toes. After ending the bending and stretching movements in the ankle, the patient sits up quickly and lets the legs hang down limply. In persons with healthy vessels there is even reactive hyperemia or filling of the vessels within 5 to 10 seconds. In persons with diseased vessels, this is delayed to an extent depending on how adequately an occlusion is compensated for by the collateral circulation.

V. Atherosclerosis of the Arterioles and Capillaries, Particularly of the Fundus of the Eye

The widely held view that a person is as old as his blood vessels has been described as incorrect by Bürger (1969) in his biomorphosis studies. He takes the contrary view, namely, that the changes in the terminal vessels determine the fate of the dependent tissue in an interaction between circulation and metabolism. He considers that the terminal vessels and their progressive thickening is a central problem in research into aging.

It is not yet certain whether the smallest vessels are involved in general atherosclerosis of the large vessels. Capillary sclerosis may be an obligate form of the general sclerosing process. On the other hand, involvement in the smaller arterial branches may be only insignificant. In general it can be said that the atherosclerotic process spares the small branches in the arterial periphery with an average diameter 500 to 50 μm if blood pressure is within the normal range. If blood pressure is high, the atherosclerotic process also extends into the smaller arteries. Hypertension not only promotes atherosclerosis, in general, but widens its distribution, carrying it deeper into the organ parenchyma. In investigations on 70 patients with severe atherosclerosis and 31 controls we found unequivocal functional and morphological changes in the terminal vessels in patients with general atherosclerosis (Mörl, 1971).

The findings in the fundus of the eye are of practical importance. Ophthalmologists have known for a long time that, with some reservations, it is possible to draw conclusions as to the situation in other vessel regions from the appearance of the fundus of the eye (Sauttner, 1960; Lieb, 1967). Lund (1964) showed that the retinal vessel system can be an indicator for atherosclerotic cardiovascular disease. There is a close relationship with the situation in the brain and also with the renal vessel system. Utermann and Klempien (1968) found in patients with atherosclerosis in the fundus of the eye, a total of 73% with simultaneous changes in the general vessel system. Patients who had visited the internist because of angiological complaints were found to have changes in the eye fundus in 72% of cases. With atherosclerosis on the vessels in the lower body region, the agreement was up to 85%. The fundus of the eye would thus be particularly suited for early detection of atherosclerosis because even the most minor changes are discernible. Goder (1968) histologically studied the temporal artery in cases of disturbed blood supply to the eye and established that these do not reflect the changes in other vessel areas, not even those of the brain. The one exception among the vessel areas studied was the ophthalmic artery, particularly, the nutritive optic vessels including the capillaries.

In summary it can be stated that the retina is a well-proved mirror of the general vessel system, allowing an early and accurate assessment of the severity of atherosclerosis. However, specific correlations must be viewed with some caution.

VI. Atherosclerosis of the Pulmonary Arteries

Primary pulmonary sclerosis (Ayerza disease) is relatively rare, as are hereditary forms. The secondary forms are more frequent; they are due to primary pulmonary hypertension of which the cause usually cannot be determined, and to active or passive lung congestion because of the cardiac defects. The increase in pulmonary hypertension in the presence of atherosclerosis of the small and intermediate pulmonary arteries because of plethora of the lungs is termed the "Eisenmenger reaction."

Atherosclerosis of the pulmonary arteries is viewed as a typical example of a high-pressure sclerosis arising without a corresponding elevation of the serum cholesterol level. This form of atherosclerosis is clearly due to wall damage caused by the high pressure; this damage then promotes subsequent lipid infiltration into the intima. The lung vessels increase in total mass due to a pronounced hyperplasia of the elastic media skeleton with a substantial modification of the original wall structure. A hypertonic overloading which has existed from birth onward is first observed as persistence of the fetal structural pattern in the intrapulmonary arteries. In pulmonary hypertension lasting for a long time, various changes develop in this persisting fetal wall structure. These changes were classified into six groups by Heath and Edwards (1958):

1. Medial hypertrophy
2. Excentric or concentric stenoses of the lumen by spreading of the intima (which has a low fiber content)
3. Increasing fibrosis of the thickened intima
4. Plexiform lesions, angiomatous changes, branches of hypertrophied' muscular arteries which are wide, resembling veins
5. Hemosiderin deposits from hemorrhages
6. Fibrin-rich wall necroses of small arteries and arterioles with inflammatory reactions

Sclerosis of the pulmonary arteries is manifested clinically by the signs of pulmonary high blood pressure, with subjective appearance of shortness of breath, and cyanosis. Objectively, an accentuation of the pulmonic component of the second heart sound can be detected. Furthermore, there is a rarefication of the pulmonary vessels in the chest X ray, chiefly in the periphery; in the ECG there is increasing right deviation of the QRS axis, and right heart hypertrophy. In the advanced phase, a chronic cor pulmonale syndrome is found.

VII. Diabetic Angiopathy

Since the introduction of insulin in therapy of diabetes, the mortality is no longer determined by diabetic coma but by cardiovascular disease. While only about 1% of all diabetics die in coma today, cardiovascular disease constitutes the most frequent cause of death among diabetics (approximately 80%). This diabetic angiopathy was first described by Bürger (1954a) (also independently by Lundbeak in 1954) and as diabetic microangiopathy by Ashton in 1958. It is readily divided into two large groups, diabetic macroangiopathy and diabetic microangiopathy.

Diabetic macroangiopathy constitutes the atherosclerotic changes in the large and medium-sized vessels of diabetic subjects. In terms of pathological anatomy, this atherosclerosis is not distinguishable from that of the nondiabetic, although there have been occasional reports on chemical differences in these atherosclerotic vessels (Hevelke, 1954, 1956; Randerath and Diezel, 1958, 1959). Autopsy findings and controls on the clinical course show clearly that generalized atherosclerosis in diabetics appears earlier, more frequently, and with a different sex distribution. In particular, it has been shown that coronary sclerosis and coronary heart disease are manifested earlier and more frequently in diabetics. Diabetic women are also affected by this. The proportion (up to 40%) of silent myocardial infarction is particularly high in diabetics, possibly due to diabetic neuropathy (Bradley and Schonfeld, 1962; Mörl, 1964; Bradley, 1971). Friedberg (1958) sees another mechanism in the chronically reduced blood flow in the intramural vessels with limited collateral circulation and continuous myocardial hypoxia; a complete occlusion does not result in prompt and complete ischemia and thus causes only slight pain or none at all. The prognosis after surviving a myocardial infarct is also very much poorer in diabetics.

Cerebral atherosclerosis is also more frequent in diabetics (1.8 times as frequent in men; twice as frequent in women). In the Joslin Clinic from 1960 to 1964, 12.4% of the diabetics died as a result of atherosclerosis of the cerebral arteries (Entmacher et al., 1964). Similar figures are quoted from Sweden, where 15.4% of male and 17.1% of female diabetics died from the results of cerebral atherosclerosis, while in the nondiabetic group the proportion was only 6.1% (Larsson, 1967). It is clinically important that hypoglycemias can evoke apoplectic manifestations which are often protracted in patients with cerebral sclerosis.

Peripheral atherosclerosis, especially of the lower extremities, is more frequent than in nondiabetics. The arteries of the lower leg and the distal sections of the femoral and popliteal arteries are particularly affected. Similar to myocardial infarction, arterial occlusion is occasionally more discrete and less symptomatic in the diabetic. The causes for the higher incidence of atherosclerotic complications in vessels of diabetics is largely obscure even today. According to Schettler and Wahl (1975), there are two mechanisms which may explain the higher incidence of atherosclerosis in diabetics: (1) diabetes favors atherogenesis directly and (2) diabetes is more frequently associated with other atherosclerosis risk factors or induces them, thus only indirectly promoting the development of atherosclerosis.

The first hypothesis would be supported if atherogenesis were related to the duration of the manifest diabetic disturbance of metabolism and how well it has been controlled. This is, indeed, valid for juvenile-type diabetes, in which the morbidity from atherosclerosis increases with the duration of the diabetic condition and the quality of metabolic control. However, such a relationship does not apply to adult-type diabetes, in which clinical manifestation of atherosclerosis (e.g., myocardial infarction or cerebral insult) often coincides with manifestation of the diabetes. The atherosclerosis must, therefore, have already begun in an early stage of diabetes. In this type of diabetes, atherosclerosis is not correlated with the duration of diabetes and probably also not with the success of metabolic control implying that this is expressed by normalization of blood sugar values. There are probably, in this type of diabetes, frequent risk factors of atherosclerosis which are associated with diabetes ("risk packet" according to Schettler, 1969).

The angiopathy which is actually specific to diabetes is diabetic microangiopathy (also termed capillaropathia diabetica). It is characterized by generalized thickening of the capillary basement membrane. The age at the time the diabetes appears obviously does not play a substantial role (Haupt and Beyer, 1974). In diabetic microangiopathy the incidence increases with increasing duration of the diabetes. The incidence is higher if long-term metabolic control is poor (Mohnike, 1959, Siege et al., 1964; Marks and Krall, 1971). It is not directly proportional to the severity of the diabetes (Marx, 1963; Anschütz, 1966). The layering of the basement membrane arises through optically empty cracks which are probably caused by infiltration of fluid. Moreover, there are increased inclusions, which are taken to be lipid particles, protein precipitates, and crystals

(Fuchs and Schwarnweber, 1968). In the walls of the smaller skin vessels of diabetics, more globulins are found than in metabolically healthy subjects (Fuchs, 1974). The basement membrane can thus probably be thickened by swelling, true incorporation of substance, or reduced degradation.

The special clinical position of specific diabetic microangiopathy is unmistakable. It is characterized by an almost exclusively acral beginning of the necroses and an increased susceptibility to infection. A further striking feature is a disproportion between the often good passability of the great arteries and the extent of peripheral necrosis, as well as the rarity of intermittent claudication as an initial sign of insufficient blood flow in the diabetic. Four specific clinical pictures can be distinguished: glomerulonecrosis (Kimmelstiel and Wilson, 1936); retinopathy; peripheral diabetic microangiopathy with gangrene of the lower extremities; and neuropathy.

References

Abrams, L., and Adams, F. (1969a). *New Engl. J. Med.* **281**, 1276.
Abrams, L., and Adams, F. (1969b). *New Engl. J. Med.* **281**, 1336.
Anschütz, F. (1966). *Deut. Med. J.* **17**, 634.
Anschütz, F. (1968). *Hippokrates* **39**, 170.
Anschütz, F. (1973). "Schmerzanalyse." Thieme, Stuttgart.
Ashton, N. (1958). *Advan. Opthalmol.* **8**, 1.
Barrett, D. L. (1968). *Ohio State Med. J.* **65**, 830.
Bartusch, M., Mörl, H., and Preuss, E.-G. (1970). *Z. Kreislaufforsch.* **59**, 60.
Bernsmeier, A., and Gottstein, U. (1965). *Internist* **5**, 207.
Bernsmeier, A., and Held, K. (1970). *Z. Kreislaufforsch.* **59**, 97.
Bradley, R. F. (1971). *In* "Cardiovascular Disease in Joslin's Diabetes Mellitus" (A. Marble, P. White, R. F. Bradley, L. P. Krall, eds.), p. 477. Lea & Febiger, Philadelphia, Pennsylvania.
Bradley, R. F., and Schonfeld, A. (1962). *Geriatrics* **17**, 322.
Büchner, F. (1960). "Spezielle Pathologie." Urban & Schwarzenberg, München-Berlin.
Bürger, M. (1954a). "Angiopathia Diabetica." Thieme, Stuttgart.
Bürger, M. (1954b). *Verh. Deut. Ges. Inn. Med.* **60**, 849.
Bürger, M. (1960). "Altern und Krankheit als Problem der Biomorphose," 4th ed. Thieme, Leipzig.
Chandler, A. B., Chapman, J. M., Erhardt, L. R., Roberts, W. C., Schwartz, C. J., Sinapius, D., Spain, D. M., Sherry, S., Ness, P. M., and Simon, T. L. (1974). *Amer. J. Cardiol.* **34**, 823.
Chapman, J. M., Loveland, D. B., Goerke, L. S., Jacobson, G., and Rotrock, W. J. (1960). *J. Chronic Dis.* **12**, 239.
Christian, P. and Hahn, P. (1972). *Internist* **13**, 421.
Classification of Atherosclerotic Lesions (1958). Report of Study Group on Definition of Terms. *WHO Tech. Rep. Ser.* **143**, 4.

Davia, J. E., Hallal, F. J., Cheitlin, M. D., Gregoratos, G., McCarty, R., and Foote, W. (1974). *Amer. Heart J.* **87**, 689.

DeBakey, M. E., and Cooley, D. A. (1953). *Surg. Gynecol. Obstet.* **97**, 257.

DeBakey, M. E. (1962). *J. Cardiovasc. Surg.* (*Torino*) **3**, 12.

DeBakey, M. E. (1974). "Operative Ergebnisse der Aortenaneurysmen." Angiologen-kongress, Ulm.

DeBakey, M. E., Crawford E. S., Cooley, D. A., Morris, G. C., Royster, F. S., and Abbott, W. P. (1964). *Ann. Surg.* **160**, 622.

Dennis, E. W., Kinard, S. A., Jr., McCall, B. W., DeBakey, M. E., Howell, J. F., and Garett, H. E. (1965). *Progr. Cardiovasc. Dis.* **7**, 544.

Derrick, J. R., and Logan, W. D. (1958). *Surgery* **44**, 823.

Deutsch, E. (1974). personal communication.

Doerr, W. (1972). *Verh. Deut. Ges. Inn. Med.* **78**, 944.

Dorndorf, W., and Gänshirt, H. (1972). "Der Hirnkreislauf," p. 512. Thieme, Stuttgart.

Drake, W. E., Jr., and Drake, M. A. L. (1966). *Circulation* **33/34** (Suppl. III), 90.

Eliot, R. S., and Edwards, V. J. (1964). *Circulation* **30**, 611.

Entmacher, P. S., Root, H. F., and Marks, H. H. (1964). *Diabetes* **13**, 373.

Erdheim, J. (1929). *Virchows Arch. Pathol. Anat.* **273**, 544.

Estes, J. E. (1950). *Circulation* **2**, 258.

Friedberg, C. K. (1958). "Diseases of the Heart." Saunders, Philadelphia, Pennsylvania.

Fuchs, U. (1974). "Angiologie." p. 638. Thieme, Stuttgart.

Fuchs, U. and Scharnweber, W. (1968). *Virchows Arch. Pathol. Anat.* **343**, 276.

Gänshirt, H. (1972). *Verh. Deut. Ges. Inn. Med.* **78**, 416.

Goder, G. (1968). "Abhandlungen aus dem Gebiet der Aupenheilkunde," Vol. 36. Thieme, Leipzig.

Goldblatt, H., Lynch, J., Hanzal, R. F., and Summerville, W. W. (1934). *J. Exp. Med.* **59**, 347.

Gore, J., and Collins, D. P. (1960). *Amer. J. Clin. Pathol.* **33**, 416.

Groom, D., McKee, E. E., Adkins, W., Pean, V., and Hudicourt, E. (1964). *Ann. Intern. Med.* **61**, 900.

Gross, R. E. (1953). *Circulation* **7**, 757.

Gsell, O. (1928). *Virchows Arch. Pathol. Anat.* **270**, 1.

Hahn, P. (1971). "Der Herzinfarkt in psychosomatischer Sicht." Vandenhock und Ruprecht, Göttingen.

Hartwich, A. (1930). *Z. Ges. Exp. Med.* **69**, 462.

Hasse, H. M. (1974). "Angiologie." Thieme, Stuttgart.

Haupt, E., and Beyer, J. (1974). In "Diabetologie in Klinik und Praxis" (von H. Mehnert and K. Schöffling, eds.). p. 357. Thieme, Stuttgart.

Heath, D., and Edwards, J. E. (1958). *Circulation* **18**, 533.

Heberer, G., Rau, G., and Löhr, H. H. (1966). "Aorta und große Arterien." Springer, Berlin-Heidelberg-New York.

Hehl, F. J. (1974). "Persönlichkeitsskalen-System 25." Beltz, Weinheim.

Hehl, F. J. and Hehl, R. (1975). *In* "Dokumentation und Information im Dienste der Gesundheitspflege" (D. Nacke, ed.), in press. Schattauer, Stuttgart.

Hering, H. E. (1917). "Der Sekundenherztod mit besonderer Berücksichtigung des Herzkammerflimmerns." Springer, Berlin.

Hess, H. (1959). "Die obliterierenden Gefäßerkrankungen." Urban & Schwarzenberg, München-Berlin.

Hevelke, G. (1954). Z. Alternsforsch. **8**, 219.

Hevelke, G. (1956). Deut. Arch. Klin. Med. **203**, 528.

Hild, R. (1966). In "Pathophysical and Clinical Aspects of Lipid Metabolism" (G. Schettler and R. Sanwald, eds.), p. 207. Thieme, Stuttgart.

Hollenhorst, R. W. (1962). Trans. Amer. Acad. Ophthalmol., p. 166.

Humphries, V., DeWolfe, G., Young, J. R., and Le Fevre F. A. (1963). In "Fundamentals of Vascular Grafting" (S. Wesolowski and C. Dennis, eds.). McGraw-Hill, New York.

Irvine, W. T., Luck, R. J., Sutton, D., and Walpita, P. R. (1963). Lancet i, 1177.

Joyce, J. W., Fairbairn, J. F., Kincaid, O. W., and Juergens, J. L. (1964). Circulation **29**, 176.

Jungblut, R., and Günther, D. (1968). Zentrabl. Chir. **93**, 329.

Kannel, W. B. (1973). Kardiol. Aktuell **1**, 4.

Kaplan, N. (1971). "Clinical Hypertension" Med. Press, New York.

Kappert, A. (1972). "Lehrbuch und Atlas der Angiologie," 6th ed. Huber, Bern-Stuttgart-Wien.

Kassirer, J. P. (1969). New Engl. J. Med. **280**, 812.

Kimmelstiel, P., and Wilson, C. (1936). Amer. J. Pathol. **12**, 83.

Kreiger, K. (1969) Dissertation, Heidelberg.

Larsson, T. (1967). Thule Int. Symp. Stroke, Stockholm, pp. 15–40. Nordiske Bokhandelus Förlag.

Leriche, R. (1923). Bull. Mem. Soc. Nat. Chir. **49**, 1404.

Leymarios, J. (1960). Thèse, No. 695, Paris.

Lieb, W. A. (1967). Med. Welt **18**, 599.

Loogen, F. (1972). Verh. Deut. Ges. Inn. Med. **78**, 984.

Lund, O.-E. (1964). "Diagnostik der Arteriosklerose." Karger, Basel.

Lundbeak, K. (1954). Schweiz. Med. Wochenschr. **84**, 538.

Marks, H. H., and Krall, L. P. (1971). In "Joslin's Diabetes Mellitus" (A. Marble, P. White, R. F. Bradley, and L. P. Krall, eds.), p. 209. Lea & Febinger, Philadelphia, Pennsylvania.

Marshall, R. J., and Mantini, E. L. (1965). Circulation **31**, 249.

Martorell, F., and Fabré, J. (1944). Med. Clin. **2**, 26.

Marx, H. (1963). Verh. Deut. Ges. Kreislaufforsch. **29**, 244.

Mathur, K. S., Kashyap, S. K., and Kumar, Y. (1963). Circulation **27**, 929.

Merill, S. L., and Pearce, M. L. (1971). Amer. Heart J. **81**, 48.

Meyer, W. W. (1968). Deut. Med. Wochenschr. **93**, 2080.

Mikkelsen, W. P. (1957). Amer. J. Surg. **94**, 262.

Mönckeberg, J. G. (1903). Virchows Arch. Pathol. Anat. **171**, 141.

Mörl, H. (1964). Virchows Arch. Pathol. Anat. **337**, 383.

Mörl, H. (1965a). Z. Kreislaufforsch. **54**, 725.

Mörl, H. (1965b). Arch. Kreislaufforsch. **45**, 1.

Mörl, H. (1971). "Atherosklerotische Gefäßerkrankungen und Mikrozirkulation." Barth, Leipzig.

Mörl, H. and Haupt, V. (1972). Zentrabl. Allg. Pathol. Pathol. Anat. **115**, 579.

Mohnike, G. (1959). Medizinische, p. 32.

Morris, G. C., and DeBakey, M. E. (1966). In "Surgical Therapy of the Carotid and Vertebral Insufficiency" (C. Kulenkampff and W. Dorndorf eds.), p. 168. Thieme, Stuttgart.

Morris, G. C., Crawford, E. S., Cooley, D. A., and DeBakey, M. E. (1962). Arch. Surg. **84**, 95.

Morris, G. C., DeBakey, M. E., and Zanger, L. C. C. (1966). *Surg. Clin. N. Amer.*, p. 931.

Münster, W., Wieny, L., and Porstmann, W. (1966). *Deut. Med. Wochenschr.* **91**, 2073.

Nüssel, E., Jahn, H., Aranasov, A. (1968). *Arch. Kreiblaufforsch.* **57**, 113.

Ortner, N. (1902). *Wien Klin. Wochenschr.* **15**, 1166.

Paul, O. (1974). *In* "The Myocardium: Failure and Infarction" (E. Braunwald ed.), p. 273. HP Publ, New York.

Porstmann, W. (1971). unpublished data.

Preuss, E. G., Häusler, M., and Seige, K. (1969). *Deut. Gesundheitsw.* **24**, 577.

Randerath, E., and Diezel, P. B. (1959) "Diabetes Mellitus," III. Thieme, Stuttgart.

Randerath, E., and Diezel, P. B. (1958). *Deut. Arch. Klin. Med.* **205**, 523.

Rasch, G. (1961). *Proc. 4th Berkeley Symp. Math. Stat. Probability, Berkeley.* Univ. of California Press, Berkeley, California.

Ratschow, M. (1936). *Zbl. Med.* p. 873.

Ratschow, M. (1965). "Handbuch der praktischen Geriatrie." Enke, Stuttgart.

Rau, G. (1970). "Lokalisierende Faktoren für Arterien- und Venenverschlüsse." (H. Hess, ed.), p. 102. Schattauer, Stuttgart.

Reifenstein, G. H., Levine, S. A., and Gross, R. E. (1947). *Amer. Heart J.* **33**, 146.

Reinich, M., Holling, H. E., Roberts, B., and Toole, J. F. (1961). *New Engl. J. Med.* **265**, 878.

Rosenman, R. H., Friedmann, M., Jenkins, D., Strauss, R., Wurm, M., and Kositschek, R. (1967). *Amer. J. Cardiol.* **19**, 776.

Ryan, E. A., Spittell, J. A., and Kincaid, O. W. (1964). *Postgrad. Med.* **36**, 77.

Sautter, H. (1960). 62 *Tag. Deut. Ophthal. Ges. Bergmann München*, p. 23.

Schettler, G. (1969). *Wien Klin. Wochenschr.* **81**, 581.

Schettler, G. (1972). *Deut. Med. Wochenschr.* **97**, 533.

Schettler, G., and Nüssel, E. (1974). *Deut. Med. Wochenschr.* **99**, 2003.

Schettler, G., and Wahl, P. (1975). "Die diabetische Angiopathie Cardiologia auf d'Oggi." in press

Schettler, G., and Wollenweber, J. (1969). "Klinik der Gegenwart" (H. E. Bock, W. Gorok, and F. Hartmann, eds). Urban & Schwarzenberg, München-Berlin.

Seige, K., Blatz, G., Schmiedel, H., Schröfel G., and Riemer, W. (1964). *Abh. Deut. Akad. Wiss. Berlin* **3**.

Siekert, R. G., Millikan, C. H., and Whisnant, J. P. (1964). *Ann. Intern. Med.* **61**, 64.

Simon, A. B., and Alonzo, A. A. (1973). *Arch. Intern. Med.* **137**, 163.

Simon, M., Rabinov, K., and Horenstein, S. (1962). *Clin. Radiol.* **13**, 201.

Sproul, G. (1963). *Circulation* **28**, 259.

Stammler, J. (1974). *In* "The Myocardium: Failure and Infarction" (E. Braunwald eds.), p. 279. HP. Publ., New York.

Stefanini, P., Fiorani, P., Benedetti-Valentini, F., Jr., Mercati, U., and Pistolese, G. R. (1964). *Angiology* **15**, 524.

Strandness, D. E., Jr. (1974). *Proc. Int. Symp. 3 Atherosclerosis* p. 707.

Strong, J. P., and McGill, H. C. (1963). *Exp. Mol. Pathol. Suppl.* **1**, 15.

Takayasu, M. (1908). *Acta Soc. Ophthalmol. Jap.* **12**, 554.

Titus, J. L., Oxman, H. A., Conolly, D. C., and Nobrega, F. F. (1973). *Singapore Med. J.* **14**, 291.

Toole, J. F. (1964). *Lancet* i, 872.

Ule, G., and Kolkmann, F. W. (1972). *In* "Der Hirnkreislauf" (H. Gänshirt, ed.), p. 47. Thieme, Stuttgart.

Utermann, D., and Klempien, E. J. (1968). *Advan. Ophthalmol.* **24,** 201.

Vollmar, J. (1974). *Deut. Med. Wochenschr.* **99,** 465.

Wenz, W. (1972). *Verh. Deut. Ges. Inn. Med.* **78,** 561.

Whisnant, J. P., Martin, M. J., and Sayre G. P. (1961). *Arch. Neurol. (Chicago)* **5,** 429.

Widmer, L. K. (1963). *Bibl. Cardiol. (Basle)* **13,** 67.

Widmer, L. K., and Duchosal, F. (1968). "Koronare Herzkrankheit und chronischer Verchluss der Gliedmaßenarterie" (A. Kapper, P. Warbel, and A. Widmer, eds.), p. 23. Huber, Bern-Stuttgart.

Widmer, L. K., Cikes, M., Kolb, P., Ludin, H., Elke, M., and Schmitt, H. E. (1967). *Schweiz. Med. Wochenschr.* **97,** 102.

Widmer, L. K., Glaus, L., and Da Silva, A. (1972). *Verh. Deut. Ges. Inn. Med.* **78,** 408.

Wilhelmsen, L. (1974). *Proc. Int. Symp. Atherosclerosis III* p. 705.

Wollenweber, J., Sheps, S. G., and Davis, G. D. (1968). *Amer. J. Cardiol.* **21,** 60.

Wood, P. (1968). "Diseases of the Heart and Circulation," 3rd ed. Eyre & Spottiswoode, London.

Wylie, E. J., and Ehrenfeld, W. K., (1970). "Extracranial Occlusive Cerebrovascular Disease. Diagnosis and Management." Saunders, Philadelphia, Pennsylvania.

Zschoch, H.-J. (1966). *Ergeb. Allg. Pathol. Pathol. Anat.* **47,** 59.

Zukel, W. J., Cohen, B. M., Mattingly T. W., and Hrvbec Z. (1969). *Amer. Heart J.* **78,** 159.

Chapter 2

ESSENTIAL FATTY ACIDS—
RECENT DEVELOPMENTS

Roslyn B. Alfin-Slater and Lilla Aftergood

School of Public Health,
University of California,
Los Angeles, California

I. Introduction

Of all the fatty acids that occur combined with glycerol as triglycerides in foods, only a small number are essential for the nutritional well being of animals, including man. These fatty acids are polyunsaturated, and have a specific structure that cannot be synthesized by animals. Linoleic acid, 18:2 ω6, is probably the most abundant of these essential polyunsaturated fatty acids. It occurs in high concentrations in many vegetable oils, e.g., corn, cottonseed, safflower, and soybean. Arachidonic acid, 20:4 ω6, occurs in small amounts in animal lipids; however, this essential fatty acid (EFA) can be synthesized from linoleic acid.

EFA deficiency has been observed in many species of animals (Alfin-Slater and Aftergood, 1968) including man (Söderhjelm *et al.*, 1970). EFA deficiency is characterized in animals by symptoms which include poor growth, impaired reproductive performance, a characteristic dermatitis, lowered caloric efficiency, decreased resistance to stress, and impairment in lipid transport. Although studies in humans are limited, borderline deficiency symptoms, including derangements in lipid transport, are probably not uncommon. The metabolic processes in which essential fatty acids participate to alleviate the observed deficiency signs are the subject of continuing research.

Some of the other often described typical deficiency symptoms have been seen in humans only in rare and special instances, e.g., in patients given prolonged intravenous feeding with fat-free solutions, and in infants administered a low fat diet free of polyunsaturated fatty acids (PUFA). Since the body stores of essential nutrients in infants are lower than in adults, and since these requirements for PUFA to support active growth are higher than in adults, there is a greater possibility of encountering PUFA deficiency in infants than in older children and/or adults. On the other hand, intravenous feeding is often necessary in patients after surgery; however, the use of fat emulsions in preparations for intravenous feeding is still in the experimental stages.

II. The Requirement for Essential Fatty Acids

The need for linoleic acid by human subjects was first demonstrated by Hansen and co-workers (1958, 1962; Adam *et al.*, 1958), and early studies indicated that the minimal requirement was approximately 1.4% of the caloric intake (Holman *et al.*, 1964). This figure is close to the requirement which has been established for a variety of mammals (Holman, 1960). As a result, the Recommended Daily Allowance (RDA) for essential fatty acids for adult humans is given as 1 to 2% of total calories.

In the normal organism, 18:2, linoleic acid, is converted into 20:4, arachidonic acid (Mead, 1961). When the animal is EFA deficient, an eicosatrienoic acid, 20:3 ω9, accumulates in the lipid fractions of various organs. This acid originates from oleic acid (Fulco and Mead, 1959), and as a result the extent of EFA deficiency may be estimated by determining the ratio of the fatty acids 20:3 ω9/20:4 ω6. This ratio is low when essential fatty acids are present in sufficient quantities and increases as the availability of EFA decreases and a state of EFA deficiency develops (Holman *et al.*, 1964). A value below 0.4 usually is indicative of EFA sufficiency. However, a low figure does not necessarily mean that the EFA levels are satisfactory; it may mean that the polyunsaturated linolenic acid (18:3) and its elongation products, 20:5 ω3 and 22:6 ω3, may have interfered with the formation of eicosatrienoic acid, even though EFA is not available. On the other hand, it has been reported that some odd-chain, unsaturated fatty acids such as 17:2 ω5 and 19:4 ω5 exhibit EFA activity (Schlenk and Sand, 1967) even though they are not converted into arachidonate; in this case, there also would be a low 20:3/20:4 ratio. In general, a ratio of total trienes to total tetraenes may be a better measure of EFA status.

III. EFA and Tissue Lipids in Infants and Children

The fatty acid composition of tissue lipids as a function of diet has been studied by various investigators. Hansen et al. (1969) studied the relationship between diet and the fatty acid composition of umbilical cord serum, infant serum, and maternal serum. They found no differences between the fatty acid compositions of the serum from premature and full-term babies when both groups were fed on breast milk. Insull et al. (1959) found that the fatty acid composition of breast milk reflected the nature of the dietary fat, without altering its volume or fat output. When the subject is in energy equilibrium, the milk fat resembles the dietary fat, but under conditions of caloric deficiency the milk fat resembles depot fat.

The fatty acid composition of depot fat has also been found to depend on the age of the child, particularly with respect to 18:2. In breast milk, levels of 18:2 in the fat vary from 7.1 to 16.0% of total fatty acids whereas in the fat in cow's milk the amount of 18:2 ranges from 2.4 to 4.3%. Lipids in umbilical cord serum had higher 20:4 and 20:3 concentrations, but lower 18:2 than did maternal serum lipids. Both 18:2 and 20:4 in serum lipids were reduced during the first 2 days of life but thereafter 18:2 increased significantly with age from the second day to the end of the first week.

Similarly, Clausen and Friis-Hansen (1971) reported that the fatty acids of serum undergo pronounced changes during the neonatal period. At birth the arachidonic acid content is higher, and the linoleic acid content is lower, than in adults. Arachidonic acid decreases during the early days after birth, whereas linoleic acid decreases during the first 2 days and then increases.

A rapid development of EFA deficiency in seven infants under 6 months of age given prolonged intravenous alimentation was recently reported by Paulsrud et al. (1972). The criterion of deficiency was the alteration in pattern of serum polyunsaturated fatty acids (PUFA). Since serum triglycerides contain only traces of acids of chain lengths longer than 18 carbons, the effects of EFA deficiency were limited to changes in the proportions of 16:1, 18:1, and 18:2. The prolonged deprivation of EFA precipitated a dermatitis in one of the patients after 3 months of intravenous feeding. Following autopsy, a large proportion of 20:3 ω9 was found in the phosphatidylcholine fraction of tissue lipids. Ratios of 20:3/20:4 varied from 1.1 (spinal cord) to 6.0 (pancreas and serum). The depletion of PUFA was least marked in nerve tissue, which might be expected to have the lowest turnover of these polyunsaturated fatty acids. These investigations suggested that to be effective in reversing the EFA deficiency, the dose level of 18:2 had to be greater than 2% of calories and the period of supplementation had to exceed 11 days, or both. The analysis of the fatty acids of the major lipid classes of serum was considered to be a good diagnostic tool as far as EFA deficiency was concerned. In one case, the triene–tetraene ratio of serum phospholipids was as high as 18.

Caldwell et al. (1972) also reported a case of severe EFA deficiency in an infant receiving prolonged fat-free alimentation. Scaly skin lesions, growth retardation, and altered plasma fatty acid pattern were observed along with thrombocytopenia (a decrease in the number of blood platelets). Parenteral administration of a soybean oil emulsion was effective in correcting the deficiency symptoms. Previously, Pensler et al. (1971) had demonstrated EFA deficiency in three children after prolonged parenteral fat-free alimentation. Changes in plasma fatty acid patterns included a reduction in 18:2, 18:3, and 20:4 and increases in 18:1, 16:1, and 20:3.

A study of 26 premature infants fed diets differing in 18:2 content for 40 days showed a close correlation between the dietary fat and serum and depot fat fatty acids. However, milk formulas containing 20% of vegetable fat produced 18:2 levels in serum and adipose tissue similar to those found in babies fed mother's milk (Ballabriga et al., 1972).

The minimal requirement of linoleate for infants seems to be about 1.4% and there seems to be no special advantage in giving linoleate at more than 4% of the daily caloric intake (Söderhjelm et al., 1970), although previously it was thought that larger amounts of linoleate (4–5% of calories) could produce optimal effects on infant feeding.

Kekomaki (1970), in discussing food requirements of normal children, mentions that although linoleic acid is considered to be an essential nutrient, it is related to energy metabolism rather than to the building up of new tissues.

Mendy et al. (1970) reported that the 20:3/20:4 ratio of total serum lipids and particularly that of serum phospholipids is a good indicator of EFA status in infants. These investigators consider that the recommended value for this ratio should be below 0.1, a value which occurs in breast-fed infants.

Olegard and Svennerholm (1971) studied the effects of diet on plasma and red cell phosphoglycerides in 3-month-old infants. Feeding linoleate-fortified formulas resulted in higher levels of 18:2 in plasma lipids. The investigators raise the question as to whether the dietary intake of 18:2 by breast-fed infants is sufficient, even though the content of EFA in human milk is approximately 7 times that found in cow's milk fat (Söder-hjelm et al., 1970).

Galli and Spagnuolo (1974) also gave increasing amounts of linoleate to lactating rats and found increased amounts of polyunsaturated fatty acids in triglycerides in the collected milk. However, milk phospholipids responded with high levels of the ω6 acids as a result of a low dietary linoleic acid intake and vice versa, suggesting a control mechanism in the secretion of polyunsaturated fatty acids in milk.

Recently, Schubert (1973) reviewed the nutritional requirements for fat, cholesterol, and EFA in infants and older children, and discussed the benefits and possible risks of alterations in intake of the dietary fat. This investigator confirmed a previous report giving the minimal requirement for linoleic acid as 1.4%; this corresponds to a concentration comparable to that in human milk (Combes et al., 1962). Schubert points out that a better screening and prospective treatment is needed for the estimated 5–7% of persons who are at risk of hyperlipoproteinemia, rather than recommending large-scale tampering with the diet of American children where there is the risk of producing dietary imbalances by making drastic changes in the usual diet.

IV. EFA Deficiency in Adults

Malabsorption of fat must inevitably lead to EFA depletion since an inadequate intake cannot be compensated for by synthesis. In support of this statement, when the fatty acids of plasma lipid fractions were determined in patients with steatorrhea, it was noted that the percentage of 18:2 was significantly lower. In fact in some of the patients, an EFA deficiency characterized by the appearance of the presence of abnormal eicosatrienoic acid in lecithin was found (Press et al., 1972). Similar findings were observed in adults with malabsorption after intestinal resection (Press et al., 1974; Wapnick et al., 1974).

In an instance where a patient had all but 40 cm of his small intestine removed and was maintained on intravenous therapy without fat for 100 days, his serum phospholipids contained 10% of Δ5,8,11-eicosatrienoic acid and he developed a skin rash (Collins *et al.*, 1971). Intravenous administration of 22.8 gm of 18:2 per day for 43 days corrected the dermatitis and increased the concentration of the 20:4 acid in serum while lowering the 20:3. These investigators suggested that an adult male requires at least 7.5 gm/day of linoleate and that the proportion of triglycerides carried by the VLDL (very low-density lipoproteins) is lowered in the absence of linoleic acid.

Recently, it was reported (Richardson and Sgoutas, 1975) that four undernourished adults developed EFA deficiency as a result of a 6- to 8-week period of fat-free total parenteral nutrition. Deficiency developed more rapidly in the two younger, more severely undernourished individuals (ages 16 and 36) than in the two older patients (ages 62 and 76). In the severe deficiency, hepatomegaly and increased liver enzyme activities were observed. Oral supplementation with linoleic acid (as safflower oil) reversed the EFA deficiency and its symptoms, i.e., the elevation of liver enzyme activity. Contrary to the results usually obtained in rats, increased rather than decreased serum triglyceride levels accompanied the development of the fatty livers. Possibly a more prolonged deficiency is required before hepatic secretion of triglyceride becomes impaired.

It has been observed that following a severe thermal injury, profound alterations similar to those seen in EFA deficiency in the lipid composition of erythrocytes may occur. Helmkamp *et al.* (1973) have described a significant reduction in the levels of linoleate, arachidonate, and docosahexaenoate in red cell membrane phospholipid in five severely burned humans. Concomitant increases in palmitate, oleate, and C24 fatty acids accompanied these decreases in EFA. Normal fatty acid patterns were reestablished following an infusion of fat emulsion.

V. EFA Transport

Under certain conditions (e.g., fasting) a considerable portion of fatty acids leaving adipose tissue as free fatty acids is incorporated in the triglycerides of the liver. Some of these subsequently leave the liver as VLDL, which in addition to the chylomicrons formed after absorption of dietary fat, are the major sources of plasma triglycerides. In studies of fatty acid patterns in human adipose tissue, liver, and plasma during starvation, it was found (Laurell and Lundquist, 1971) that the level of linoleic acid in plasma tended to rise during its passage through the transport system. The authors suggest that this finding confirms the observation of Fredrickson and Gordon (1958), who reported that in fasting men 18:2 was oxidized less and recycled more as free fatty acid than were palmitic and oleic acids.

A lack of a dietary source of EFA tends to initiate their release from tissue stores. This may explain why both dietary saturated fat and/or cholesterol, which preferentially uses EFA for esterification (Aftergood and Alfin-Salter, 1967), accelerate EFA deficiency. The growth of rats fed diets deficient in EFA is suppressed even more by addition of saturated fat to the fat-free diet (Deuel et al., 1955; Peifer and Holman, 1959). It is also possible that EFA are required for proper metabolism or utilization of saturated acids.

VI. EFA and Vitamin E Interrelationships

The double bonds of polyunsaturated fatty acids are extremely susceptible to oxidation. The trend toward the increased ingestion of polyunsaturated fatty acids as a factor in the regulation of serum cholesterol levels and in providing protection against coronary heart disease has provoked studies to measure the effectiveness of antioxidants in the prevention of lipid oxidation. The antioxidants most effective at the cellular level are the tocopherols, and investigations on polyunsaturated fatty acid–tocopherol interrelationships have led to elucidation of mechanisms by which the most common of the tocopherols, α-tocopherol, or vitamin E, exerts its activity. Vitamin E has been shown to be an obligatory cofactor in a variety of dehydrogenation reactions, particularly those reactions in which heme enzymes are active. However, since the discovery of vitamin E (Evans and Bishop, 1922), it was considered that probably one of the most important functions of vitamin E is its ability to prevent oxidation of essential, easily oxidizable substances in cellular systems and to prevent the formation of lipoperoxides which may be toxic (Schwarz, 1961).

Tappel (1965) has presented evidence for the occurrence in vivo of damaging lipid peroxidation products in the vitamin E-deficient animals, and Pritchard and Singh (1960) found decreased levels of PUFA in heart, liver, adrenals, and plasma of vitamin E-deficient rats, which they attributed to lipid peroxidation. These findings, however, were not confirmed by Bunyan et al. (1967).

In 1969, exploring his free radical theory of aging, Harman indicated that in some cases increasing the unsaturation of dietary fat without also increasing antioxidants tended to shorten the mean life span of mice; this could be prevented by giving antioxidants. This investigator concluded that the oxidation products of the PUFA may contribute to the development of age-associated degenerative diseases. This theory needs further confirmation.

In support of this theory, Chen (1974) found increased liver lipid peroxidation with age which could only partially be alleviated by "normal" vitamin E supplementation suggesting an increased requirement for vitamin E by the older animals.

A possibly related finding is that of Harland et al. (1973) in which cholesteryl linoleate hydroperoxides isolated from atheromatous plaques appeared to be similar to those produced by autoxidation.

As a result of these and other experiments, it has been suggested (Horwitt, 1959, 1962) that as the intake of PUFA increases so should the intake of vitamin E. Fortunately most of the natural dietary sources of PUFA are also excellent sources of active forms of vitamin E.

Holman (1970) has also concluded that the need for vitamin E is related to the amount of PUFA in the diet and tissues, and that the vitamin E/PUFA relationships must be considered when the nutritive adequacy of diets high in PUFA is being evaluated.

A ratio in which milligrams of α-tocopherol to grams of polyunsaturated fatty acid was 0.6 was originally suggested as that necessary to ensure adequate vitamin E nutrition (Harris and Embree, 1963). Some experiments with rats have indicated that the vitamin E requirement does not increase linearly with the intake of linoleic acid and may be as low as 0.1 mg α-tocopherol per gram linoleic acid at low PUFA intake (Jager and Houtsmiller, 1970). In other studies with rats, Alfin-Slater et al. (1969a,b) found no evidence of an increased vitamin E requirement when vegetable seed oils rich in PUFA were fed at a level of 30% of the diet without an additional source of vitamin E. More recently the same authors (Alfin-Slater et al., 1972) have reported that a ratio of milligrams of vitamin E to grams of PUFA of 0.4 was satisfactory for the rat's well being. In human studies, Ahrens et al. (1959) found no evidence of creatinuria, myopathy, encephalopathy or other vitamin E deficiency symptoms in patients fed highly unsaturated menhaden oil (low in tocopherol) or corn oil at 40% of calories for 5½ months. Furthermore, Herting and Drury (1963) had also previously concluded that cottonseed oil contained sufficient vitamin E to protect its PUFA content.

These conclusions have been confirmed by the report of Christiansen and Wilcox (1973) who observed that a high dietary polyunsaturated/saturated (P/S) fatty acid ratio of ~1:1 did not result in a decrease in serum tocopherol. These workers reiterated the fact that increased amounts of highly unsaturated oils also contributed significant amounts of tocopherol which were reflected by the serum tocopherol levels.

In a recent publication, Bieri and Evarts (1973) pointed out that tocopherols other than α-tocopherol, γ-tocopherol in particular, are present in foods in substantial amounts and also are capable of providing significant antioxidant activity. These authors indicated that the 1968 RDA for adults for vitamin E (25–30 IU) was probably much too high based on both experimental and epidemiological evidence. As a result, in 1973 the RDA recommendation for vitamin E was reduced to 15 mg. It was suggested, however, that the permissible variation of intake of vitamin E should depend on the type and amount of dietary fat ingested.

Particularly susceptible to symptoms of vitamin E insufficiency are newborn babies who usually have low reserves of tocopherol (Dju et al., 1958). In fact, newborn infants deprived of colostrum and mother's milk require a longer time than do breast-fed infants to attain normal levels of serum tocopherol (Woodruff et al., 1964). Feeding PUFA-rich formulas to premature infants has resulted in lowered plasma tocopherol levels, accompanied by undesirable clinical and hematological symptoms (Hassan et al., 1966). However, in a 3-year study of premature infants, Panos et al. (1968) concluded that the anemia of prematurity was not actually a vitamin E or a linoleic acid related condition, but was a result of other factors. However, since the intestinal absorption of vitamin E is impaired in premature infants and an additional need for vitamin E occurs as a result of feeding diets rich in PUFA, the provision of adequate vitamin E supplements becomes important.

Linoleate-rich diets such as that provided by a cottonseed formula may promote tocopherol depletion of infants during the first month of life and induce certain transient hematological changes similar to those described in premature infants maintained on PUFA-rich diets (Hashim and Asfour, 1968). Recently, Davis (1972) tested the vitamin E adequacy of infant diets and concluded that even though many formulas contain 5 IU or more of vitamin E, the ratio of milligrams vitamin E to grams PUFA falls below 0.6. However, Clausen and Friis-Hansen (1971) concluded that normally there is no risk of diet-induced vitamin E deficiency in normal newborns. Furthermore, Lewis et al. (1973) reported that high intakes of polyunsaturated fat over many years did not alter the plasma vitamin E levels of children and that the vitamin E:PUFA ratio of approximately 0.37 was sufficient for maintaining normal plasma levels of tocopherol.

Witting (1972) has discussed the limits of applicability of the E:PUFA ratio. It is known that EFA-deficient animals synthesize eicosatrienoic acid, C20:3, which is also an easily oxidizable polyunsaturated fatty acid and therefore these EFA-deficient animals may become vitamin E-deficient even when the diet is free of PUFA. It has also been shown that the nonessential polyenoic acids interfere with the conversion of linoleate to arachidonate (Century and Horwitt, 1964) and in this way reduce the incidence of chick encephalomalacia, which is caused by an essential fatty acid (particularly arachidonate)-induced vitamin E deficiency. It appears that the vitamin E requirement is related to tissue PUFA content and this, in turn, may bear little relation to the dietary PUFA level. Therefore an individual's E:PUFA status should be determined by the average intake over a period of months or years. Witting suggests the following formula for the recommended vitamin E intake: milligrams a-d-tocopherol per day/percentage linoleate in adipose tissue fatty acids should equal 0.6. However, this measurement is not very practical in humans.

Witting and Lee (1975) have studied the relationship of dietary PUFA and vitamin E to the linoleate content of adipose tissue of female student volunteers. They observed a general increase in tissue linoleate (approximately 13%) as compared with the results obtained a decade earlier (approximately 8%). They interpreted this as being a reflection of the general trend toward increased consumption of PUFA. The investigators expect further increases in tissue linoleate which may occur as a result of still further dietary increases, thereby increasing the requirement for vitamin E. They suggest that adipose tissue linoleate levels in the general population be used as a baseline for the periodic evaluation and revision of the recommended dietary allowance for vitamin E which should be related to the linoleate content of adipose tissue fatty acids.

For at least several years after discontinuing a high PUFA diet, the tocopherol requirement should be determined by the stores of adipose tissue linoleate rather than by the diet currently ingested, and therefore in cases where a high PUFA diet is discontinued, the elevated level of dietary vitamin E should be maintained to prevent oxidation of the stored PUFA.

Both human and animal studies have shown that at equilibrium the linoleate content of adipose tissue triglycerides will be the same as the linoleate content of the dietary fat, if the species EFA requirement is exceeded (Witting, 1970). The half-life ($T_{1/2}$) of linoleic acid in human adipose tissue is approximately 26 months.

VII. EFA and LCAT Enzyme

An enzyme that affects the content of unesterified cholesterol in plasma and tissues and the content of cholesteryl ester in plasma is lecithin-cholesterol acyltransferase (LCAT) (Glomset and Norum, 1973). Since this enzyme transfers the fatty acid from the C-2 position of lecithin to the hydroxyl group on the cholesterol molecule, and since the fatty acid involved in the transfer is usually either linoleic or arachidonic acid, LCAT obviously plays an important role in PUFA metabolism.

LCAT is secreted into plasma by the liver (Osuga and Portman, 1971). It circulates in plasma complexed to high-density lipoproteins. As a result of the activity of LCAT, free cholesterol and lecithin of the plasma lipoproteins decrease while cholesteryl esters increase. Although LCAT activity has been found to be increased in experimental diabetes, no elevations have been seen in cases of diabetes mellitus. Furthermore, no effect of insulin on the activity of this enzyme was observed (Misra et al., 1974). However, in various cases of hepatic disease, low LCAT activity accounted for impaired cholesterol esterification (Simon et al., 1974). In fact, a possible association between an abnormal low-density lipoprotein and nephropathy in LCAT deficiency has been suggested by Gjone et al. (1974). It may also be of interest that a 53% increase in LCAT activity was observed in patients undergoing Colestipol resin therapy as a means of decreasing serum cholesterol levels (Clifton-Bligh et al., 1974). This occurred despite the decrease in the serum concentrations of cholesterol.

The recent discovery of a familial LCAT deficiency has provided an unique opportunity to study the metabolic role of the LCAT reaction (Gjone and Norum, 1968). The main characteristics of this disease associated with the metabolic defect have been: an anemia associated with a twofold increase in erythrocyte lecithin and free cholesterol, a very high free cholesterol, and a very low cholesteryl ester level in plasma, with a fatty acid composition resembling human chylomicron cholesteryl esters formed in the intestinal mucosa rather than that found in HDL or LDL (low-density lipoprotein) cholesteryl esters normally formed by the LCAT reaction. Plasma phospholipids are also abnormal—there is an extremely high content of lecithin and an unusually low level of lysolecithin. In addition, there is a hyperlipemia with particularly elevated triglycerides. Other accompanying abnormalities include corneal opacities, lipid vacuoles in the cells of bone marrow and spleen, proteinuria and, in the most advanced cases, renal failure.

An interesting observation is provided by the fact that metabolic abnormalities similar to these described in familial LCAT deficiency have been seen in cholesterol-fed guinea pigs (Ostwald et al., 1970) and in patients (and animals) with obstructions of the bile duct (Quarfordt et al., 1972). Although it is probable that different mechanisms are responsible for the observed accumulation of free cholesterol in these three different conditions, the resulting abnormal composition of the lipoproteins is similar. It is assumed that the role of the LCAT reaction is to prevent unesterified cholesterol derived from the surfaces of chylomicrons and VLDL from accumulating in the plasma.

The type of cholesteryl esters formed by the LCAT reaction is related to the species specificity of the enzyme, differing in man from that of the rat. The human enzyme preferentially transfers linoleate, while in the rat it is arachidonate that changes its location (Portman and Sugano, 1964). The rat also possesses an active hepatic acyl-CoA:cholesterol transferase which contributes a certain portion of plasma cholesteryl esters, whereas man is solely dependent for its plasma cholesteryl esters on LCAT. In EFA deficiency, cholesterol esterification in plasma is enhanced, presumably because of an increased enzyme concentration (Sugano and Portman, 1965).

Gustafson et al. (1974) has reported that during the estrogen-dominated follicular phase of the menstrual cycle, a reduction of linoleate and an increase in arachidonate takes place in the HDL lecithin, suggesting the influence of gonadal hormones on lipoprotein lecithin synthesis in the liver. Linoleate in the β position is the preferred substrate for the LCAT reaction. Thus a decrease in linoleate may influence the LCAT enzyme activity.

The physiological role of the LCAT reaction has not been completely elucidated. It has been suggested (Glomset, 1968; Scherer, 1974) that it plays a role in the transport of unesterified cholesterol from peripheral tissues to the liver. It also stabilizes lipoprotein structure. Hyperlipidemia is often associated with an increased LCAT activity. The most recent research on LCAT has been presented in a symposium proceedings edited by Gjone and Norum (1974).

VIII. EFA and Brain Development

Recently, considerable evidence has accumulated indicating that the maturation of the brain may be affected by malnutrition, resulting in reversible and/or irreversible damage to the central nervous system.

Selivonchik and Johnston (1975) reported that fat deficiency in rats during the development period (from day 14 of gestation) leads to increased susceptibility to experimental allergic encephalomyelitis induced in animals at 54 days of age. A supplement of linoleic acid had a marked protective effect against this disease.

Changes in brain lipids occurring during EFA deficiency are usually less pronounced and slower developing than those in other tissues. There is a relative metabolic stability of brain lipids in comparison with lipids of other tissues. In addition, the major portion of lipids in the brain is formed before birth or in the neonatal period.

The investigations of the effects of EFA deficiency on the central nervous system in weaning animals (at a late stage in the development of nervous tissue) led to observàtions of rather limited alterations in brain fatty acid composition (Walker, 1967; Rathbone, 1965).

However, as a result of studies with rats, Galli (1973) concluded that the effects of EFA deficiency on the brain of growing rats consist mainly of a reduction of ω6 (20:4 and 22:4) acids and an increase in ω9 (20:3 and 22:3) acids of brain phospholipids. These changes are reflected by an elevation of the ratio of 20:3/20:4. Although the amount of unsaturation in each lipid class remains constant, the newly formed ω9 trienes substituting for ω6 tetraenes may not have the same physiological significance. The observed alteŕations in brain lipids are most pronounced in myelin, and even in this tissue, changes can be almost completely reversed by improving the nutritional status of the animal.

Sun et al. (1974) observed that as time progressed, mice derived from prenatally fat-deficient mothers showed an increase of 20:3 ω9 and a decrease in 20:4 ω6 in the phosphoglycerides of subcellular brain fractions. Elongated products of 20:3 ω9 such as 22:3 ω9 and 22:5 ω9 were also present. These acids were preferentially linked to the alkenylacyl glycerophosphorylethanolamines. Similar changes in fatty acid composition, but at a slower rate, occurred in both myelin and synaptosome-rich fractions even when the deficient diet was initiated after maturation.

Alling et al. (1974), on the other hand, described a relative inertness of rat cerebrum to the wide differences in the dietary EFA level. The ratio between the lowest and the highest level of EFA used in the diets of two generations of rats was 1:40. In the third generation, there were only slight differences in the total PUFA content of the cerebrum among the different groups. The amount of 22:6 ω3 was lower in the group fed the low EFA whereas the relative concentrations of 22:5 ω6, 20:3 ω9, and 22:3 ω9 were increased.

In a recent review dealing with the uptake and transport of fatty acids into the brain, Dhopeshwarkar and Mead (1973) concluded that the rat brain does not have to depend on the supply of EFA received during the early period of development and growth but can take up these essential nutrients throughout adult life. It has been observed by White et al. (1971) that the return to normal of EFA-deficient animals after shifting to a balanced diet is quite rapid. It seems, however, that EFA deficiency could have some effect on the permeability of brain capillaries affecting the blood–brain barrier.

According to Svennerholm (1968) changes in the distribution and fatty acid composition of phosphoglycerides of normal human brains are rather small after early childhood. There is a difference between the fatty acids of the lipids from the cerebral cortex and from the adjacent white matter. The phospholipids of cerebral white matter contain more monoenoic but less PUFA than those of cerebral cortex. With increasing age, the concentration of the fatty acids of the linoleate family diminish while that of the linolenate family increases. A large concentration of 20:4 is evident in inositol phosphoglycerides; this is nearly as high for white as it is for gray matter and shows only small changes with age.

Disease of the biliary tract of mothers has been associated with the incidence of mental retardation in the offspring (Churchill *et al.*, 1967). It has been postulated that avoidance or malabsorption of fat as a consequence of gallbladder disease diminishes the supply of lipids essential for brain growth of the fetus. When the effect of maternal diet in rats during pregnancy on the learning ability of offspring was tested (Caldwell and Churchill, 1966), the impairment of fetal brain development by a fat-free diet became obvious. Linoleic acid appeared to be required for normal brain development (Churchill *et al.*, 1967).

IX. EFA and Plasma Lipids

Since the risk of developing coronary heart disease is greater in persons whose serum cholesterol levels are elevated (Keys, 1971), there is a great deal of interest in finding ways to lower serum cholesterol levels in hypercholesterolemic subjects. Kinsell and his group (1952) were among the first investigators to suggest that linoleic acid was the active component of fats which could decrease serum cholesterol levels. This finding has been confirmed many times in experimental animals, humans, and in epidemiological studies. In one of the recent epidemiological studies, Keys (1970) demonstrated that 80% of the serum cholesterol variability among population groups could be explained by the different amount of saturated fat in the diet.

The plasma cholesterol and triglyceride concentrations in men were lowered when PUFA were substituted for saturated fatty acids in the diet (Nestel *et al.*, 1970). These confirmatory findings came from studies in which fatty acids derived from animal sources were compared with those derived from vegetable sources or with synthetic glycerides.

When hypercholesterolemic subjects were placed on a low cholesterol–high PUFA diet and were followed for up to 3 years, both plasma cholesterol and triglyceride concentration decreased substantially (Evans *et al.*, 1972). The authors have also observed reductions in cholesterol levels on this regimen in individuals with initial serum cholesterol values considered normal.

Although the hypocholesterolemic effect of vegetable oils is generally considered a function of their linoleate content, fish oils, rich in more highly polyunsaturated fatty acids and of longer chain length, also exhibit a hypocholesterolemic effect in spite of the fact that they possess negligible EFA activity (Thomasson, 1953). It is probable that EFA activity is not an absolute requirement for this additional property of PUFA. In one of the studies, however, in which 51 male subjects ingested capsules of fish oils for several weeks, no effect on serum cholesterol was observed, although the average triglyceride level was lowered in 21 individuals (Engelberg, 1966).

Other investigators have suggested that the hypocholesterolemic effect of fatty acids, such as linoleic and linolenic, is due to their ease of conversion to very highly unsaturated fatty acids, rather than to an inherent property of these acids themselves. Eskimos are the only people who are believed habitually to consume a diet rich in PUFA (Bang et al., 1971) and the incidence of coronary heart disease is considered to be very low in this population.

The American Heart Association recommends reducing the total fat content of the diet to no more than 30% of total calories, with saturated fat restricted to approximately 10% of total calories. The available data indicate that this type of cholesterol-lowering diet could substantially reduce the risk of coronary heart disease (Anderson et al., 1973; Mueller, 1973). According to Jolliffe (1961) the P/S ratio might be considered a measure of the effectiveness of the hypocholesterolemic effect of fat.

The observation that elevated plasma lipids could be reduced by a diet relatively rich in unsaturated fat, led to an interest in the possible differences in the metabolism of different fatty acids. Kirkeby et al. (1972) performed a study of the fatty acid composition of serum lipids in men with myocardial infarctions. Of the 73 patients, those who had had heart attacks in the past, had a higher percentage of 18:2 than those with the more recent coronary episode. However this difference was believed to be due to an increased dietary content of essential fatty acids as a consequence of treatment of this disease. Higher values for pentaenoic and hexaenoic fatty acids were also obtained in the patients with acute myocardial infarction, thus indicating that these fatty acids are not of significance in preventing coronary heart disease. Linoleic acid levels, however, were lower than in controls. Essential fatty acids appear to be more important in the prevention of coronary heart disease than are the highly unsaturated fish oils, even though the reverse had been previously proposed (Notevarp, 1966). In a parallel study, serum lipids of the wives of these patients were investigated (Kirkeby et al., 1972) and no differences were demonstrated between these women and the wives of healthy men; therefore, particular dietary habits of families did not seem to be involved in the differences previously obtained.

In several cases of familial hypercholesterolemia, however, no significant lowering of serum cholesterol levels was found after treatment with corn oil (Strisower et al., 1968; Segall et al., 1971). Although results of this type are not usual, particular resistance to PUFA could be due to a specific inherited metabolic abnormality. In general, there are many dietary trials using polyunsaturated fatty acids aimed at prevention of coronary heart disease which have yielded consistent reductions in the incidence of coronary heart disease (Rinzler, 1968; Bierenbaum et al., 1970; Turpeinen, 1971).

When 59 nonobese and hyperlipidemic men ingested a fat-controlled (AHA recommended) diet, a follow-up study after 5 months revealed a fall in serum cholesterol and lipoprotein levels, and in triglyceride levels as well. Increases in the linoleate content of plasma phospholipid and adipose tissue reflected the increases in the ingestion of PUFA. It appears that these increased levels of 18:2 may be used instead of dietary interviews in assessing PUFA intake as early as 6 months after initiating the modified diet (Wilson et al., 1971).

An epidemiological study of serum lipids performed on nine groups of widely differing male adults in South Africa (Truswell and Mann, 1972) indicated that the triglyceride fatty acid patterns reflected the dietary fatty acids. The increased percentage of 18:2 correlated well with decreased serum cholesterol and even with lowered serum triglycerides. Apparently 18:2 intake is one of the determinants of serum cholesterol levels in a free-living population as well as in the metabolic ward.

Although the lowering effect of unsaturated fats on high serum cholesterol levels has been repeatedly demonstrated in men, the biochemical mechanism responsible for this phenomenon has not been established to the satisfaction of all investigators. Wood et al. (1966) reported that the reduction of plasma cholesterol levels resulting from the substitution of PUFA-containing fats for more saturated dietary fat was accompanied by an increased rate of fecal steroid excretion. Claims for increased cholesterol elimination (Haust and Beveridge, 1963) and total bile acid excretion (Roels and Hashim, 1962) observed in individuals given diets rich in unsaturated fats have been questioned with respect to the methodology used in these investigations (Miettinen et al., 1965; Avigan and Steinberg, 1965). More recently, Ali et al. (1966) studied the excretion of fecal bile acids in three adult males during several dietary periods where 35–60% of total calories was provided by corn oil or butterfat. Because of the large individual variations the significance of differences in the fecal bile acid excretion with these diets again remains in doubt.

Campbell *et al.* (1972) and Sodhi *et al.* (1964) suggest that the increased turnover of bile salts seen after feeding corn oil, together with the resultant acceleration in cholesterol catabolism, may partly explain the hypocholesterolemic effect of PUFA. Moore *et al.* (1968) observed that changing from saturated to unsaturated fat in the diet resulted in a 28% decrease in serum cholesterol together with an increase in fecal sterol excretion, which was equally distributed between neutral sterol and bile acids. On the other hand, Grundy and Ahrens (1970) while studying the effects of unsaturated dietary fats on absorption, excretion, synthesis, and distribution of cholesterol in patients with familial hypercholesterolemia, observed that decreased cholesterol levels were not accompanied by marked increases in fecal neutral sterols and suggested that more likely, cholesterol was redistributed from plasma into various tissues. However, in a more recent study Grundy (1975) observed that the feeding of polyunsaturated fats, as compared with saturated fats, to eleven patients with hypertriglyceridemia caused an increased excretion of endogenous neutral steroids, acidic steroids, or both.

Reductions in plasma cholesterol during the feeding of polyunsaturated fats were seen in most patients and these changes were usually associated with a decrease in concentration of plasma triglycerides. It is possible that in hypertriglyceridemic patients, polyunsaturated fats may contribute to cholesterol reduction by changing the metabolism of triglycerides or VLD lipoproteins. In a previous study, the same authors (Grundy and Ahrens, 1966) using both chemical and isotopic methods to study fecal steroid excretion when corn oil was substituted for butter oil in the diet of *one* patient, had found that as a result of the dietary treatment, plasma lipid concentration, cholesterol, and serum triglycerides decreased sharply; this was accompanied by an increased fecal sterol excretion which was not due to increased cholesterol synthesis. The daily mean cholesterol decrease amounted to 227 mg whereas fecal sterol excretion increased by 748 mg which was equally divided between the acidic and neutral steroids of endogenous origin. A net outflux of cholesterol therefore occurred from tissue, as well as from plasma, into the gut lumen. Connor *et al.* (1969) did steroid balance studies in six healthy adult males and observed that the consumption of corn oil resulted in an increased excretion both of bile acids and of neutral sterols. This increase more than accounted for the concomitant disappearance of cholesterol from the plasma pool.

Nestel *et al.* (1975) also reported lowered plasma cholesterol levels in men on a high polyunsaturated fat intake both with high and low dietary cholesterol in their diet. Sterol excretion was greater on a high PUFA, high cholesterol diet than on a high PUFA, low cholesterol diet.

Both neutral sterol and bile excretion were higher on the polyunsaturated fat as compared to the saturated fat diet.

The results of Bhattacharyya *et al.* (1969), using eleven students as experimental subjects, also support the hypothesis that the serum cholesterol-lowering effect of dietary polyunsaturated fat is mediated by increased elimination of neutral steroids in the feces. When dietary essential fatty acids are not available, the more saturated cholesteryl esters accumulate in the liver. In turn, an optimum availability of essential fatty acids assures formation of phospholipids and cholesterol esters which are structurally necessary for lipoprotein synthesis and subsequently for more efficient transport of lipids (Söderhjelm *et al.*, 1970). It has been suggested that the amount of essential fatty acids in phospholipids governs the carrying capacity of VLDL for triglycerides. Phospholipids containing arachidonate are more firmly bound to protein (De Pury and Collins, 1966). In essential fatty acid deficiency, less phospholipid is bound to protein, thus less triglyceride can be transported, leading to an accumulation of fat in the liver.

Chait *et al.* (1974) reported that the greatest reduction of serum triglycerides and cholesterol as a result of unsaturated fat feeding occurred in subjects with the highest baseline levels. The triglyceride reduction was due to a decrease in VLDL triglycerides and cholesterol in decreased VLDL and LDL. These investigators measured rates of incorporation of palmitic and linoleic acids into VLDL triglycerides and showed that palmitic acid was preferentially incorporated. Thus polyunsaturated fat may decrease the rate of secretion of VLDL triglycerides into plasma by substituting for a preferred substrate.

Spritz and Mishkel (1969) proposed that for steric reasons less lipid can be accommodated by the lipoprotein molecules when saturated and monounsaturated acyl chains are substituted by polyunsaturated ones.

It has been reported that the fractional turnover rate of linoleic acid exceeds that of palmitic acid when they are administered as free fatty acids (Nestel, 1965). Injected linoleic acid turns over relatively faster in hepatic glycerides, and is incorporated more extensively into plasma triglycerides. In dogs there is an inverse correlation between the fractional turnover of linoleate (but not of palmitate) and the concentration of plasma triglyceride (Nestel *et al.*, 1962). When the metabolisms of palmitic acid and linoleic acid were compared in seven subjects during constant infusion of radioactive tracers, it was observed that the fractional turnover of linoleic acid was nearly always greater than that of palmitic, although its total turnover rate was less (Nestel and Barter, 1971). Since a smaller proportion of linoleate than of palmitate was incorporated into plasma triglycerides, the investigators felt that this might be a factor in the lowering of plasma triglyceride concentrations by diets rich in PUFA.

Bollinger and Reiser (1965), in reviewing the metabolic fate of fatty acids derived from dietary triglycerides, concluded that linoleic acid is preferentially catabolized. On the other hand, dietary linoleate has a far greater inhibitory effect on lipogenesis than does saturated fat (Allman and Gibson, 1965). In another study of hepatic lipogenesis in mice, only diets high in 18:2 were found to suppress lipogenesis (Sabine et al., 1969). Similarly whether or not fat is present in the diet affects the rate of chain elongation and desaturation (Marcel et al., 1968). The microsomes from essential fatty acid-fed rats appear to be less active than those from fat-deficient animals. This observation is related to the well-documented hypothesis that there exists a competition among oleic, linoleic, and linolenic acids in reactions leading to the synthesis of more highly unsaturated fatty acids (Mohrhauer and Holman, 1963, 1967; Peluffo et al., 1963; Brenner and Peluffo, 1966). The more unsaturated fatty acids have greater affinity for desaturating enzymes than do the more saturated fatty acids; i.e., there is more of a tendency to effect linoleic acid conversions to more polyunsaturated fatty acids than to bring about the dehydrogenation of oleic acid (Gidez, 1964). Similarly it has been shown that high dietary unsaturated fat stimulated desaturation in the 6 position but inhibited desaturation in position 9 (Inkpen et al., 1969). Considerable information on the elongation and desaturation of polyunsaturated fatty acids has been provided by Brenner and his group (1969).

PUFA exert a specific effect in vivo in effecting the lowering of the activities of at least five enzymes concerned with the synthesis of saturated and monounsaturated fatty acids in liver of rats. This coincides with a rapid reciprocal change in the relative amounts of ω6 and ω9 fatty acids in the liver lipids (Muto and Gibson, 1970). Free fatty acids inhibit enzyme activity and the PUFA are more active in this respect than are saturated fatty acids.

According to Sinclair (1964), polyunsaturated fatty acids should be antiketogenic since carbon atoms involved in the methylene interrupted double bond system should not contribute to the acetate pool from which ketone bodies are formed. In confirmation of this theory, it has been shown by Chung and Dupont (1968), that high corn oil diets do not elevate plasma acetoacetate. Also Mathias and Dupont (1969) have reported that linoleate is partially oxidized via a pathway similar to that in which propionic acid is metabolized, i.e., through methylmalonyl-CoA and succinyl-CoA.

It has been reported recently (Kasper et al., 1973) that with a low carbohydrate and protein intake, increasing amounts of fat high in linoleic acid increases the metabolic rate. Weight gain did not correlate with the caloric intake and a marked sensation of heat was experienced. In addition, cholesterol and triglyceride levels in serum were found to decrease. On the other hand, triglyceride synthesis was stimulated by essential fatty acid deficiency (Solyom et al., 1967).

Long-term ingestion of a high PUFA diet is usually reflected by characteristic changes in depot fat composition where a significant accumulation of linoleate takes place at the expense of more saturated fatty acids. This fact may be considered to be a useful index of adherence to a modified diet (Christakis et al., 1965; Dayton et al., 1967). Kaunitz et al. (1961) have even proposed that the linoleate deposited in adipose tissue depots decreases the relative proportion of neutral fat to body weight and, in fact, induces the deposition of a fat whose composition is more representative of the dietary fat.

In an 8-year study relating a diet high in unsaturated fats with serum cholesterol levels and fatty acid composition of adipose tissue of older men (Dayton et al., 1969), it was found that the high linoleate (38% of fat) diet induced a prompt and sustained drop in serum cholesterol (12.7% of starting level). The linoleate content of adipose tissue rose from 10.9 to 33.7%. At the same time, fatal atherosclerotic events in the control group reached 70 as compared with 48 in the group fed the high dietary linoleate.

A comparative study of serum lipids of 40 healthy individuals with 35 atherosclerotic patients revealed markedly less 18:2 and 20:4 in serum lipids, accompanied by an elevated 20:3 in the atherosclerotic group (Alimova and Endakova, 1970). An increased amount of odd-chain fatty acids was also noted in the patients. Triglycerides of patients contained a higher proportion of saturated and monoenoic acids.

Determinations of total serum fatty acids in obese and normal men and women have indicated a higher proportion of linoleate, and a lower amount of oleate and palmitoleate, in controls as compared to the obese (Reuter et al., 1969). These subjects, however, were not subdivided according to age and sex. The observation of Insull et al. (1970) may be of interest in this connection. These workers compared the fatty acid composition of adipose tissue lipids in 1962 and in 1966 in urban American men who died suddenly and unexpectedly. Men were matched by race, age, and extent of coronary arteriosclerosis. It appears that in 1966 the proportion of linoleic acid was higher (+1.7%) and stearic acid lower (−1.1%) in adipose tissue lipids than it was in 1962. It is possible that these differences are due primarily to the increased use of dietary polyunsaturated fat since 1960 as a result of nutrition education programs, especially those promoted by the American Heart Association (Diet and Heart Disease Statements; 1961, 1965, 1968, 1973).

Comparative studies of the fatty acid distribution in the cholesteryl ester fractions of LDL and VLDL have been performed in human atherosclerotic subjects. More oleic and less palmitic and linoleic acids were found in the VLDL fraction. Although these differences were not very significant it was suggested that these results (Jaillard et al., 1972) support the hypothesis by which LCAT promotes exchanges of HDL cholesteryl ester for VLDL triglycerides and free cholesterol (Nichols and Smith, 1965; Rehnborg and Nichols, 1964). Jaillard et al. also relate the decrease in plasma 18:2 observed in severe hypertriglyceridemias to the presence of plasma chylomicrons which are low in linoleic acid.

On the other hand, no clear-cut influence of the dietary fat on the concentration of serum triglycerides was observed by Arvidson and Malmros (1972) in healthy male Swedes during dietary-induced changes in serum cholesterol.

Kingsbury and Brett (1974) reported an abnormal serum fatty acid pattern which permits the distinguishing between the patients with and without myocardial infarction. This pattern suggests EFA deficiency with apparent increased synthesis of monoenoic acid coupled with a high C20:3 ω9 and a relative inadequacy of linoleic acid. Previously, however, Hagenfeldt et al. (1973) suggested that atherosclerotic subjects might exhibit a block in the conversion of 20:3 ω6 to 20:4 ω6 accompanied by a low 18:3 ω6. This may be due to the fact that there is an inhibition of the desaturation of 18:2 ω6 to 18:3 ω6 caused by the accumulation of 20:3 ω6. These authors report that the accumulation of 20:3 ω6 correlates positively with the linoleate level contrary to the situation in EFA deficiency where 20:3 ω9 accumulates as a result of decreasing levels of 18:2. Obviously further studies are needed to clarify the inconsistency between these two studies.

Attempts to provide foods with higher linoleic acid content have resulted in several interesting developments. Recently, it was found that the linoleic acid content of ruminant meat and milk fat could be substantially elevated, thus producing polyunsaturated animal fats. The hydrogenation of linoleic acid in seed oils by the microorganisms in the rumen can be prevented if the oil droplets are coated with formaldehyde-treated casein. The 18:2 content of the milk and milk fats can then be raised in proportion to that contained in the feed, principally at the expense of the saturated fatty acid, palmitic (Cook et al., 1970, 1972). Recently, significant reductions in plasma cholesterol levels were observed in 5 of 6 subjects, and later in 8 men, on diets which included dairy and meat products derived from these polyunsaturated fat-fed animals. The total excretion of sterols increased by an average of 18% when polyunsaturated ruminant fats were ingested (Nestel et al., 1973, 1974). A stable level was reached after 20 days after which time bile acid excretion was similar both on the control and the modified fat diet.

Even though many investigators feel that the hypocholesterolemic role of polyunsaturated fat and the hypercholesterolemic influence of saturated fat have been demonstrated in sufficient detail to warrant acceptance of the recommendation to increase polyunsaturated fat in the diet for better health and for the prevention or retardation of coronary heart disease, others (Reiser, 1973) feel that a more critical scrutiny of the results obtained so far is necessary. Criticisms include the fact that these results were not obtained in significant numbers, particularly when normal individuals were studied, and that the experimental design in a number of cases could lead to an erroneous interpretation of results. It is felt that studies should be done where comparisons of saturated and polyunsaturated fat are made with a "neutral diet" rather than comparing two extremes with each other. Evidently polyunsaturated fats may result in a lowering of serum cholesterol through multiple mechanisms and it seems unlikely that a single mechanism may be responsible for all the effects of polyunsaturated fats on plasma lipids.

X. EFA and Platelet Aggregation

The effect of the replacement of dietary saturated fat by polyunsaturated fat on lowering the incidence of ischemic heart disease, in addition to its effect on cholesterol metabolism, may be related to the observed reduction of platelet aggregability by polyunsaturated fatty acids (Hampton and Mitchell, 1966). There is evidence that malfunctioning blood platelets play a role in the pathogenesis of occlusive arterial disease in men (Haerem, 1971). Recently, Hornstra et al. (1973) studied the effect of a diet low in saturated fat and high in polyunsaturated fat on platelet function in 63 men compared with 73 normally fed controls. Using a new method which assesses platelet aggregation in flowing blood, a pronounced and significant reduction in aggregability was found in the experimental group.

These results were confirmed by Fleischman et al. (1974) in studies in which the same subjects were placed on either a saturated fat diet or a linoleic acid-enriched diet so individuals served as their own controls rather than having comparisons between independent groups. A 2-week period was sufficient to significantly affect both platelet aggregation and disaggregation. Subsequently (Fleischman et al., 1975), it was reported that platelet aggregation time significantly increases within 48 hours in response to an increase in dietary linoleate of 4% while disaggregation time decreases significantly in 96 hours. A change as small as 0.5% of calories as linoleate was associated with significant increases in aggregation time within 4 days.

When platelets aggregate, the phospholipids contained therein are made available for plasma coagulation (Nordoy and Lund, 1968). These phospholipids are thus involved in an important step in thrombus formation (Nordoy and Rodset, 1970). Experimental studies indicate that there is a relationship between fatty acid composition of platelet phospholipids and platelet function. Nordoy and Rodset (1971) studied two groups of 10 subjects, each ingesting for 21 days a diet containing medium-chain triglycerides (MCT). In subjects given the soybean oil diet, a significant reduction in serum total cholesterol and serum phospholipids occurred. Some of the platelet function tests indicated changes which could possibly lead to an increased susceptibility to aggregation of platelets in the MCT-fed individuals and a decrease of susceptibility of the same fraction in the soybean oil-fed subjects.

Although the increased tendency to thrombosis of atherosclerotic patients appeared to be reduced by linoleic acid (Owren, 1965), it was shown later that actually it is the 18:3 acid, linolenate, which counteracts platelet aggregation enhanced by saturated fatty acids in vitro (Mahadevan et al., 1966). An apparent difference in response of platelet adhesiveness to the treatment with either 18:2 or 18:3 acids suggested that the mechanism of action of these acids in this respect may be different (Bentzen et al., 1972).

Vergroesen (1972) has suggested that one of the possible modes of action of linoleic acid as related to cardiovascular disease may be through stimulation of synthesis of prostaglandins (discussed later) which, in turn, are inhibitors of platelet aggregation.

Rose et al. (1974), on the other hand, suspect that arachidonic acid exerts its effects on systemic arterial pressure, myocardial activity, and on platelets in dogs through its conversion to an intermediate in the biosynthesis of PGE_2 rather than as PGE_2 itself.

The inhibition of the collagen-induced platelet aggregation by PGE, was found to be considerably reduced in the plasma of EFA-deficient rats (Vincent et al., 1974).

XI. EFA in Disease States

Changes in essential fatty acid levels and distribution have been observed in various disease states. Whether these alterations precede or result from the disease remain to be established. Bernsohn and Stephanides (1967) have suggested that a lack of EFA in the diet might be related to the development of multiple sclerosis. Belin *et al.* (1971) have shown that the percentage of 18:2 in the serum of some patients with multiple sclerosis is lower than that in the controls while that of the oleate is higher. Following the administration of linoleate to patients, the content of 18:2 increased but was not maintained as well as in controls, when the supplementation ceased. The rate of loss of linoleate in these patients seems to be abnormally high. Previous biochemical studies of the brain lipids in multiple sclerosis patients had shown that the relative proportions of saturated and unsaturated fatty acids were different from that found in healthy people (Baker *et al.*, 1963). In addition, rats bred on a diet deficient in unsaturated fatty acids showed an increased susceptibility to the development of experimental allergic encephalitis, which is regarded by some investigators as a disorder similar to multiple sclerosis (Clausen and Moller, 1967).

Attempts to treat multiple sclerosis patients with polyunsaturated fatty acids have met with only limited success. When 75 patients with multiple sclerosis were given daily supplements of a vegetable oil mixture containing either linoleate or oleate for 2 years in a double-blind control trial (Millar *et al.*, 1973), relapses tended to be less frequent and were significantly less severe and of shorter duration in the 18:2-supplemented group than in those receiving the 18:1 mixture, but clear evidence that treatment affected the overall rate of clinical deterioration was not obtained.

On the other hand, Love *et al.* (1974) pointed out that a reduction of linoleate in serum lipids is not specific to multiple sclerosis since it has been observed in patients with acute nonneurological illnesses as well. The resemblance of the serum fatty acid patterns to those seen in EFA deficiency in these cases could be a general phenomenon.

Patients with Laennec's cirrhosis were studied by Sullivan *et al.* (1969). It was observed that each of the 15 individuals, regardless of their serum triglyceride level, showed a relative decrease in serum 18:2. The relative deficiency of this fatty acid may play a role in abnormalities of serum and hepatic triglycerides found in alcoholic cirrhotics.

When alcoholic patients were fed a high PUFA diet as a part of their treatment, it was observed through liver biopsy that, in a majority of cases, retrogression of fatty changes in the liver took place (Irsigler *et al.*, 1970). Kyriakides *et al.* (1970) performed lipid analyses on tissues of two siblings with Wolman's disease, a primary familial xanthomatosis with involvement and calcification of the adrenals. They found that linoleic acid was greatly reduced in the tissue lipids in one patient who received a skim milk diet but was present in normal concentrations in the second receiving a soybean formula. In both patients massive accumulations of triglycerides and cholesterol in liver and spleen were found. Previously, Konno *et al.* (1966) reported low levels of linoleic acid in the tissue lipids in their patients with Wolman's disease. There is a possibility that these lower levels are actually a secondary reflection of the diet and are not related to the disease process itself.

One of the genetic diseases which possibly involves an EFA deficiency is abetalipoproteinemia (Phillips and Dodge, 1968) where decreased 18:2 and elevated ω9 acids are found in red blood cells. A possible long-term EFA deficiency secondary to malabsorption and low fat intake has been offered as a possible explanation for these findings and this disease. However, this could also be a genetically induced metabolic aberration (Söderhjelm *et al.*, 1970).

Another disease state involving EFA is the familial LCAT (lecithin-cholesterol acyltransferase) deficiency which was discussed previously. This LCAT enzyme is important in regulating the EFA content of lipids in transport.

The dominating biochemical abnormality in Refsum's disease (here-dopathia atactica polyneuritiformis) is the accumulation of phytanic acid in the tissues, secondary to defective α-hydroxylation. In addition, extremely low levels of 18:2 have been reported in lipids isolated from different organs (Kahlke, 1964). When the incorporation of 18:2 (labeled) was followed in a patient, it was observed that it was recycled to a lesser extent in the patient than in normal subjects and that its incorporation into plasma triglycerides was lower. The oxidation of 18:2 was not substantially increased (Laurell *et al.*, 1972).

Even though an increase in esterified saturated and unsaturated fatty acids has been reported in diabetic patients with hyperlipidemia (Schrade *et al.*, 1963), Albutt and Chance (1969) found no difference between the fatty acid composition of the fasting plasma cholesteryl ester fraction in normal children and that in diabetic children on a standard diet.

In kwashiorkor the high levels of plasma monoenoic and polyenoic acids can be attributed to the excess demand for the synthesis of fat and/or preferential utilization of the unsaturated fatty acids to help mobilize other lipid components (Hafiez *et al.*, 1971). Other investigators, however, reported normal serum PUFA levels in kwashiorkor patients (Schendel and Hansen, 1961), but during the recovery from kwashiorkor, the pattern of serum fatty acids was similar to that of EFA deficiency, i.e., the di-, tetra-, penta-, and hexaenoic acids were decreased and trienoic acids increased. It is probable that tissue reserves of EFA are minimal in kwashiorkor, and that the mobilization of liver and depot fat places a great demand on the limited supply of these acids required for lipid transport. Therefore, a relative EFA deficiency is induced. Taylor

(1971) also reported that the main abnormality observed in serum total fatty acids of children suffering from kwashiorkor was a marked decrease in the percentage and concentrations of 18:2 and 20:4.

Lack of dietary linoleate has been suggested as the cause of phryno-derma (skin eruptions) in older children and adults (Bagchi *et al.*, 1959). Increased susceptibility to infections has also been observed in fat-deficient animals and infants (Hansen *et al.*, 1963).

Finally, an apparent deficiency in EFA accompanied by a marked reduction in serum vitamin E observed in cystic fibrosis may be at least partially responsible for the defects in membrane structure and stability and also for decreased production of prostaglandins (Rosenlund *et al.*, 1974).

XII. EFA and Cholelithiasis

Recently, controversial reports on the relationship of dietary PUFA to the formation of gallstones have appeared in the literature. In general, the formation of gallstones depends on the precipitation of cholesterol-containing particles from a supersaturated solution of bile, and the retention of some of these particles which then serve as nuclei for further precipitation. Thus, formation of cholesterol gallstones presupposes that the bile contains more cholesterol than it can hold in solution (Dam, 1971). The components normally holding cholesterol in solution in the bile are primarily the bile salts and the phospholipids. Animal studies have shown that diets that contain both cholesterol and fat generally increase the cholesterol content in serum and liver. In some species, e.g., squirrel monkeys (Osuga and Portman, 1971), such diets produce cholesterol gallstones. It is also relatively easy to produce gallstones in hamsters by feeding a diet containing 10% butterfat. In this species, the addition of cholesterol to the butterfat-containing diet actually *reduces* the incidence of cholesterol-containing gallstones (Dam *et al.*, 1968). On the other hand, fats containing PUFA (with or without cholesterol) counteract the formation of cholesterol gallstones in hamsters.

In humans the relative concentrations of cholesterol, phospholipids, and bile salts in the bile appear to be the determinants of cholesterol solubility. A variety of dietary interventions have been attempted in order to modify these ratios. Thistle and Schoenfield (1969), for example, reported an improvement in the cholesterol-solubilizing power of human bile by the daily ingestion of chenodeoxycholic acid.

Dam *et al.* (1967) studied human bile composition following the ingestion of diets containing either butterfat or margarine (containing

40% of linoleic acid). Ratios of total bile acids to cholesterol, and of lipid soluble phosphorus to cholesterol, did not change in the direction favoring increased solubility of cholesterol when margarine was substituted for butter. On the other hand, bile is considered to be lithogenic at low bile acid output (Northfield and Hofmann, 1973; Danziger et al., 1972) and corn oil was found to promote higher fecal bile excretion (Hagerman and Schneider, 1973) than did a diet containing medium-chain triglycerides.

Quite recently, Sturdevant et al. (1973) suggested the possibility that a high P/S fatty acid ratio promotes, rather than counteracts, gallstone formation. Their conclusion was based on a postmortem comparison of gallstone incidence in elderly men who underwent long-term feeding of a diet high in PUFA, low in saturated fat, and low in cholesterol with control subjects fed a diet higher in saturated fat and cholesterol and lower in polyunsaturated fatty acids. Hofmann et al. (1973) pointed out that such results may be due in part to a stimulation of cholesterol biosynthesis as a result of the high plant sterol ingestion on the high PUFA diet. Also the lowering of serum cholesterol resulting from feeding diets containing fats with a high P/S ratio may increase cholesterol excretion in bile which, in turn, may increase bile lithogenicity. This was shown to be the case in the squirrel monkey where the addition of unsaturated fat to a cholesterol-containing diet appeared to increase bile lithogenicity (Melchior et al., 1972).

Grundy (1975) also observed that in several fasting, hypertriglyceridemic patients, gallbladder bile became more lithogenic after the administration of polyunsaturated fats. It has been suggested that polyunsaturated fats may increase bile lithogenicity in some patients through mobilization of cholesterol into bile.

XIII. EFA in Tumor Lipids

Many metabolic studies of lipid turnover indicate that there is a very active metabolism of all lipids in malignant tissue. Cholesterol, phospholipid, and fatty acids appear in higher concentrations in malignant neoplasms than in benign tumors or normal tissues (Yasuda and Bloor, 1932). In addition, a positive correlation between death rates from certain neoplasms and consumption of dietary fats has been demonstrated (Lea, 1966). When the lipid in biopsied livers of patients with malignant neoplastic diseases was analyzed it was observed that this lipid had higher 20:4 and lower 18:1 than the lipids from normal liver (Nakazawa and Yamagata, 1971). Similarly it was observed that sphingolipid and ceramide fatty acids in tumor lipids contained as much as 9% of 24:2 (Δ15, 18), a fatty acid normally not present in nontumorous tissue (Wood and Harlow, 1970).

De Alvarez et al. (1967) performed a very extended study of the fatty acid composition of serum of patients with gynecological malignancy and found that the cancer patients showed a generalized reduction in serum linoleic acid, accompanied by elevations in palmitic and stearic acids. In a later study (1969), these authors found a progressive rise in long-chain, highly unsaturated fatty acids in malignant tumor tissues as compared to benign gynecological tumors and normal tissue. Most consistently there was an elevation in the amount of 22:6. Similarly, Wells et al. (1973) observed an accumulation of 20:4 in liver tumors resulting from aflatoxin administration in rats. The authors speculate that the selective use of 18:2 by the tumor might deplete liver stores and lead to the reductions in serum linoleate observed in the cancer patients.

Kabara et al. (1972) suggest that the effect of dietary fat is exerted at the induction of carcinogenesis. A high fat diet when fed to mice after exposure to carcinogens enhanced tumor production. Although tumor-bearing rats quite often develop elevated plasma fatty acids, this is a rather infrequent finding in humans (Mays, 1969). This difference could possibly be explained by a difference in the proportion of tumor tissue to body weight.

Pearce and Dayton (1971) have reported excess deaths from carcinoma among elderly men ingesting a cholesterol-lowering, i.e., a high P/S ratio diet in a controlled clinical trial. However, several additional trials in various locations failed to confirm this finding (Ederer et al., 1971).

XIV. EFA and Prostaglandins

One of the more exciting developments in the field of unsaturated fatty acids has been the demonstration that EFA serve as precursors of prostaglandins (PG) (Bergstrom et al., 1964; Van Dorp et al., 1964a,b). Although prostaglandins were discovered some 40 years ago, it is only comparatively recently that their importance in metabolic systems has been recognized, and actually in 1966, Von Euler received a Nobel Prize Award for elucidating some of the biological properties of these physiologically active substances.

Prostaglandins are a group of closely related derivatives of prostanoic acid, a C-20 fatty acid. Even though their activity was first detected in the human semen, it is now known that these compounds affect many biological processes, e.g., blood pressure, heart rate, muscle contraction, lipolysis, and fertility. One of the interesting characteristics of PG activity is that they are active in minute amounts, and, depending on the situation and the type of prostaglandin, may effect opposing reactions (Horton, 1969; Bergstrom et al., 1968; Weeks, 1972; Ramwell et al., 1968).

In earlier work it was found that incubation of radioactive arachidonate with homogenates of vesicular glands from sheep resulted in the formation of radioactive PGE_2. Later it was established (Van Dorp et al., 1964b) that Δ8,11,14-eicosatrienoic acid is a precursor of PGE_1 and that 4,7,10,13-nonadecatetraenoic acid gives rise to PGE_2. These precursor–product relationships reconfirmed the essentiality of linoleic acid from which most of these more highly polyunsaturated fatty acids are derived. Beerthuis et al. (1968) pointed out striking correlations between the structure and EFA activity of the prostaglandin precursor acids; the efficiency of the PUFA as precursors for prostaglandins is dependent on their biological activity. Evidently the enzymes involved in the conversion are sensitive to the location of the double bond in the reacting substances.

Thus PGE_1 and PGF_1a are formed by the enzymatic oxidation and cyclization of a C-20:2 acid, dihomo-γ-linolenic, and PGE_2 and PGF_2 are formed from arachidonic.

The E and F series differ in that the group attached to C-9 is either a keto or an a hydroxy group. The 1 and 2 series differ in that there are 1 or 2 double bonds, respectively, on the side chains. PGA_1 and PGA_2 are dehydrated derivatives of PGE in which there is an additional double bond between C-10 and C-11.

Samuelsson (1972) has recently reviewed the mechanism of PG biosynthesis and catabolism. He suggested that since the enzymatic synthesis of these compounds involves cyclization of unesterified precursor acids and since the concentration of free fatty acids is generally very low in tissues, the possibility exists that esterified fatty acids might undergo oxidative cyclization either prior to or following hydrolysis. However, it was found that no cyclization occurred when the precursor was esterified to phospholipid (Vonkeman and Van Dorp, 1968) and therefore, PG synthesis is probably regulated by the action of phospholipases which release certain fatty acids from membranes (Pace-Asciak and Wolfe, 1970). In fact, Kunze and Vogt (1971) have suggested that since the precursors of PG are mainly esterified to phospholipids, it is phospholipase A which regulates PG synthesis.

There is a rapid conversion of the free precursor fatty acid to PG indicating that active PG-forming enzymes are constantly available in the intact tissue (Kunze, 1970). It is the amount of free substrate present which governs the production of prostaglandins.

It has been shown (Christ and Nugteren, 1970) that epididymal fat pads and isolated fat cells are able to release arachidonic acid as well as PGE_2 during lipolysis. Enzyme systems responsible for converting EFA into PG are present in adipose tissue. Very low concentrations of PG efficiently inhibit lipolysis. The addition of epinephrine to tissues promotes the activity of cAMP which, in turn, stimulates lipolysis. During lipolysis, free fatty acids, which are precursors of PG, are released and the prostaglandins, by inhibiting lipolysis, assure a continuous feedback control of excessive lipolysis in adipose tissue. In EFA-deficient adipose tissue there is a decreased release of PG and therefore an increase in lipolysis. It has been shown that in PUFA-fed rats there is less glycerol and less free fatty acids released from adipose tissue (Pawar and Tidwell, 1968; DePury and Collins, 1965).

However, the release of free fatty acids during noradrenaline infusion in man is not influenced by PGE_1. It is possible that PGE_1 is not transported to adipose tissue (Carlson et al., 1970) under these conditions.

Christ and Van Dorp (1972) compared PG biosynthetic activity in a variety of animal tissues ranging from lung to renal medulla to bladder. Synthetase activity was found in all tissues investigated. Although Jonsson and Anggard (1972) have pointed out that, normally, the level of endogenous PG in human skin is low, efficient enzyme systems are available for both prostaglandin formation and their metabolism.

The fact that the mechanism of activity of PG is quite complex is due in part to their great specificity in regulating metabolic processes. Even though PGE_1 prevents the hormone-stimulated increase in cAMP in isolated fat cells (Butcher and Baird, 1968), it neither inhibits isolated adenyl cyclase (Vaughan and Murad, 1968) nor does it stimulate phosphodiesterase (Christ and Nugteren, 1970).

An impairment of PG biosynthesis in sheep vesicular glands after hypophysectomy was considered by Privett et al. (1972) to be an indication that certain endocrine secretions are involved in the relationship between EFA and PG synthesis.

In view of the competition existing between unsaturated fatty acids in chain elongation and desaturation (Holman and Mohrhauer, 1963) it was expected that this competition might also be observed in PG formation. Thus Pace-Asciak and Wolfe (1968) reported that the conversion of arachidonic acid to PGE_2 by seminal vesicle (sheep) or stomach acetone powders (rat) was inhibited by oleic, linoleic, and linolenic acids. Similarly, PG synthesis was competitively inhibited by the Δ8-*cis*, 12-*trans*, 14-*cis*, 20:3 and Δ5-*cis*, 8-*cis*, 12-*trans*, 14-*cis*, 20:4 acids (Nugteren, 1970) and by Δ5,8,11,14-tetraenoic acid (Ahern and Downing, 1970).

Another type of inhibition of formation of PG is effected by corticosteroids. Greaves and McDonald-Gibson (1972) showed that in skin homogenates corticosteroids inhibited the formation of PGE_2 and $PGF_2\alpha$ from arachidonate. It may be that at least part of the antiinflammatory properties of corticosteroids is due to this inhibition of PG biosynthesis.

Biological inactivation of PG includes dehydrogenation, β and ω oxidation, and reduction of keto groups; these metabolites are excreted in the urine (Samuelsson et al., 1971). In general, men excrete greater amounts of prostaglandin metabolites than do women (Hamberg, 1972). The excretion rate of PG metabolites was found to be suppressed following oral aspirin or sodium salicylate ingestion; these substances are also antiinflammatory agents.

Prostaglandins have been found to affect ovarian steroids. In pregnant animals, treatment with PG results in a depressed progesterone level (Behrman et al., 1971). It has been conclusively established that PG administration during pregnancy results in abortion (Karim, 1971).

The discovery that the prostaglandins are biosynthesized from EFA led to the speculation that the symptoms of EFA deficiency might be attributed to lack of PG. When low concentrations of PG were added to isolated fat cells from rats deficient in EFA, it was found that the inhibiting effect on lipolysis was very pronounced, indicating that EFA-deficient cells were still sensitive to exogenous PG. This suggests that the hypersensitivity of lipolytic agents might be due to the impairment of the synthesis of endogenous PG from the PUFA. Another interpretation might include the possibility that higher levels of cAMP may be produced in EFA-deficient rats, or perhaps that the lipolytic system is subject to enhanced activation of cAMP (Paoletti et al., 1968).

In a recent study (Hafiez, 1974), administration of PGE_2 to male rats fed an EFA-deficient diet resulted in improvement in fertility from 45 to 100%. This may be compared with a similar improvement after methyl arachidonate administration. In female rats, treatment with PGE_2 or methyl arachidonate was equally effective in preventing irregular estrous cycles and prolongation of gestation but not the delayed vaginal opening caused by the EFA deficiency. In addition, lactation in the PGE_2 group was actually impaired rather than improved.

However, not all EFA deficiency symptoms resulting from a dietary deficiency of EFA could be corrected by a parenteral administration of PG. Apparently dermal signs of EFA deficiency are not caused by a secondary deficiency of PG. It was concluded that the role of EFA must be more than just to serve as precursors for prostaglandin synthesis (Kupiecki et al., 1968). However since approximately 9% of the fatty acids in human skin are 20:4 (Vroman et al., 1969), experiments which were undertaken to test whether 20:4 in the skin is metabolized to PG showed that biosynthesis of PG did occur, with PGE_2 being the major metabolic product

(Ziboh and Hsia, 1971). Understandably then, the skin of EFA-deficient animals is virtually void of PG (Van Dorp, 1970).

On the other hand, Ziboh and Hsia (1972) reported that EFA-deficient rats accumulated sterol esters in skin, that PGE_2 markedly inhibited the biosynthesis of these sterol esters and, at the same time, PGE_2 exhibited a curative action on the scaliness of skin associated with EFA deficiency. Also cholesteryl ester synthetase in the liver had been previously found to be inhibited by prostaglandins (Schweppe and Jungmann, 1970).

Cardiovascular effects of PGE_2 in man are quite spectacular. The rapid intravenous injection of 100 mg of PGE_2 or a slow continuous injection of 0.8 mg of PGE_2/kg of body weight increases heart rate and produces marked decreases in blood pressure. Similar effects can be produced by intramuscular doses of PGE_2 albeit at a higher level (Karim et al., 1971).

Adenosine diphosphate (ADP) induces aggregation of blood platelets and increases their adhesiveness to glass. This effect is inhibited by PGE_1 (Irion and Blombäck, 1969). Platelet thrombus formation in injured vessels in vivo is at least partially suppressed by PGE_1, either applied topically or injected intravenously into rabbits (Emmons et al., 1967).

Thomasson (1969) has related certain properties of prostaglandins to the effect of polyunsaturated fat on the development of atherosclerosis. Since saturated fats appear to be more atherogenic than unsaturated fats, one of the early theories of atherogenesis (Sinclair, 1956) was that this disease might be due to a chronic deficiency of EFA, resulting in increased capillary permeability. According to another theory ("hemostasis") (Owren, 1964), in the initial stages of atherosclerosis and thrombosis, a trauma in the arterial wall causes adhesion of blood platelets and eventually a thrombus is formed since the adhering platelets release ADP and thus perpetuate platelet aggregation. It is possible that PGE_1 is involved in the development and disappearance of the initial thrombus. Acute formation of thrombi have been demonstrated in arteriosclerotic subjects who have a relatively low supply of linoleic acid (Hornstra, 1970) and who, therefore, potentially synthesize less PG.

In addition, PGE_1 can also cause vasodilatation and decrease blood pressure. High blood pressure is one of the risk factors involved in the increased susceptibility to coronary heart disease. On the other hand, the contractable force of the perfused heart was shown to be stimulated by PGF_2 (Bergstrom et al., 1968). It is obvious that prostaglandins are a complex group of compounds which need considerably more study to elucidate their mode of action.

Understanding the mechanisms by which EFA exert their activity has been stimulated by the discovery of these extremely potent prostaglandins. However, when one considers the different concentrations of these substances which are required for activity in various processes, i.e., considerable amounts of EFA as compared with minute amounts of prostaglandins, either prostaglandins have an extremely high turnover rate and require a constant source of available EFA, or essential fatty acids, in addition to acting as precursors of PG, have other functions which are still to be elucidated.

REFERENCES

Adam, D. J. D., Hansen, A. E., and Wiese, H. F. (1958). *J. Nutr.* **66,** 555.
Aftergood, L., and Alfin-Slater, R. B. (1967). *J. Lipid Res.* **8,** 126.
Ahern, D. G., and Downing, D. T. (1970). *Biochim. Biophys. Acta* **210,** 456.
Ahrens, E. H., Jr., Hirsch, J., Peterson, M. L., Insull, W., Jr., Stoffel, W., Farquhar, J. W., Miller, T., and Thomasson, H J. (1959). *Lancet* i, 115.
Albutt, E. C., and Chance, G. W. (1969). *Amer. J. Clin. Nutr.* **22,** 1552.
Alfin-Slater, R. B., and Aftergood, L. (1968). *Physiol. Rev.* **48,** 758.
Alfin-Slater, R. B., and Aftergood, L. (1971). *Progr. Biochem. Pharmacol.* **6,** 214.
Alfin-Slater, R. B., Morris, R. S. Hansen, H., and Proctor, J. F. (1965). *J. Nutr.* **87,** 168.
Alfin-Slater, R. B., Hansen, H., Morris, R. S., and Melnick, D. (1969a). *J. Amer. Oil Chem. Soc.* **16,** 563.
Alfin-Slater, R. B., Morris, R. S., Aftergood, L., and Melnick, D. (1969b). *J. Amer. Oil Chem. Soc.* **16,** 657.
Alfin-Slater, R. B., Shimma, Y., Hansen, H., Wells, P., Aftergood, L., and Melnick, D. (1972). *J. Amer. Oil Chem. Soc.* **49,** 395.
Ali, S. S., Kuksis, A., and Beveridge, J. M. R. (1966). *Can. J. Biochem.* **44,** 1377.
Alimova, E. K., and Endakova, E. A. (1970). *Vop. Med. Khim.* **16,** 310.
Alling, C., Bruce, A., Karlsson, F., and Svennerholm, L. (1974). *Nutr. Metab.* **16,** 181.
Allman, D. W., and Gibson, D. M. (1965). *J. Lipid Res.* **6,** 51.
Anderson, J. T., Grande, F., and Keys, A. (1973). *J. Amer. Diet. Ass.* **62,** 133.
Arvidson, G., and Malmros, H. (1972). *Z. Ernaehrungswiss,* **11,** 105.
Avigan, J., and Steinberg, D. (1965). *J. Clin. Invest.* **44,** 1845.
Bagchi, K., Halder, K., and Chowdhury, S. R. (1959). *Amer. J. Clin. Nutr.* **7,** 251.
Baker, R. W. R., Thompson, R. H. S., and Zilkha, K. J. (1963). *Lancet* i, 26.
Ballabriga, A., Martinez, A., and Gallart-Catala, A. (1972). *Helv. Paediat. Acta* **27,** 91.
Bang, H. O., Dyerberg, J., and Nielsen, A. B. (1971). *Lancet* i, 1143.
Beerthuis, R. K., Nugteren, D. H., Pabon, H. J. J., and Van Dorp, D. A. (1968). *Rec. Trav. Chim. Pays-Bas* **87,** 461.
Behrman, H. R., Yashinaga, K., Wyman, H., and Greep, R. O. (1971). *Amer. J. Physiol.* **221,** 189.
Belin, J., Petter, N., Smith, A. D., Thompson, R. H. S., and Zilkha, K. J. (1971). *J. Neurol. Neurosurg. Psychiat.* **34,** 25.
Bentzen, A. J., Jacobsen, P. A., and Munch-Petersen, S. (1972). *Gerontol. Clin.* **14,** 217.
Bergstrom, S., Danielsson, H., and Samuelsson, B. (1964). *Biochim. Biophys. Acta* **90,** 207.

Bergstrom, S., Carlsson, L. A., and Weeks, J. R. (1968). *Pharmacol. Rev.* **20**, 1.

Bernsohn, J., and Stephanides, L. M. (1967). *Nature (London)* **215**, 821.

Bhattacharyya, A. K., Thera, C., Anderson, J. T., Grande, F., and Keys, A. (1969). *Amer. J. Clin. Nutr.* **22**, 1161.

Bierenbaum, M. L., Fleischman, A. I., and Green, D. P. (1970). *Circulation* **42**, 943.

Bieri, J. G., and Evarts, R. P. (1973). *J. Amer. Diet. Ass.* **62**, 147.

Bollinger, J. N., and Reiser, R. (1965). *J. Amer. Oil Chem. Soc.* **42**:1130.

Brenner, R. R., and Peluffo, R. O. (1966). *J. Biol. Chem.* **241**, 5213.

Brenner, R. R., Peluffo, R. O., Nervi, A. M., and DeTomas, M. E. (1969). *Biochim. Biophys. Acta* **176**, 420.

Bunyan, J., Diplock, A. T., and Green, J. (1967). *Brit. J. Nutr.* **21**, 217.

Butcher, R. W., and Baird, C. E. (1968). *J. Biol. Chem.* **243**, 1713.

Caldwell, D. F., and Churchill, J. A. (1966). *Psychol. Rep.* **19**, 99.

Caldwell, M. D., Jonsson, H. T., and Othersen, H. B., Jr. (1972). *J. Pediat.* **81**, 894.

Campbell, C. B., Cowley, D. J., and Dawling, R. H. (1972). *Eur. J. Clin. Invest.* **2**, 332.

Carlson, L. A., Ekelund, L. G., and Orö, L. (1970). *Acta Med. Scand.* **188**, 379.

Century, B., and Horwitt, M. K. (1964). *Arch. Biochem. Biophys.* **104**, 416.

Chait, A., Onitiri, A., Nicoll, A., Rabaya, E., Davies, J., and Lewis, B. (1974). *Atherosclerosis* **20**, 347.

Chen, L. H. (1974). *Nutr. Rep. Int.* **10**, 339.

Christ, E. J., and Nugteren, D. H. (1970). *Biochim. Biophys. Acta* **218**, 296.

Christ, E. J., and Van Dorp, D. A. (1972). *Biochim. Biophys. Acta* **270**, 537.

Christakis, G. J., Rinzler, S. H., Archer, M., Hashim, S. A., and Van Itallie, T. B. (1965). *Amer. J. Clin. Nutr.* **16**, 243.

Christiansen, M. M., and Wilcox, E. B. (1973). *J. Amer. Diet. Ass.* **63**, 138.

Chung, L. H., and Dupont, J. (1968). *Lipids* **3**, 545.

Churchill, J. A., Ayers, M. A., and Caldwell, D. F. (1967). *J. Amer. Med. Ass.* **201**, 482.

Clausen, J., and Friis-Hansen, B. (1971). *Z. Ernaehrungswiss.* **10**, 264.

Clausen, J., and Moller, J. (1967). *Acta Neurol. Scand.* **43**, 375.

Clifton-Bligh, P., Miller, N. E., and Nestel, P. J. (1974). *Metabolism* **23**, 437.

Collins, F. D., and Sinclair, A. J. (1969). *Proc. Australian Biochem. Soc.* **2**, 19.

Collins, F. D., Sinclair, A. J., Royle, J. P., Coats, D. A., Maynard, A. T., and Leonard, R. F. (1971). *Nutr. Metab.* **13**, 150.

Combes, M. A., Pratt, E. L., and Wises, H. F. (1962). *Pediatrics* **30**, 136.

Connor, W. E., Witiak, D. T., Stone, D. B., and Armstrong, M. L. (1969). *J. Clin. Invest.* **48**, 1363.

Cook, L. J., Scott, J. W., and Ferguson, K. A. (1970). *Nature (London)* **228**, 178.

Cook, L. J., Scott, T. W., and Faichney, G. J. (1972). *Lipids,* **7**, 83.

Dam, H. (1971). *Amer. J. Med.* **51**, 596.

Dam, H., Kruse, I., Jensen, M. K., and Kallehauge, H. E. (1967). *Scand. J. Clin. Lab. Invest.* **19**, 367.

Dam, H., Frange, I., and Sondergaard, E. (1968). *Z. Ernaehrungswiss.* **9**, 43.

Danziger, R. G., Hofman, A. I., Schoenfield, L. J., and Thistle, J. L. (1972). *New Engl. J. Med.* **286**, 1.

Davis, K. C. (1972). *Amer. J. Clin. Nutr.* **25**, 933.

Dayton, S., Hashimoto, S., and Pearce, M. L. (1967). *J. Lipid Res.* **8**, 508.

Dayton, S., Pearce, M. L., Hashimoto, S., Dixon, W. J., and Tomiyasu, U. (1969). *Circulation* **40** (Suppl. 2), 1.

De Alvarez, R. R. (1969). *Amer. J. Obstet. Gynecol.* **104**, 230.
De Alvarez, R. R., Goodel, B. W., and Zighelboim, I. (1967). *Amer. J. Obstet. Gynecol.* **97**, 419.
De Gomez Dumm, I. N. T., Peluffo, R. O., and Brenner, R. R. (1972). *Lipids* **7**, 590.
De Pury, G. G., and Collins, F. D. (1965). *Biochim. Biophys. Acta* **106**, 213.
De Pury, G. G., and Collins, F. D. (1966). *Chem. Phys. Lipids* **1**, 1.
Deuel, H. J., Jr., Alfin-Slater, R. B., Wells, A. F., Kryder, G. D., and Aftergood, L. (1955). *J. Nutr.* **55**, 337.
Dhopeshwarkar, G. A., and Mead, J. F. (1973). *Advan. Lipid Res.* **11**, 109.
Dju, M. Y., Mason, K. E., and Filer, L. J., Jr. (1958). *Amer. J. Clin. Nutr.* **6**, 50.
Ederer, F., Leren, P., Turpeinen, O., and Frantz, I. D., Jr. (1971). *Lancet* **ii**, 203.
Emmons, P. R., Hampton, J. R., Harrison, M. J. G., Honour, A. J., and Mitchell, J. R. A. (1967). *Brit. Med. J.* **2**, 468.
Engelberg, H. (1966). *Metabolism* **15**, 26.
Evans, D. W., Turner, S. M., and Ghosh, P. (1972). *Lancet* **i**, 72.
Evans, H. M., and Bishop, K. S. (1922). *J. Metab. Res.* **1**, 335.
Fleischman, A. I., Bierenbaum, M. L., Justice, D., Stier, A., Sullivan, A., and Fleischman, M. (1975). *Amer. J. Nutr.* **28**, 601.
Fleischman, A. I., Justice, D., Watson, P., and Bierenbaum, M. L. (1974). *Fed. Amer. Soc. Exp. Biol.* **33**, 244.
Fredrickson, D. S., and Gordon, R. S. (1958). *J. Clin. Invest.* **37**, 1504.
Fukazawa, T., and Privett, O. S. (1972). *Lipids* **7**, 387.
Fukazawa, T., Privett, O. S., and Takahashi, Y. (1971). *Lipids* **6**, 388.
Fulco, A. J., and Mead, J. F. (1959). *J. Biol. Chem.* **234**, 1411.
Galli, C. (1973). *In* "Dietary Lipids and Postnatal Development" (C. Galli, G. Jacini, and A. Pecile, eds.), p. 191. Raven Press, New York.
Galli, C., and Spagnuolo, C. (1974). *Lipids* **9**, 450.
Gidez, L. E. (1964). *Biochem. Biophys. Research Commun.* **14**, 413.
Gjone, E., and Norum, K. R. (1968). *Acta Med. Scand.* **183**, 107.
Gjone, E., Blomhoff, J. P., and Skarbovik, A. J. (1974). *Clin. Chim* **A54**, 11.
Gjone, E., and Norum, K. R. (1974). *Scand. J. Clin. Lab. Invest.* **33** (Suppl. 137), 1.
Glomset, J. A. (1968). *J. Lipid Res.* **9**, 155.
Glomset, J. A., and Norum, K. R. (1973). *Advan. Lipid Res.* **11**, 1.
Greaves, M. W., and McDonald-Gibson, W. (1972). *Brit. Med. J.* **2**, 83.
Grundy, S. M. (1975). *J. Clin. Invest.* **55**, 269.
Grundy, S. M., and Ahrens, E. H., Jr. (1966). *J. Clin. Invest.* **45**, 1503.
Grundy, S. M., and Ahrens, E. H., Jr. (1970). *J. Clin. Invest.* **49**, 1135.
Guarnieri, M., and Johnson, R. M. (1970). *Advan. Lipid Res.* **8**, 115.
Gustafson, A., Lilienburg, L., and Svanborg, A. (1974). *Scand. J. Clin. Lab. Invest.* **33** (Suppl. 137), 63.
Haerem, J. W. (1971). *Atherosclerosis* **14**, 417.
Hafiez, A. A. (1974). *J. Reprod. Fert.* **38**, 273.
Hafiez, A. A., Khalifa, K., Soliman, L., Fayad, I., Fayek, K., and Abdel-Wahab, F. (1971). *Lipids* **6**, 208.
Hagenfeldt, L., Paasikivi, J., and Sjögren, A. (1973). *Metabolism* **22**, 1349.
Hagerman, L. M., and Schneider, D. L. (1973). *Proc. Soc. Exp. Biol. Med.* **143**, 93.
Hamberg, M. (1972). *Biochem. Biophys. Research Commun.* **49**, 720.
Hampton, J. R., and Mitchell, J. R. A. (1966). *Lancet* **ii**, 764.
Hansen, A. E., Haggard, M. E., Boelsche, A. N., Adam, D. J. D., and Wiese, H. F. (1958). *J. Nutr.* **66**, 565.

78   ROSLYN B. ALFIN-SLATER AND LILLA AFTERGOOD

Hansen, A. E., Stewart, R. A., Hughes, G., and Söderhjelm, L. (1962). *Acta Paediat. Suppl.* **51**, 137.
Hansen, A. E., Wiese, H. F., Boelsche, A. N., Haggard, M. E., Adam, D. J. D., and Davis, H. (1963). *Pediatrics* **31** (Suppl. 1), 171.
Hansen, I. B., Friis-Hansen, B., and Clausen, J. (1969). *Z. Ernaehrungswiss.* **9**, 352.
Harland, W. A., Gilbert, J. D., and Brooks, C. J. W. (1973). *Biochim. Biophys. Acta* **316**, 378.
Harman, D. (1969). *J. Amer. Geriat. Soc.* **17**, 721.
Harris, P. L., and Embree, N. D. (1963). *Amer. J. Clin. Nutr.* **13**, 385.
Hashim, S. A., and Asfour, R. H. (1968). *Amer. J. Clin. Nutr.* **21**, 7.
Hassan, H., Hashim, S. A., Van Itallie, T. B., and Sebrell, W. H. (1966). *Amer. J. Clin. Nutr.* **19**, 147.
Haust, H. L., and Beveridge, J. M. R. (1963). *J. Nutr.* **81**, 13.
Helmkamp, G. M. Jr., Wilmore, D. W., Johnson, A. A., and Pruitt, B. A., Jr. (1973). *Amer. J. Clin. Nutr.* **26**, 1331.
Herting, D. C., and Drury, E. E. (1963). *J. Nutr.* **81**, 335.
Hofman, A. F., Northfield, T. C., and Thistle, J. L. (1973). *New Engl. J. Med.* **288**, 46.
Holman, R. T. (1960). *J. Nutr.* **70**, 405.
Holman, R. T. (1968). *Progr. Chem. Fats* **9**, 275.
Holman, R. T., Mohrhauer, H. (1963). *Acta Chem. Scand.* **17**, 584.
Holman, T. T. (1970). *Progr. Chem. Fats* **9**, 607.
Holman, T. T., Caster, W. O., and Wiese, H. F. (1964). *Amer. J. Clin. Nutr.* **14**, 70.
Hornstra, G. (1970). *Fette Seifen Anstrichm.* **72**, 960.
Hornstra, G., Lewis, B., Chait, A., Turpeinen, O., Karvonen, M. J., and Vergroesen, A. J. (1973). *Lancet* **i**, 1155.
Horton, E. W. (1969). *Physiol. Rev.* **49**, 122.
Horwitt, M. K. (1959). *Fed. Proc. Fed. Amer. Soc. Exp. Biol.* **18**, 520.
Horwitt, M. K. (1962). *Vit. Horm.* (*N. Y.*) **20**, 541.
Inkpen, C. A., Harris, R. A., and Quackenbush, F. W. (1969). *J. Lipid Res.* **10**, 277.
Insull, W., Hirsch, J., James, T., and Ahrens, E. H., Jr. (1959). *J. Clin. Invest.* **38**, 443.
Insull, W., Lang, P. D., and Hsi, B. P. (1970). *Amer. J. Clin. Nutr.* **23**, 17.
Irion, E., and Blombäck, M. (1969). *Scand. J. Clin. Invest.* **24**, 141.
Irsigler, K., Kryspin-Exner, K., and Mildschuh, W. (1970). *Acta Hepatosplenol.* **17**, 103.
Jager, F. C., and Houtsmiller, U. M. T. (1970). *Nutr. Metab.* **12**, 3.
Jaillard, J., Sezille, G., Scherpered, P., Fruchart, J. C., and Biserte, G. (1972). *Pathol. Biol.* **20**, 51.
Jolliffe, N. (1961). *Metabolism* **10**, 497.
Jonsson, C. E., and Anggard, E., (1972). *Scand. J. Clin. Lab. Invest.* **29**, 289.
Kabara, J. J., Chapman, B. B., and Borin, B. M. (1972). *Proc. Soc. Exp. Biol. Med.* **139**, 100.
Kahlke, W. (1964). *Klin Wochenschr.* **42**, 1011.
Karim, S. (1971). *Ann. N.Y. Acad. Sci.* **180**, 482.
Karim, S., Somers, K., and Hillier, K. (1971). *Cardiovasc. Res.* **5**, 255.
Kasper, H., Thiel, H., and Ehl, M. (1973). *Amer. J. Clin. Nutr.* **26**, 197.
Kaunitz, H., Slanetz, C. A., Johnson, R. E., and Babayan, V. K. (1961). *J. Nutr.* **72**, 386.
Kekomaki, M. P. (1970). *Acta Anaesthesiol. Scand.* (*Suppl.*) **37**, 18.

Keys, A. (1970). *Circulation* (*Suppl.*) **41**, 1.

Keys, A. (1971). *In* "Coronary Heart Disease" (H. I. Russek, and B. L. Zohman, eds.), p. 59. Lippincott, Philadelphia, Pennsylvania.

Kingery, F. A. J., and Kellum, R. E. (1965). *Arch. Dermatatol.* **91**, 272.

Kingsbury, K. J., and Brett, C. (1974). *Postgrad. Med. J.* **50**, 425.

Kinsell, L. W., Partridge, J., Boling, L., Margen, S., and Michaels, G. D. (1952). *J. Clin. Endocrinol. Metab.* **12**, 909.

Kirkeby, K., Ingvaldsen, P., and Bjerkedal, I. (1972). *Acta Med. Scand.* **192**, 521.

Konno, T., Fujii, M., Watanuki, T., and Kaizumi, K. (1966). *Tohoku J. Exp. Med.* **90**, 375.

Kunze, H. (1970). *Biochim. Biophys. Acta* **202**, 180.

Kunze, H., and Vogt, W. (1971). *Ann. N.Y. Acad. Sci.* **190**, 123.

Kupiecki, F. P., Sekhar, N. C., and Weeks, J. R. (1968). *J. Lipid Res.* **9**, 602.

Kyriakides, E. C., Fillipone, N., Paul, B., Grattan, W., and Balint, J. A. (1970). *Pediatrics* **46**, 431.

Laurell, S., and Lundquist, A. (1971). *Scand. J. Clin. Invest.* **27**, 29.

Laurell, S., Nilsen, R., and Norden, A. (1972). *Clin. Chim. Acta* **36**, 169.

Lea, A. J. (1966). *Lancet* **ii**, 332.

Lewis, J. S., Pian, A. K., Baer, M. T., Acosta, P. B., and Emerson, G. A. (1973). *Amer. J. Clin. Nutr.* **26**, 136.

Love, W. C., Cashell, A., Reynolds, M., and Callaghan, N. (1974). *Brit. Med. J.* **2**, 18.

Mahadevan, V., Singh, H., and Lundberg, W. O. (1966). *Proc. Soc. Exp. Biol. Med.* **121**, 82.

Marcel, Y. L., Christiansen, K., and Holman, R. T. (1968). *Biochim. Biophys. Acta* **164**, 25.

Mathias, M. M., and Dupont, J. (1969). *Fed. Proc., Fed. Amer. Soc. Exp. Biol.* **28**, 370.

Mays, E. T. (1969). *J. Surg. Res.* **9**, 273.

Mead, J. F. (1961). *Fed. Proc., Fed. Amer. Soc. Exp. Biol.* **20**, 952.

Mendy, F., Hirtz, J., Berret, R., Rio, B., Serville, F., and Verger, P. (1970). *Arch. Sci. Physiol.* **24**, 279.

Melchior, G. W., Clarkson, T. B., and Bullock, B. C. (1972). *Circulation* **46** (Suppl. 2), 19.

Miettinen, T. A., Ahrens, E. H., Jr., and Grundy, S. M. (1965). *J. Lipid Res.* **6**, 411.

Millar, J. H. D., Zilkha, K. J., Langman, M. J. S., Wright, H. P., Smith, A. D., Belin, J., and Thompson, R. H. S. (1973). *Brit. Med. J.* **1**, 765.

Misra, D. P., Staddon, G., Powell, N., Misra, J., and Crook, D. (1974). *Clin. Chim. Acta* **56**, 83.

Mohrhauer, H., and Holman, R. T. (1963). *J. Nutr.* **81**, 67.

Mohrhauer, H., and Holman, R. T. (1967). *J. Nutr.* **91**, 528.

Moore, R. B., Anderson, J. T., Taylor, H. L., Keys, A., and Frantz, I. D., Jr. (1968). *J. Clin. Invest.* **47**, 1517.

Mueller, J. F. (1973). *J. Amer. Diet. Ass.* **62**, 613.

Muto, Y., and Gibson, D. M. (1970). *Biochem. Biophys. Research Commun.* **38**, 9.

Nakazawa, I., and Yamagata, S. (1971). *Tohoku J. Exp. Med.* **103**, 129.

Nestel, P. J. (1965). *Metabolism* **14**, 1.

Nestel, P. J., and Barter, P. (1971). *Clin. Sci.* **40**, 345.

Nestel, P. J., Bezman, A., and Havel, R. J. (1962). *Amer. J. Physiol.* **203**, 914.

Nestel, P. J., Carroll, K. F., and Havenstein, N. (1970). *Metabolism* **19**, 1.

Nestel, P. J., Havenstein, N., Whyte, H. M., Scott, T. J., and Cook, L. J. (1973). *New Engl. J. Med.* **288**, 379.

Nestel, P. J., Havenstein, N., Scott, T. W., and Cook, L. J. (1974). *Aust. N. Z. J. Med.* **4**, 497.

Nestel, P. J., Havenstein, N., Homma, Y., Scott, T. W., and Cook, L. J. (1975). *Metabolism* **24**, 189.

Nichols, A. V., and Smith, L. (1965). *J. Lipid Res.* **6**, 206.

Nordoy, A., and Lund, S. (1968). *Scand. J. Clin. Lab. Invest.* 22:328.

Nordoy, A., and Rodset, J. M. (1970). *Acta Med. Scand.* **188**, 133.

Nordoy, A., and Rodset, J. M. (1971). *Acta Med. Scand.* **190**, 27.

Northfield, T. C., and Hofmann, A. (1973). *Lancet* i, 747.

Notevarp, O. (1966). *Tek. Ukebl.* **113**, 465.

Nugteren, D. H. (1970). *Biochim. Biophys. Acta* **210**, 171.

Olegard, R., and Svennerholm, L. (1971). *Acta Paediat. Scand.* **60**, 505.

Ostwald, R., Yamanaka, W., and Light, M. (1970). *Proc. Soc. Exp. Biol. Med.* **134**, 814.

Osuga, T., and Portman, O. W. (1971). *Proc. Soc. Exp. Biol. Med.* **136**, 722.

Owren, P. A. (1964). *Nutr. Dieta Suppl.* **6**, 156.

Owren, P. A. (1965). *Ann. Intern. Med.* **63**, 167.

Pace-Asciak, C., and Wolfe, L. S. (1968). *Biochim. Biophys. Acta* **152**, 784.

Pace-Asciak, C., and Wolfe, L. S. (1970). *Biochim. Biophys. Acta* 218:539.

Panos, T. C., Stinnett, B., Zapata, G., Eminians, J., Marasigan, B. V., and Beard, A. G. (1968). *Amer. J. Clin. Nutr.* **21**, 15.

Paoletti, R., Puglisi, L., and Usardi, M. M. (1968). *Advan. Exp. Med. Biol.* **2**, 425.

Paulsrud, J. R., Pensler, L., Whitten, C. F., Stewart, S., and Holman, R. T. (1972). *Amer. J. Clin. Nutr.* **25**, 897.

Pawar, S. S., and Tidwell, H. C. (1968). *Biochim. Biophys. Acta* 164:167.

Pearce, M. L., and Dayton, S. (1971). *Lancet* ii, 464.

Peifer, J. J., and Holman, R. T. (1959). *J. Nutr.* **68**, 155.

Peluffo, R. O., Brenner, R. R., and Mercuri, O. (1963). *J. Nutr.* **81**, 110.

Peluffo, R. O., De Gomez Dumm, I. N. T., and Brenner, R. R. (1972). *Lipids* **7**, 363.

Pensler, L., Whitten, C., Paulsrud, J., and Holman, R. T. (1971). *J. Pediat.* **78**, 1067.

Phillips, G. B., and Dodge, J. T. (1968). *J. Lab. Clin. Med.* **71**, 629.

Portman, O. W., and Sugano, M. (1964). *Arch. Biochem. Biophys.* **105**, 532.

Press, M., Kikuchi, H., and Thompson, G. R. (1972). *Gut* **13**, 836.

Press, M., Kikuchi, H., Shimoyama, T., and Thompson, G. R. (1974). *Brit. Med. J.* **2**, 247.

Pritchard, E. T., and Singh, H. (1960). *Biochem. Biophys. Res. Commun.* **2**, 184.

Privett, O. S., Phillips, F., Fukazawa, T., Kaltenbach, C. C., and Sprecher, H. W. (1972). *Biochim. Biophys. Acta* **280**, 348.

Quarfordt, S. H., Oelschlager, H., and Krigbaum, W. R. (1972). *J. Clin. Invest.* **51**, 1979.

Ramwell, P. W., Shaw, J. E., Clarke, G. B., Grostic, M. F., Kaiser, D. G., and Pike, J. E. (1968). *Progr. Chem. Fats* **9**, 233.

Rathbone, L. (1965). *Biochem. J.* **97**, 629.

Rehnborg, C. S., and Nichols, A. V. (1964). *Biochim. Biophys. Acta* **84**, 596.

Reiser, R. (1973). *Amer. J. Clin. Nutr.* **26**, 524.

Reuter, W., Ries, W., and Klinger, H. (1969). *Deut. Z. Verdau. Stoffwechselk.* **29**, 149.

Richardson, T. J., and Sgoutas, D. (1975). *Amer. J. Clin. Nutr.* **28**, 258.

Rinzler, S. H. (1968). *Bull. N.Y. Acad. Med.* **44**, 936.

Roels, O. A., and Hashim, S. A. (1962). *Fed. Proc., Fed. Amer. Soc. Exp. Biol.* **21** (Suppl. 2), 71.

Rose, J. C., Johnson, M., Ramwell, P. W., and Kot, P. A. (1974). *Proc. Soc. Exp. Biol. Med.* **147**, 652.

Rosenlund, M. L., Kim, H. K., and Kritchevsky, D. (1974). *Nature (London)* **251**, 719.

Sabine, J. R., McGrath, H., and Abraham, S. (1969). *J. Nutr.* 98:312.

Samuelsson, B. (1972). *Fed. Proc., Fed. Amer. Soc. Exp. Biol.* **31**, 1442.

Samuelsson, B., Granstrom, E., Green, K., and Hamberg, M. (1971). *Ann. N.Y. Acad. Sci.* **180**, 138.

Schendel, H. E., and Hansen, J. D. L. (1961). *Amer. J. Clin. Nutr.* **9**, 735.

Scherer, R. (1974). *Klin. Wochenschr.* **52**, 203.

Schlenk, H., and Sand, D. M. (1967). *Biochim. Biophys. Acta* **144**, 305.

Schrade, W., Boehle, E., Biegler, R., and Harmuth, E. (1963). *Lancet* **i**, 285.

Schubert, W. K. (1973). *Amer. J. Cardiol.* **31**, 581.

Schwarz, K. (1961). *Amer. J. Clin. Nutr.* **9**, 71.

Schweppe, J. S., and Jungmann, R. A. (1970). *Proc. Soc. Exp. Biol. Med.* **133**, 1307.

Segall, M. M., Fosbrooke, A. S., Lloyd, J. K., and Wolff, O. H. (1971). *Amer. Heart J.* **82**, 707.

Selivonchik, D. P., and Johnston, P. V. (1975). *J. Nutr.* **105**, 288.

Simon, J. B., Kepkey, D. L., and Poon, R. (1974). *Gastroenterology* **66**, 539.

Sinclair, H. M. (1956). *Lancet* **i**, 381.

Sinclair, H. M. (1964). In "Nutrition" (G. H. Beaton and E. W. McHenry, eds.), Vol. 1, p. 93. Academic Press, New York.

Smith, W. L., and Lands, W. E. M. (1972). *Biochemistry* **11**, 3276.

Sodhi, H., Wood, P., and Kinsell, L. (1964). *Clin. Res.* **12**, 108.

Söderhjelm, L., and Hansen, A. E. (1962). *Pediat. Clin. N. Amer.* **9**, 927.

Söderhjelm, L., Wiese, H. F., and Holman, R. T. (1970). *Progr. Chem. Fats* **9**, 555.

Solyom, A., Muhlbachova, E., and Puglisi, L. (1967). *Biochim. Biophys. Acta* **137**, 427.

Spritz, N., and Mishkel, M. A. (1969). *J. Clin. Invest.* **48**, 78.

Strisower, E. H., Adamson, G., and Strisower, B. (1968). *Amer. J. Med.* **45**, 488.

Sturdevant, R. A. L., Pearce, M. L., and Dayton, S. (1973). *New Engl. J. Med.* **288**, 24.

Sugano, M., and Portman, O. (1965). *Arch. Biochem. Biophys.* **109**, 302.

Sullivan, J. F., Kelly, M., and Suchy, N. (1969). *Amer. J. Dig. Dis.* **14**, 864.

Sun, G. Y., Go, J., and Sun, A. Y. (1974). *Lipids* **9**, 450.

Svennerholm, L. (1968). *J. Lipid Res.* **9**, 570.

Tappel, A. L. (1965). *Fed. Proc. Fed. Amer. Soc. Exp. Biol.* **24**, 73.

Taylor, G. O. (1971). *Amer. J. Clin. Nutr.* **24**, 1212.

Thistle, J. L., and Schoenfield, L. J. (1969). *J. Lab. Clin. Med.* **74**, 1020.

Thomasson, H. J. (1953). *Int. Z. Vitaminforsch.* **25**, 62.

Thomasson, H. J. (1969). *Nutr. Dieta* **11**, 228.

Truswell, A. S., and Mann, J. I. (1972). *Atherosclerosis* **16**, 15.

Turpeinen, O. (1971). *Ann. Clin. Res.* **3**, 433.

Van Dorp, D. A. (1970). *Ann. N.Y. Acad. Sci.* **180**, 181.

Van Dorp, D. A., Beerthuis, R. K., Nugteren, D. H., and Vonkeman, H. (1964a). *Biochim. Biophys. Acta* **90**, 204.

Van Dorp, D. A., Beerthuis, R. K., Nugteren, D. H., and Vonkeman, H. (1964b). *Nature (London)* **203**, 839.

Vaughan, M., and Murad, F. (1968). *Biochemistry* **8**, 3092.

Vergroesen, A. J. (1972). *Proc. Nutr. Soc.* **31,** 323.

Vincent, J. E., Melai, A., and Bonta, I. L. (1974). *Prostaglandins* **5,** 369.

Vonkeman, H., and Van Dorp, D. A. (1968). *Biochim. Biophys. Acta* **110,** 430.

Vroman, H. E., Nemecek, R. A., and Hsia, S. L. (1969). *J. Lipid Res.* **10,** 507.

Walker, B. L. (1967). *Lipids* **2,** 497.

Wapnick, S., Norden, D. A., and Venturas, D. J. (1974). *Gut* **15,** 367.

Weeks, J. R. (1972). *Annu. Rev. Pharmacol.* **12,** 317.

Wells, P., Aftergood, L., and Alfin-Slater, R. B. (1973). Unpublished data.

White, H. B., Jr., Galli, C., and Paoletti, R. (1971). *J. Neurochem.* **18,** 869.

Wilson, W. S., Hulley, S. B., Burrows, M. I., and Nichaman, M. Z. (1971). *Amer. J. Med.* **51,** 491.

Witting, L. A. (1970). *Progr. Chem. Fats* **9,** 517.

Witting, L. A. (1972). *Amer. J. Clin. Nutr.* **25,** 257.

Witting, L. A., and Lee, L. (1975). *Amer. J. Clin. Nutr.* **28,** 577.

Wood, R., and Harlow, R. D. (1970). *Arch. Biochem. Biophys.* **141,** 183.

Wood, P., Shioda, R., and Kinsell, L. W. (1966). *Lancet* **ii,** 604.

Woodruff, C. W., Bailey, M. C., Davis, J. T., Rogers, N., and Coniglio, J. G. (1964). *Amer. J. Clin. Nutr.* **14,** 83.

Yasuda, M., and Bloor, W. R. (1932). *J. Clin. Invest.* **11,** 677.

Ziboh, V. A., and Hsia, S. L. (1971). *Arch. Biochem. Biophys.* **146,** 100.

Ziboh, V. A., and Hsia, S. L. (1972). *J. Lipid Res.* **13,** 458.

Chapter 3

METHODS FOR EVALUATION

OF HYPOLIPIDEMIC DRUGS

IN MAN: MECHANISMS

OF THEIR ACTION

Tatu A. Miettinen

Second Department of Medicine, University of Helsinki,
Helsinki, Finland

I. Introduction

The basic lesion of atherosclerotic disease is an arterial cholesterol-rich atheroma, the development of which is believed to be markedly enhanced by the concomitant existence of hyperlipoproteinemia, in particular, β-hyperlipoproteinemia and also pre-β-hyperlipoproteinemia and a combination of the two. On the other hand, reduction of an elevated plasma lipid level can be expected to prevent development and slow down

83

progression of already existing atheromatosis. As a matter of fact, dietary studies have demonstrated rather convincingly that the use of plasma cholesterol-lowering diets reduces the occurrence of fatal and nonfatal coronary events in subjects free of or with manifest coronary heart disease (Leren, 1966; Christakis *et al.*, 1966; Dayton *et al.*, 1969; Miettinen *et al.*, 1972). Similar results have been obtained with clofibrate (*Brit. Med. J.*, 1971; Krasno and Kidera, 1972), the most frequently used hypolipidemic drug and with a bile acid sequestering agent Colestipol (Dorr *et al.*, 1974). The favorable results of hypolipidemic therapy on coronary artery disease have further strengthened the view that hyperlipoproteinemias should be treated with all available means, primarily with diet and, in diet-resistant cases, with hypolipidemic drugs. Though the evaluation of effectiveness of these agents on lipids can be best determined by serial determinations of the blood lipid levels, possibly with ultracentrifugal separation and analysis of different lipoprotein fractions, it would be important to know their effects on lipid synthesis, absorption and elimination, turnover of lipoproteins, and especially on the tissue contents of lipids. A reliable measurement of the tissue contents of lipids, if possible *in vivo*, can be considered extremely valuable for cholesterol, the major lipid constituent of arterial atheromas, particularly because in some individuals even a negligible change in plasma lipid levels can be expected to cause a significant mobilization of tissue cholesterol (Grundy *et al.*, 1972). In practical clinical work hyperlipoproteinemias are usually detected at a fairly late stage when clinical manifestations reveal the apparent presence of atheromatosis in peripheral, coronary, and/or cerebral arteries. Therefore, reduction in size of these atheromas becomes one of the basic objectives of the hypolipidemic therapy, which ultimately would then improve the prognosis of the patient.

This chapter deals chiefly with methods which can be used to study lipid metabolism in man. In addition, it considers the drug-induced changes in lipid metabolism. Since different forms of hypercholesterolemia have been considered to be the most important risk factor in atheromatous coronary heart disease and since atheromas are rich in cholesterol, primary attention will be paid to the measurement of different parameters in cholesterol metabolism and to the detection of changes in them. Methods dealing with plasma triglycerides will also be dealt with briefly. Finally, a short survey will be presented on the mechanism of action of different hypolipidemic drugs.

II. Outlines of Cholesterol Metabolism

The cholesterol of the body originates from two different sources, diet and endogenous synthesis. The latter is usually the predominant factor and takes place in practically every cell, but especially in the liver and intestinal mucosa (cf. Dietschy and Wilson, 1970). Dietary cholesterol is mixed with biliary cholesterol and forms micelles for absorption with the aid of bile acids, phospholipids, free fatty acids, and monoglycerides. About 30 to 50% of the cholesterol (Borgström, 1969; Quintao et al., 1971b; Kudchodkar et al., 1973) is absorbed and transported primarily as chylomicra and to a small extent as very low-density lipoproteins (VLDL, pre-β-lipoprotein in electrophoresis) via the thoracic duct into the blood stream. According to current knowledge (cf. Steinberg, 1974), with the contribution of different lipases (lipoprotein lipase and liver lipase), chylomicrons are converted to VLDL and further to low-density lipoprotein (LDL, β-lipoprotein in electrophoresis), the major plasma cholesterol-carrying lipoprotein, to which endogenously synthesized cholesterol is incorporated homogeneously with the dietary one.

Cholesterol in plasma lipoproteins is fairly rapidly equilibrated with a certain portion of tissue cholesterol (rapidly equilibrating cholesterol pool, called pool A or pool I, located mainly in the blood, liver, and intestine), whereas mixing with the major part of tissue cholesterol takes place very slowly (slowly equilibrating pool, called pool B or pool II, located mainly in muscle, adipose, and connective tissue and in lipid deposits, atheromas, and xanthomas) (Goodman and Noble, 1968; Goodman et al., 1973b; Nestal et al., 1969; Samuel and Lieberman, 1973). Though the amount of exchangeable tissue cholesterol may not be correlated with the serum cholesterol concentration (Samuel and Lieberman, 1973), the connective tissue cholesterol increases with age (Crouse et al., 1972), and lipid deposits are formed more frequently at high rather than at low or normal serum cholesterol concentrations. Studies with labeled cholesterol have suggested that reduction of plasma cholesterol in man is associated with mobilization of tissue cholesterol (Miettinen, 1968ab; Grundy et al., 1972; Sodhi et al., 1973), the exact origin of which is unknown. On the other hand, increased entry of dietary cholesterol into the plasma lipoproteins may result in a net flux into the tissue cholesterol pool(s) even without any clear-cut increase of plasma cholesterol concentration (Quintao et al., 1971b).

Removal of cholesterol from the body takes place primarily via the gastrointestinal route into the stools as bile acids and cholesterol and its bacterial transformation products, coprostanol and coprostanone. In the liver, cholesterol is converted to primary bile acids, the cholic and chenodeoxycholic acids, via several steps initiated by 7α-hydroxylation of cholesterol (cf. Danielsson and Einarsson, 1969). They are secreted into the bile and about 95 to 97% of them is reabsorbed primarily from the terminal ileum and to a small extent from the colon back to the liver. This forms the enterohepatic circulation of the bile acid pool, about 2 to 5% of which is lost into stools during each cycle. In man about two-thirds of the overall fecal elimination of cholesterol takes place as neutral sterols and one-third as bile acids (cf. Miettinen, 1970a, 1973a). If reabsorption of bile acids is inhibited, even to a small extent, fecal excretion increases rapidly and the size of the bile salt pool is decreased. This enhances hepatic bile acid synthesis by activating the rate-limiting step, 7α-hydroxylation of cholesterol and releases finally the negative feedback inhibition of cholesterol synthesis between hydroxy-methylglutaryl (HMG)-CoA and mevalonate (HMG-CoA reductase is activated) so that hepatic cholesterol production may increase several-fold. Lack of intestinal bile acids also stimulates the mucosal cholesterol synthesis (Dietschy, 1968, Dietschy and Gamel, 1971). Increased fecal bile acid excretion is virtually always associated with a reduction of the plasma cholesterol level although an excessive compensatory increase in cholesterol synthesis prevents this occasionally. The same appears to be the case with cholesterol itself. If the amount of dietary cholesterol is reduced or especially if cholesterol absorption is inhibited (fecal excre-tion of neutral sterols increases), HMG-CoA reductase is activated. However, serum cholesterol tends to decrease, despite increased choles-terol synthesis. It is interesting to note that though HMG-CoA reductase activity is the primary rate-limiting step of cholesterol synthesis, later steps, including cyclization of squalene and demethylation and rearrangements of double bonds of lanosterol, may also become rate-limiting. A visible sign of this is an increase of cholesterol precursors in plasma when cho-lesterol synthesis is increased, e.g., by interrupted enterohepatic circula-tion of bile acids (Miettinen, 1970a).

Plasma cholesterol is also eliminated to a small extent via the skin (Bhattacharyya et al., 1972). It corresponds to about 10% of the fecal cholesterol output and is not influenced by the plasma cholesterol con-centration. A small portion of cholesterol is also excreted into the urine and feces as steroid hormones (cf. Borkowski et al., 1967). However, the skin and steroid hormone routes of cholesterol elimination are apparently quite stable and not greatly affected by changes in cholesterol metab-olism; therefore, they have usually been neglected in studies of choles-terol metabolism. Normally urinary cholesterol and bile acid excretion is negligible; the former can increase somewhat in the nephrotic syn-drome, while the urinary output of bile acids is markedly enhanced in many hepatobiliary diseases (cf. van Berge Henegouwen, 1974). Accord-ingly, cholesterol elimination can be measured quantitatively by deter-mining fecal neutral sterols of cholesterol origin and fecal bile acids, and to which urinary bile acids can be added in patients with liver disease (Miettinen, 1972a).

The different aspects of cholesterol synthesis and metabolism that may be needed when evaluating the mechanism of action of a hypolipidemic drug include the following: overall cholesterol production and measurement of the activity of the primary regulatory step between HMG-CoA and mevalonate in different tissues (if possible), fecal elimination of cholesterol as bile acids and neutral steroids, bile acid kinetics, amount of dietary sterols, absorption of dietary and endogenous cholesterol, biliary secretion of cholesterol and bile acids, pool size of bile acids, pool sizes and turnover of plasma lipoproteins especially LDL, size of pool I and especially of pool II, and the fluxes between the two pools.

III. Methods for Measurement of Different Parameters in Cholesterol Metabolism

A. Sterol Balance Techniques

1. CHEMICAL BALANCE METHOD

The sterol balance value is obtained as the difference between the dietary cholesterol and the cholesterol eliminated from the body. As already stated, the elimination is quantitated by fecal measurement of total bile acids and neutral sterols of cholesterol origin (Grundy et al., 1965; Miettinen et al., 1965). In man the sterol balance value is almost always negative, indicating that the net loss of cholesterol from the body is offset by synthesis provided that the sterol metabolism is in a steady state. Thus, in the latter circumstance, the sterol balance value equals the cholesterol synthesis. At present, the sterol balance technique appears to be the fastest and most convenient way to measure cholesterol synthesis in the steady state. The latter means that the daily influx of cholesterol from diet and synthesis and the fecal elimination of cholesterol are stable over a fairly long period of time when the amounts of exchangeable cholesterol in the different pools remain constant.

From the practical point of view, detection of the steady state is difficult. If the diet (including dietary cholesterol), body weight, plasma cholesterol level, and cholesterol elimination into feces (all measurable parameters) have been stable over a considerably long period of time, cholesterol metabolism can be considered to be in the steady state (Grundy and Ahrens, 1969). However, if plasma cholesterol has decreased recently, as usually is the case when a patient is put on a special diet, is institutionalized, or is on a hypolipidemic drug, there may be a net flux of tissue cholesterol to plasma and feces over an unmeasurable long period, especially in cases with marked tissue deposits of cholesterol.

A possibility exists under those nonsteady state conditions that a drug, as may be the case with clofibrate (Grundy *et al.*, 1972), reduces cholesterol synthesis and plasma cholesterol and yet simultaneously increases fecal elimination of cholesterol as a consequence of enhanced mobilization of tissue cholesterol. Following discontinuation of a hypolipidemic drug the plasma cholesterol level usually rises and may be accompanied by a net flux of cholesterol from the blood compartment to tissues for an undetectable long period of time so that the sterol balance value is less than the synthesis. Thus the usual experimental design of drug trials, first a control period, followed by the drug period, and finally another control period, reveals that with the sterol balance technique the cholesterol synthesis can be determined only during the first period, the two other periods being possibly influenced by changes in the tissue pools.

The technical performance of the sterol balance studies requires determination of the dietary cholesterol intake, degradation of the steroid ring during intestinal passage, fecal flow, and fecal steroids. To make the daily intake of dietary components and especially that of cholesterol and plant sterols as constant as possible, a liquid formula diet has been widely used in the sterol balance studies (Ahrens, 1970). A solid food diet is preferable, however, because it mimicks the more normal dietary habits of man. The action of intestinal bacteria on neutral sterols and perhaps also on acidic steroids appears to be stronger on solid rather than liquid formula diets (Denbesten *et al.*, 1970). Under metabolic ward conditions the daily fluctuation in cholesterol intake on the solid food diet remains fairly small. In ambulatory subjects, living under home conditions, the cholesterol intake varies considerably and the validity of sterol balance studies becomes questionable unless the dietary habits are carefully recorded or the cholesterol-containing major dietary ingredients are standardized. The disadvantages in using high cholesterol diets in sterol balance studies of drug trials are that quantitatively larger errors may arise in the cholesterol intake, and that a drug-induced further decrease in synthesis and changes in a basically high fecal neutral steroid output may become difficult to detect.

Dietary cholesterol and plant sterols can be measured by gas–liquid chromatography from the homogenate of daily food collections (Miettinen *et al.*, 1965). It should be borne in mind that the usual alkaline ethanolic hydrolysis splits only ester linkages of dietary sterols. In our experience, from 5 to 40% of the plant sterols in a solid food diet are in an alkaline-resistant form, apparently as an ether, and can be split by acidic hydrolysis. If this is not taken into consideration the fecal output of plant sterols may exceed the dietary intake because most sterols are deconjugated by bacteria during intestinal passage. This may result in confusion if unlabeled plant sterols (dietary plus added) are used for the detection and correction of the sterol ring degradation known to take place during the intestinal transit on the liquid formula diet, in particular (Grundy et al., 1968). On a solid food diet this type of loss is less marked (in my own experience) or even absent (Kottke and Subbiah, 1972; Kudchodkar et al., 1972a). Since neutral sterols of cholesterol and plant sterol origin are equally lost, at least during formula feeding, and since plant sterols are practically unabsorbable, cholesterol elimination as neutral sterols should in all sterol balance studies be corrected with the fecal recovery of daily administered (dietary plus added or labeled) β-sitosterol (major component of plant sterols). It is not known whether on a solid food diet degradation of the plant sterol nucleus equals that of cholesterol. According to our experience the fraction of dietary plant sterols converted to secondary products (coprostanol and coprostanone derivatives) during intestinal passage is, on a normal and modified solid food diet, consistently lower than that of cholesterol even at an equal fecal output levels. Fecal flow for bile acid determination can be corrected, e.g., with unabsorbable chromic oxide (Davignon et al., 1968).

Fecal bile acids and neutral sterols of cholesterol and plant sterol origin, respectively, are usually measured with the thin-layer chromatographic (TLC) gas–liquid chromatographic (GLC) technique from fecal homogenates of 1- to 3- or several-day collections (Grundy et al., 1965, Miettinen et al., 1965). In our experience, GLC determination of fecal bile acids with or without previous TLC purification gives identical results provided that the amount of internal standard, 5α-cholestane, is large enough. Therefore, in most of our studies on bile acids the TLC step has been omitted.

Briefly, in our laboratory the technical performance of the sterol balance studies for drug trials is as follows: When the patient is hospitalized he is given Cr_2O_3 200 mg t.i.d. and β-sitosterol (85% pure) 200 mg t.i.d. (unlabeled β-sitosterol can be replaced by [³H]β-sitosterol) and placed on an isocaloric solid food diet. This provides about 30% of calories from fat (mixture of lard and soybean oil) and contains on an average of 125 mg of cholesterol and β-sitosterol per 2400 kcal. After 5 to 10 days on the markers and diet, 3-day (sometimes 2-day) stool collections are started for the baseline determination of fecal steroid excretion. As a sign of an apparent steady state of sterol metabolism the steroid excretion of these collections is reasonably constant provided that the body weight has been stable and the serum cholesterol has become stabilized early enough. Depending on experimental design, the drug can be administered after two to four collections, using, e.g., carmine to differentiate the stools of the control and the drug periods from each other. Collections during the transition periods when plasma cholesterol is changing after the start and discontinuation of the drug are usually pooled. The second control period has been frequently omitted because of excessively long hospitalization and a nonsteady state.

An increase in fecal bile acids and/or neutral sterols of cholesterol origin caused by a drug can be due to inhibited absorption or enhanced biliary secretion. There appears to be no drug, however, which could consistently enhance fecal bile acid excretion by stimulating primarily the biliary secretion of bile acids. Fecal steroid analysis, as such, can not distinguish between the two mechanisms. Since dietary cholesterol is mixed with the endogenous one within the intestinal lumen, any increase in fecal neutral sterols indicates enhanced excretion of both dietary and endogenous cholesterol, provided that the absorption percentage of cholesterol decreases. If the biliary secretion of cholesterol increases and the absorption percentage is unchanged, an increment of fecal neutral sterols originates solely from endogenous sources of cholesterol, the absolute amount of unabsorbed dietary cholesterol remaining unaltered. To reveal changes in absorption or biliary secretion, fecal sterol analysis should be combined with measurement of cholesterol absorption, or biliary lipid secretion should be quantitated directly (Grundy and Metzger, 1972).

2. ISOTOPIC BALANCE METHOD

In this method radioactive cholesterol or a labeled precursor of cholesterol is given as a single dose to label the exchangeable pool of cholesterol in the body. After reasonable equilibration is achieved, usually attainable for isotopic balance studies after 3 to 4 weeks, the specific activity of serum cholesterol can be considered to equal that of biliary cholesterol and bile acids in the bile acid pool. Under these conditions, division of daily radioactivity in acidic and neutral sterol fractions of stools by the specific activity of serum cholesterol equals the fecal elimination of endogenous cholesterol as bile acids and neutral sterols (Grundy and Ahrens, 1966, 1969). The method measures the daily turnover of serum cholesterol irrespective of the amount of dietary cholesterol but it measures synthesis on zero intake of cholesterol only. After baseline determinations the drug to be tested can be administered, and by using continuous serial measurements the effect of the drug on the fecal steroids of endogenous origin (turnover) and on the specific activity of serum cholesterol can be recorded.

As compared to the chemical balance method, the major advantage of the isotopic method is that the amount of dietary cholesterol need not be known. Thus it is more suitable than the chemical balance method for ambulatory use in which the cholesterol intake varies, provided that the fecal collection can be made quantitative. Furthermore, some conclusions can be drawn from drug-induced changes in the specific activity-time curve of serum cholesterol provide that an isotopic steady state is achieved. An increase in the endogenous fecal steroid output associated with an increased slope of the curve suggests that cholesterol synthesis is increased while a decrease of the slope has been interpreted to indicate reduced synthesis (Nestel *et al.*, 1965; Grundy and Ahrens, 1966), especially if the fecal output is decreased. However, many absorbable hypocholesterolemic drugs, e.g., thyroxine (Miettinen, 1968a), clofibrate (Nestel *et al.*, 1965; Horlick *et al.*, 1971; Grundy *et al.*, 1972), nicotinic acid (Miettinen, 1968b; Sodhi *et al.*, 1973), and probucol (Miettinen, 1972a), may result in an enhanced fecal steroid output and a clear-cut concomitant decrease in the slope, sometimes visible as a marked hump in the curve during the transition period. These changes are mostly explainable by a mobilization of radioactive tissue cholesterol, while synthesis may be decreased, unchanged, or enhanced depending on the amount of increase in fecal steroids and the amount and specific activity of mobilized tissue cholesterol (Miettinen, 1968a,b). The amount of mobilized tissue cholesterol has been calculated from the fecal data and specific activity-time curve of serum cholesterol (Sodhi *et al.*, 1973).

Several disadvantages of the isotopic balance technique can be visualized. A relatively long time is needed to attain a reasonable isotopic equilibration and the procedure measures synthesis of cholesterol on a cholesterol-free or very low cholesterol diet only. As compared to the chemical balance method, the analytical technique is more simple but possible degradation of neutral sterols to undetectable products during intestinal transit is not encountered in the isotope method. The use of radioactivity may not be desirable in young individuals, in particular. Difficulties may arise with the intestinal transit time, i.e., with the time passing between the biliary secretion of cholesterol and bile acids and their fecal excretion. Usually this is about 1 day and reasonably reliable results can be obtained if the radioactivity in a 3-day stool collection is divided by the specific activity of serum cholesterol at the beginning of the collection period. Because the half-life of the bile acid pool is several days, the radioactivity in the bile acid fraction should be divided actually by the specific activity of the serum cholesterol 2 or 3 days earlier than in the case of neutral sterols. It should be borne in mind that the transit time may vary from one subject to another and that it may be modified by drugs (e.g., bile acid sequestrants may cause constipation and neomycin diarrhea). The transit time should, therefore, be measured in every subject on and off the drug, e.g., with carmine. A marked compensatory increase in cholesterol synthesis by drugs may result in incomplete equilibration of labeled cholesterol with newly synthesized cholesterol. Thus, during treatment with bile acid sequestrants the specific activities of biliary bile acids and cholesterol and of mucosal cholesterol are lower than that of serum cholesterol (Grundy et al., 1971; Miller et al., 1973). Under those conditions the isotopic balance technique underestimates the overall cholesterol synthesis. Similar results may be obtained in obese and hyperglyceridemic subjects even under normal conditions (Miettinen, 1968a).

A combination of the chemical balance and isotopic methods is practical for drug trials because it allows measurements to be made of cholesterol synthesis (only before drug administration), cholesterol turnover, and cholesterol absorption. Dietary sterols, fecal bile acids, and neutral sterols are then quantitated directly with GLC. The difference between the GLC and the isotopic measurement of fecal neutral sterols reveals the amount of unabsorbed cholesterol, the remainder of the dietary cholesterol having been absorbed. Any secretion of unlabeled cholesterol via the bile or with sloughed mucosal cells could cause an underestimation of absorption. The methods used for the measurement of absorption using intravenous label, oral label, or isotopic steady state have been discussed (Borgström, 1969; Grundy and Ahrens, 1969; Quintao et al., 1971a; Kudchodkar et al., 1973; Connor and Lin, 1974).

The most simple way to measure absorption, though perhaps not the most reliable one (Quintao et al., 1971a), is to give in a test meal [^{14}C]-cholesterol and unabsorbable [^{3}H] β-sitosterol (ratio ^{14}C/^{3}H is known) with unlabeled carriers and to measure from the stools the change in the ratio during intestinal passage (Sodhi et al., 1974). The absorption percentage is calculated from this change. A disadvantage of this method, as of those using a single oral dose, is that it measures absorption during only a short period of time when the composition of and absorption from small intestinal contents may differ from the average of the whole day. However, "off" and "on" the drug conditions can be standardized, the change between the two tests revealing the effect of the drug on absorption. Even then difficulties are encountered with the unabsorbable drugs, which apparently disturb absorption only when the drug is present in the small intestine, absorption being unaffected during the rest of the day. The radioactivity of the control test may disturb the second one, especially if the tests are performed close to each other. Correction for this error can be made by determining the resting specific activity of plasma cholesterol at the beginning of the retest and subtracting the corresponding amount of radioactivity from the radioactivity of fecal sterols originating from cholesterol, provided that fecal neutral sterols are quantitated. The possible secretion into the gut of cholesterol not equilibrated with plasma cholesterol will not underestimate the absorption in this method because the labeled cholesterol of the meal is apparently mixed homogeneously with the endogenous cholesterol delivered into the intestinal lumen during digestion and absorption of the test meal.

In experimental animals cholesterol absorption has also been measured by the simultaneous administration of known amounts of one isotopic cholesterol orally and another one intravenously (Zilversmit, 1972; Zilversmit and Hughes, 1974). Since both isotopes are apparently taken up rapidly by hepatic cells and released in the identical lipoprotein form back into the blood stream the specific activity-time curves of the two isotopes parallel within a few days. The absorption percentage can be calculated from the plasma isotopic ratio when the administered doses are known. Our preliminary studies on man have indicated that this method gives variable results from those obtained with the true fecal excretion method in about 20% of the cases.

B. Kinetic Methods

The first attempts to measure the rate of turnover of cholesterol with the aid of the specific activity-time curve of plasma cholesterol after a single dose of radioactive cholesterol were based on the assumption that the exponential decline of the curve reflects the turnover of one readily exchangeable cholesterol pool (Chobanian and Hollander, 1962; Chobanian et al., 1962). Later, however, Goodman and Noble (1968) interpreted the behavior of the curve differently: the intravenously administered cholesterol first mixes rapidly with pool A, which simultaneously mixes slowly with the other, pool B. The mixing between the two pools was considered to be complete in about 5 to 6 weeks. About a 10-weeks experiment appeared to be long enough for the measurement of the turnover of cholesterol with the two-pool model. Cholesterol production measured with the sterol balance technique appeared to be slightly lower (15%) than that obtained with the two-pool compartmental analysis (Grundy and Ahrens, 1969). Samuel and Perl (1970) observed, however, that in some of the patients the slope of the curve deviated from linearity in their very long-term (up to 63 weeks) studies, suggesting the presence of a third component. They subjected the data to computerized input-output analysis (Perl and Samuel, 1969) and observed that a three-exponential curve fit was frequently better than the two-exponential curve. Furthermore, the input rate (absorbed dietary plus biosynthesized cholesterol) was smaller, and the total exchangeable and the slowly exchangeable cholesterol masses were greater in the three- than the two-

exponential fit, while the fourth measurable parameter, rapidly exchangeable mass of cholesterol, was the same (Samuel and Perl, 1970; Samuel
and Lieberman, 1973). Goodman et al. (1973a) also analyzed their data
with the two-pool and, in long-term (up to 41 weeks) studies, with three-
pool models. With this method the production rate, the size of pool A,
the rate constants for transfer between pools, and the upper and lower
limiting values of the two more slowly exchangeable pools of body cholesterol can be calculated. The results indicated that the medium-term
data may be invalid for the production rate (8–9% too high) and for the
sizes of the more slowly exchangeable pools.

An advantage of the kinetic methods is that a more or less definitive
picture can be obtained from the tissue pools of cholesterol. The major
aim of the hypolipidemic therapy is to prevent increase of or even to
reduce tissue cholesterol of a hyperlipidemic subject, primarily in
atheromas and other lipid deposits. By carrying out kinetic analyses
before and during a long enough period of hypocholesterolemic treatment it could be possible to reveal changes in tissue cholesterol. As a
matter of fact, kinetic studies made after the ileal bypass operation, the
most powerful measure to reduce serum cholesterol, have demonstrated
shrinking of the slowly exchangeable cholesterol pool as compared to the
preoperative status (Moore et al., 1969). However, it is probably because
of a too short follow-up of the specific activity-time curves that the cholesterol turnover studies have failed to show reduction in tissue cholesterol pools by ion exchangers (Goodman and Noble, 1968; Goodman et
al., 1973a; Moutafis and Myant, 1969; Miller et al., 1973), even though re-
gression in the size of the visible lipid deposits, xanthelasmas and tendon
xanthomas, can be detected clinically in many of the patients treated with
these drugs. However, neomycin (Samuel et al., 1968) and clofibrate
(Grundy et al., 1972), which usually reduce the serum cholesterol level
less effectively than cholestyramine, seem to shrink tissue cholesterol
pools.

A disadvantage of the turnover studies is the long period of time
required for the drug trials; two adequate long-term studies performed
before and during the drug would require more than 2 years provided
the drug is "on" for 6 months only before the relabeling is performed.
Therefore shortening of the procedure has been attempted by using
tritiated water (specific activity of the precursor can be determined in
the body water) and labeled mevalonate as precursors of cholesterol
and determining the production rate and synthesis by chemical fecal
steroid analysis (Kekki et al., 1973). Computer analysis of the data obtained with the two-pool model is based on a 6- to 8-weeks follow-up of
the specific activity-time curves of plasma cholesterol. This method

measures the size of the exchangeable cholesterol pool, indirectly the size of pool B (size of pool A is calculated from blood, liver, and gut cholesterol; Moore et al., 1970), fluxes between the pools, and the production rate and synthesis of cholesterol in pools A and B separately. In contrast to what has been assumed previously (Goodman and Noble, 1968), analysis of the data revealed that a considerable proportion of cholesterol is synthesized in pool B. This is seen directly when [^{14}C]cholesterol and [^{3}H]mevalonate are used simultaneously. The specific activity-time curve of plasma [^{3}H]cholesterol falls slower than that of [^{14}C] cholesterol. This can be expected if [^{3}H]cholesterol synthesized in pool B returns with a high specific activity to the plasma compartment. The combination of heavy water with labeled cholesterol minimizes the amount of radioactivity, yet a complete kinetic analysis of cholesterol metabolism can be obtained within a fairly short period of time. London and Rittenberg (1950) administered heavy water to a human subject continuously to keep deuterium enrichment constant in body water, and measured the rate of newly synthesized cholesterol by means of deuterium. Since then, this method has been used by Taylor et al. (1966) for detecting changes in cholesterol synthesis by dietary cholesterol in man.

C. Quantitation of Bile Acid Metabolism

The effects of drugs on the elimination of cholesterol as bile acids can be measured in connection with either the chemical or the isotopic balance technique. However, the kinetics of bile acids can be analyzed by the method of Lindstedt (1957) when the pool size, half-lives, and turnover rates of the two primary bile acids are obtained, provided that the administered cholic acid and chenodeoxycholic acid are differently labeled. A rough estimate of the synthesis of the two bile acids can be obtained by using one isotope only and applying its kinetic parameters to the calculation of the turnover of the other from their mass ratio in the bile (Lindstedt, 1970). However, since, e.g., cholestyramine markedly reduces the relative amount of chenodeoxycholate in the bile (Dam et al., 1971; Garbutt and Kenney, 1972; Wood et al., 1972), though its fecal output may increase almost proportionately to that of cholic acid (Miettinen, 1974a), use of the double label technique is preferable when a drug effect on cholesterol elimination as bile acids is measured with the method of Lindstedt. With this technique, cholestyramine has been shown to increase cholesterol elimination as cholic and chenodeoxycholic acids differently in different hyperlipoproteinemias (Einarsson et al., 1974); clofibrate has no effect on bile acid synthesis in patients with type II abnormality, while a decreased production of cholic acid is observed in subjects with type IV hyperlipoproteinemia (Einarsson *et al.*, 1973). The kinetics of cholesterol and cholesterol catabolism, as neutral sterols and bile acids, can also be measured by sequential determination of biliary cholesterol and bile acid-specific activities after a single injection of labeled cholesterol (Quarfordt and Greenfield, 1973). The procedure also allows measurement of the kinetic parameters of individual bile acids.

D. Use of Labeled Precursors

Most of the methods presented above are time-consuming and require a steady state, a criterion not usually fulfilled in drug trials. From a practical point of view it would be valuable to know rapidly after institution of a hypolipidemic measure whether it actually has any effect on cholesterol synthesis independent of a possible nonsteady state. Different isotopic precursor methods have been used for this purpose.

1. LABELED WATER

The use of tritiated water was investigated by Loud (1955) and compared with [^{14}C]acetate in experimental animals. Moore *et al.* (1970) studied the incorporation of tritium from body water into serum cholesterol during the 2 days following the administration of an oral dose of tritiated water. The ratio of the specific activity of the serum cholesterol to that of body water was determined and, as a sign of increased cholesterol synthesis, was shown to be increased by a factor of 5.7 by the ileal bypass operation. The cholesterol synthesis rate was measured by multiplying the amount of cholesterol in pool A by the tritium incorporation ratio. Calculated in this way, surgery increased the synthesis 4.2-fold. The disadvantage of using a relative large amount of tritium can be avoided by replacing ^3H_2O by heavy water.

2. LABELED ACETATE

Labeled acetate alone has been used *in vivo* studies of cholesterol synthesis in man. These investigations have actually revealed that the incorporation of acetate to plasma cholesterol is stimulated by thyroid hormones (Gould, 1959) and suppressed by dietary cholesterol in a considerable proportion of normo- and hypercholesterolemic subjects (Pawliger and Shipp, 1965) but not in a hyperlipidemic child (Fujiwara *et al.*, 1965).

3. THE ACETATE-MEVALONATE TEST

The use of labeled [^{14}C]acetate alone has several disadvantages in drug trials, in particular: the cholesterol synthesis cannot be quantitated, the size of the acetate pool may vary from one experiment to another and thus dilute the label differently, and the pool sizes of cholesterol precursors after mevalonate, including methyl sterols and cholesterol itself, may change, the result being an altered dilution of ^{14}C products. Since cholesterol synthesis is regulated between HMG-CoA and mevalonate (cf. Gould and Swyrud, 1966), the simultaneous use of the two precursors would measure activity and changes in this rate-limiting step, and correct errors arising from changes in pool sizes after HMG-CoA, particularly if the cholesterol formed from HMG-CoA is related to that produced from mevalonate. For practical reason HMG-CoA can be replaced by acetate. Thus, after an intravenous injection of a mixture of the two precursors (e.g., 10–20 μCi of [^{3}H]mevalonate) the ratio of the two isotopes (^{14}C/^{3}H) can be measured serially in plasma-free, esterified, or total cholesterol (Miettinen, 1966, 1970a). This ratio of the acetate-mevalonate test relays the activity of cholesterol synthesis between HMG-CoA and mevalonate. The possibility of errors due to changes in the acetate, acetoacetate, and HMG-CoA pools still exists. Therefore, if the effect of a hypocholesterolemic drug is to be studied, the diet and other experimental conditions should be rigidly the same during the two experimental periods. Even then the average deviation of the ^{14}C/^{3}H ratio from the mean of double determinations, carried out in the same subjects under basal conditions 1 to 3 days apart, is almost 10%, indicating that, because of apparent variation in acetate metabolism, small changes in cholesterol synthesis can not be detected with a single acetate-mevalonate test. However, reliability and sensitivity are greatly increased with relatively little additional laboratory work if a small dose of the labeled mixture is administered twice or even three times during each experimental period. Some of the pool size problems can be overcome by replacing acetate by octanoate which, in vitro at least, is known to generate effectively acetyl-CoA not only to the hepatic intra- but also to the extramitochondrial compartment where cholesterol synthesis takes place (Dietschy and Brown, 1974).

Although the acetate-mevalonate test is correlated with the basic cholesterol synthesis measured with the sterol balance technique (Miettinen, 1970a; Sodhi and Kudchodkar, 1973b), this association can be lacking in many conditions for several reasons. The test measures synthesis during a short period of time only because of rapid acetate and mevalonate metabolism. Thus the average daily flux of acetate to mevalonate and further to cholesterol in relation to that of mevalonate is not measured because the activity of HMG-CoA reductase shows a marked diurnal variation (Hamprecht et al., 1969). This probably also occurs in human subjects. Furthermore, the uptake and utilization of the precursors varies in different tissues, the ratio probably reflecting cholesterol synthesis in a few organs only but by no means in the body as a whole. Thus the relative increase in synthesis indicated by the acetate-mevalonate test can be markedly greater, e.g., after ileal bypass, than that in the overall production measured by the sterol balance technique. In addition, the overall conversion of mevalonate to cholesterol is not measured; in a heavy subject it is quite apparently much greater than in a small one, yet the ratio can be the same; in liver damage the overall cholesterol production may be low, yet the utilization of acetate for cholesterol synthesis can be effective in the few remaining parenchymal cells of the liver, resulting in a high $^{14}C/^{3}H$ ratio (Miettinen, 1972b).

Despite the limitations mentioned above, changes in cholesterol synthesis between acetate and mevalonate induced by lipid-lowering agents, particularly by unabsorbable ones that affect acetate metabolism very little, can be easily discovered with the acetate-mevalonate test within the limits of experimental error. Thus, cholestyramine, Colestipol, and neomycin increase the synthesis (Table I, Miettinen, 1970a, 1974a), clofibrate depresses it (Sodhi et al., 1971a), while the change appears to vary during nicotinic acid therapy (Miettinen, 1971a).

E. Cholesterol Precursors in Serum

Stimulation and inhibition of HMG-CoA reductase activity would increase and diminish, respectively, the intracellular amount of at least some of the numerous intermediates of cholesterol synthesis after that step. Accordingly, the release of these compounds from the cell and then into the blood stream would parallel the rate of cholesterol synthesis. Therefore serum cholesterol precursors could be used as an indicator of cholesterol synthesis. The first experiments (Miettinen, 1966) were made with triparanol-treated rats, using changes in the elevated desmosterol level to relay alterations in cholesterol synthesis at the earlier step. As a matter of fact, cholesterol feeding, which in the rat is known to suppress hepatic cholesterol synthesis effectively, significantly decreased the plasma desmosterol level. Similar studies performed later by Bricker et al. (1972) confirmed the finding and showed that cholesterol feeding almost totally inhibited the triparanol-induced increase of plasma desmosterol. In cor-

TABLE I

CHOLESTEROL METABOLISM IN HYPERCHOLESTEROLEMIC SUBJECTS TREATED WITH CHOLESTYRAMINE AND NEOMYCIN ALONE AND IN COMBINATION

| Treatment | Serum cholesterol (mg%) | Fecal steroids (mg/day) | | | $^{14}C/^{3}H \times 100$ in cholesterol[e] | DMS[i] | |
		Bile acids	Neutral sterols	Total sterols		μg/100 ml	μg/mg FCH × 10²
Cholestyramine–neomycin study (6 patients)[j]							
None	365 ± 29	185 ± 23	640 ± 93	825 ± 113	5.5 ± 0.9	22 ± 5	26 ± 5
Resin[a]	265 ± 18	1353 ± 231	766 ± 148	2119 ± 339	15.0 ± 1.8	40 ± 8	60 ± 9
Resin[a] plus neomycin	216 ± 16[f,g]	1455 ± 373[f]	1043 ± 130[g]	2498 ± 457[f,g]	13.6 ± 2.6[f]	29 ± 7[f,g]	47 ± 10[f]
Neomycin study (8 subjects)[k]							
None	375 ± 34	205 ± 26	643 ± 60	848 ± 68	9.0 ± 1.9	29 ± 7	34 ± 6
Neomycin[b]	284 ± 32[f]	238 ± 55	871 ± 92[f]	1109 ± 108[f]	13.7 ± 3.3[f]	21 ± 5	32 ± 8
Neomycin[c] (1.5 gm)	293 ± 24[f]	277 ± 39	731 ± 84[f]	1008 ± 106[f]	6.0 ± 2.0	15 ± 4[f]	19 ± 6[f]
Neomycin[d] (2–6 gm)	268 ± 20[h]	386 ± 99	817 ± 89[h]	1203 ± 121[h]	7.0 ± 1.9	18 ± 4	25 ± 7

[a] Cholestyramine administered at a dosage of 32 gm/day for 10 to 12 days each.

[b] Neomycin administered at a dosage of 1.5 gm/day for 10 to 12 days. In one subject administration was for 135 days.

[c] Five patients were restudied after 1 to 2 years, using neomycin at a dosage of 1.5 gm/day.

[d] Values obtained after an additional 12 days on a dosage of neomycin of 2–6 gm/day.

[e] Ratio of $^{14}C/^{3}H$ in serum total cholesterol obtained 8 hours after i.v. administration of a [^{14}C]acetate-[^{3}H]mevalonate mixture (10 μCi/5 μCi).

[f] Statistically significant ($p < 0.05$) change from basal values.

[g] Statistically significant change from resin values ($p < 0.05$).

[h] Statistically significant change from small neomycin dose ($p < 0.05$).

[i] DMS, dimethyl sterol from serum unesterified methyl sterol mixture; FCH, unesterified cholesterol.

[j] The diet contained on an average 125 mg of cholesterol/2400 kcal, indicating that the change in fecal neutral sterols was only to a minor extent accounted for by dietary cholesterol. The changes caused by cholestyramine in fecal bile acids and total steroids correlated with those in the acetate-mevalonate test (at 8 hours $r = 0.824$ and 0.839, respectively) and methyl sterols (for DMS, $r = 0.918$ and 0.890, respectively), while the correlation with the changes in total and unesterified serum cholesterol did not reach a significant level.

[k] In the neomycin study the change in fecal steroids correlated with the decrease in serum cholesterol ($r = 0.736$).

responding experiments on human subjects, cholesterol feeding did not reduce the plasma demosterol level, though accumulation of this sterol appeared to occur at a higher rate after the cholesterol feeding was discontinued (Miettinen, 1966, 1970a).

Since triparanol causes well-known side effects and its use in man seemed impractical, possibilities to quantitate basal sterol precursors in plasma were investigated particularly because earlier studies (Goodman, 1964) suggested that squalene and lanosterol were present in the human blood and that they could be in equilibrium with their site of origin. The presence of lanosterol and other methyl sterols in human serum was verified by the TLC-GLC mass spectrometric methods (Miettinen, 1968c). Quantitation of those compounds revealed, as was expected, that in man total fasting, known to decrease cholesterol synthesis, reduced plasma concentrations, while cholestyramine treatment, known to enhance synthesis, increased plasma concentrations (Miettinen, 1968c, 1970a, 1971b). Since changes in cholesterol synthesis appeared to be reflected more sensitively in the unesterified than the esterified methyl sterols, quantitation of the former is preferable. Of the mixture of free methyl sterols, the dimethyl sterols, especially diunsaturated dimethylsterol (probably $\Delta^{8,24}$-$4\alpha,4\beta$-dimethyl cholestan-3β-ol), and the trimethyl sterols, especially lanosterol and dihydrolanosterol, were in general increased most sensitively, as if their further demethylation had been rate-limiting or their release into and removal from the blood had been different. The few determinations made of plasma C-27 precursor sterols showed that lathosterol was significantly increased by cholestyramine, suggesting that demethylated precursor sterols of plasma may also relay cholesterol synthesis. On the other hand, although the plasma and liver squalene were increased in cholestyramine-treated rats parallel with methyl sterols, hepatic cholesterol synthesis *in vitro,* and fecal elimination of cholesterol, in man, with a differently altered cholesterol synthesis, the parallelism between serum squalene, on the one hand, and methyl sterols, the acetate-mevalonate test, and fecal steroid excretion, on the other hand, was inconsistent (Miettinen, 1969, 1970a). Thus, cyclization of squalene may not always be a greatly rate-limiting step in cholesterol synthesis in man.

Under certain basal conditions, e.g., in varying body size, hypercholesterolemia, gluten enteropathy, hypertriglyceridemia, and hyperthyroidism, the plasma methyl sterols are not consistently correlated with cholesterol synthesis as measured by the sterol balance technique. Since the greater part of the free methyl sterols, as of cholesterol, are normally transported in plasma LDL (Miettinen, 1971b), any change in that lipoprotein level can change the methyl sterol concentration without actually being related to cholesterol synthesis. Thus, in hypercholesterolemia the increased LDL can bind more methyl sterols increasing their concentration, though the sterol balance values indicate that synthesis is low normal. On the other hand, reduction of the LDL level by dietary means or drugs may decrease plasma level of methyl sterols even though the concentration of methyl sterols within the LDL molecule may remain unchanged as a sign of unaltered cholesterol synthesis. The determination of methyl sterols in the different serum lipoprotein fractions indicated that changes in cholesterol synthesis were most consistently seen in the methyl sterols of LDL, particularly when expressed per free cholesterol, inconsistently and from the quantitative point of view insignificantly in VLDL, provided its amount was within normal limits, and quantitatively insignificantly in HDL (Miettinen, 1971b). In hypertriglyceridemic subjects, however, a fairly large amount of the free methyl sterols, as of free cholesterol, was transported in VLDL; the methyl sterol content of the latter, however, appeared to parallel the changes in cholesterol production. The findings indicated that when the free methyl sterol level without lipoprotein fractionation is used as an index of alterations in cholesterol synthesis the values should be expressed per free total cholesterol.

Serum methyl sterols showed parallel changes with the results of both the acetate-mevalonate test and the sterol balance technique when cholesterol synthesis was inhibited by total fasting or when it was stimulated with cholestyramine; under the latter condition, the three parameters were significantly correlated with each other (Miettinen, 1971b). The drug-induced changes in cholesterol synthesis revealed by serum methyl sterols are, however, only qualitative, as in the case of the acetate-mevalonate test, and cannot be expressed in terms of milligrams per day. Furthermore, though the decreased cholesterol synthesis during total fasting was associated with a fall in serum methyl sterols within a few days, the opposite change after stimulation of the synthesis, which, e.g., after interruption of the enterohepatic circulation of bile acids by the ileal bypass operation or by cholestyramine reached its maximum within the first 3–4 days according to the acetate-mevalonate test, took place fairly slowly; usually it was not detectable until after several days and sometimes after 2 weeks; occasionally the values rose gradually during several months (Miettinen, 1970a, 1971b). Under the latter conditions the relative increase of methyl sterols was usually disproportionate to that of fecal steroids, although somewhat lower than that shown by the acetate-mevalonate test. Several theoretical explanations can be visualized for these observations. The activity of the cholesterol-synthesizing enzymes is normally much higher after mevalonate than the activity of the rate-limiting step HMG-CoA reductase (Gould and Swyrud, 1966). Since this may apply also to the enzymes converting lanosterol to cholesterol, the increase in cholesterol synthesis between acetate and mevalonate should be considerable before methyl sterols accumulate in cells and serum. Furthermore, enzymes oxidizing methyl sterols may be induced under these conditions as is suggested by a fast turnover of serum methyl sterols during cholestyramine (half-life time about 2 hours in contrast to 5 hours normally) treatment despite their increased serum concentration (Miettinen, 1970a). Demethylation of methyl sterols has been actually shown to be increased in the rat liver during cholestyramine treatment (Moir *et al.*, 1970). Methyl sterols may also be distributed extravascularly within the exchangeable cholesterol pool, including cholesterol deposits. Particularly in hypercholesterolemic subjects with a large pool, the amount of precursor sterols removed from the serum in this way may be considerable. Red cell methyl sterols, at least, appear to run parallel with serum methyl sterols (Miettinen, 1974b). Removal of the latter via the bile may also be altered by hypolipidemic drugs. If methyl sterols are absorbable, the drugs interfering with cholesterol absorption, such as bile acid sequestrants, neomycin, β-sitosterol, and probucol, would interrupt the enterolymphohepatic circulation of methyl sterols, enhancing their removal from the serum into feces. This is one possible explanation for the finding (Table I) that, though the acetate-mevalonate test indicated an increased cholesterol synthesis by neomycin, the serum methyl sterols remained unchanged. No information is available on whether methyl sterols *in vivo* can be converted to bile acid derivatives and whether cholestyramine can enhance this process.

F. Quantitation of Cholesterol Synthesis with Squalene

Since apparently all the carbons of the cholesterol molecule have flowed during cholesterol biosynthesis through squalene and since the latter is present in small amounts in plasma, its turnover and subsequent cholesterol synthesis can be quantitated after intravenous administration of labeled mevalonate (Liu *et al.*, 1974). For this purpose a known amount of labeled cholesterol and mevalonate is given intravenously. The percentage of mevalonate converted to cholesterol and, thus, the initial dose of labeled squalene can be calculated from the specific activity-time curves (followed up to the fifth week) of labeled cholesterols. The specific activity-time curve and the dose of squalene determine the rate of cholesterol synthesis, the results of the whole process being available in less than 6 weeks. Advantages of the method are that the synthesis can be measured in terms of milligrams per day, it is an outpatient procedure, it does not require controlled diets, stool collections or metabolic steady state, and it can be repeated at frequent intervals to evaluate, e.g., drug effects.

G. Turnover of Low-Density Lipoproteins

Since plasma cholesterol is mainly transported by LDL it can be assumed that the major end products of plasma cholesterol, i.e., fecal bile acids and neutral sterols, actually are also the major measurable end products of LDL. Consequently, any drug-induced changes in cholesterol turnover should be associated with parallel changes in the turnover of LDL, a subject which has not been studied very carefully. In view of the present findings that LDL may be catabolized on the cell membranes of extrahepatic tissue (Goldstein and Brown, 1973; Brown and Goldstein, 1974) and that catabolism of LDL may be negligible in the liver (Steinberg, 1974), this type of association, though apparent in many clinical conditions, becomes difficult to understand. Certain drugs apparently decrease the turnover of LDL, yet they can increase fecal elimination of cholesterol (see Section V,B).

The turnover of LDL can be measured by labeling the protein moiety of the isolated LDL with radioiodine (Gitlin et al., 1958; Walton et al., 1963, 1965). After intravenous administration of the label a kinetic analysis of LDL can be obtained and the half-time, pool size, and absolute (corresponding to synthesis) and fractional catabolic rates calculated. The study can be repeated "off" and "on" the drug. It has been shown that, though not consistently (Walton et al., 1963; Scott and Hurley, 1969), catabolism of LDL is impaired in hypercholesterolemia (Langer et al., 1972), thyroid hormones enhance the turnover of LDL (Walton et al., 1965), cholestyramine decreases LDL concentration and increases the fractional catabolic rate of LDL in type II patients (Langer et al., 1969), while nicotinic acid (Langer and Levy, 1971) and clofibrate (Walton et al., 1963, Scott and Hurley, 1969) lower the plasma LDL level but has no effect on the fractional catabolic rate.

IV. Methods for Measurement of Triglyceride Metabolism

As compared to cholesterol the overall plasma triglyceride turnover and drug-induced changes in it are difficult to quantitate and detect, respectively, because of the lack of measurable end products. Furthermore, the transport lipoprotein of triglycerides, VLDL, is known to be metabolically and chemically heterogeneous, the different lipid and protein moieties having different turnover rates in relation to each other and in different densities of VLDL, the molecule being finally converted to LDL (cf. Steinberg, 1974). The regulation of the further catabolism of VLDL to LDL and the exact role of plasma lipases in this process, especially of those activated by heparin, are not known. Drug-induced decrease in

plasma triglycerides can be obtained by increasing the removal of triglycerides from the plasma, i.e., by enhancing the conversion of VLDL to
LDL, by decreasing the production rate, or by combinations of the two,
the opposite changes leading to an increase in triglycerides. In the
steady state of plasma triglyceride concentration the two factors, removal
and production, are in balance. Results obtained by different techniques
of evaluation of plasma triglyceride kinetics (Kekki and Nikkilä, 1975)
are conflicting as to whether endogenous hyperglyceridemias are primarily due to enhanced production or decreased removal (cf. Nikkilä and
Kekki, 1971; Olefsky et al., 1974). Since this discrepancy may be partly
methodological and partly due to heterogeneity of the hyperglyceridemic
state, conflicting results may be obtained on the mechanisms of hypotriglyceridemic drugs.

The methods defined to quantitate plasma triglyceride or VLDL triglyceride transport are based on the use of radioactive precursors, free
fatty acids or glycerol, or are nonisotopic. The latter includes the procedure in which the influx of triglycerides via the hepatic vein into plasma
is measured directly by multiplying the hepatic venous-arterial difference
of triglyceride concentration by the hepatic plasma flow (Carlson and
Ekelund, 1963; Havel et al., 1970; Boberg, 1971). Though this method
is feasible at low plasma triglyceride levels, large measurement errors
may result at high triglyceride concentrations because, in view of determination errors of plasma triglycerides, the concentration gradient across
the splanchnic region is small. Furthermore, the procedure misses the
triglyceride flow taking place via the thoracic duct.

Another nonisotopic method for the determination of plasma triglyceride turnover has been presented by Porte and Bierman (1969). They
infused heparin to enhance the plasma lipolytic activity to its maximum
and to reduce the plasma triglycerides to a constant lower level. The
determination of triglyceride removal was based on the plasma triglyceride lowering and on the in vitro assay of the plasma postheparin lipolytic
activity. This method is easy to perform and repeated determinations
can be carried out easily in a single subject. Recent studies have indicated
that plasma postheparin lipase activity consists of lipoprotein lipase and
hepatic lipase activities which can be quantitated separately (Krauss et al.,
1974). Of these, the lipoprotein lipase activity is absent or markedly low
in type I hypertriglyceridemia and plays an important role at least in
the removal of plasma triglycerides of chylomicrons, while the role of
hepatic lipase in the removal is unknown in vivo. Oxandronolone, which
is a moderate plasma triglyceride-lowering agent enhancing postheparin
lipolytic activity (Glueck, 1971; Glueck et al., 1973), has no effect on
heparin-induced lipoprotein lipase activity but increases hepatic lipase

activity, changes which have, however, no correlation with the reduction of the plasma triglyceride level (Ehnholm et al., 1975).

Estimation of the rate of incorporation of labeled free fatty acids into plasma triglycerides or VLDL triglycerides has been most popularly used for the measurement of the plasma triglyceride turnover according to different modifications of the original Ryan and Schwartz (1965) method. In the latter method the labeled free fatty acid is infused at a constant rate. The production rate of triglycerides is calculated from the increase of labeled triglycerides, infusion rate, and specific activity of free fatty acid. Since the transport data obtained with this procedure are erroneously too low, the method has been modified by using specific activity measurements of the hepatic vein (Havel et al., 1970; Boberg, 1971) or by computerizing the specific activity-time curves according to multicompartmental models (Eaton et al., 1969; Quarfordt et al., 1970; Eaton, 1971). The clearance rate of plasma triglycerides, which in the steady state should be equal to the production, can be obtained as a difference between the net splanchnic secretion rate (measured as the hepatic venous-arterial difference times the hepatic flow) and increase of labeled triglycerides in plasma (Boberg et al., 1972a,b). The studies using labeled fatty acids as precursors of triglycerides indicate that type IV hyperglyceridemia is, in most cases, due to impaired removal of plasma triglycerides (Ryan and Schwartz, 1965; Sailer et al., 1966; Havel et al., 1970; Eaton, 1971; Boberg et al., 1972a).

The transport of plasma triglycerides can be determined from the disappearance rate of radioactivity from plasma triglycerides or VLDL triglycerides. Most of the studies have been made with the glycerol labeling technique (Reaven et al., 1965; Nikkilä and Kekki, 1971a) because this precursor has been suggested to recirculate less than when fatty acids are used for this purpose (Farguhar et al., 1965). Quite recently the labeling has been performed with fatty acids and it was suggested that the errors caused by the recirculation were small (Adams et al., 1974). Using decay of radioactivity, triglyceride concentration, and plasma triglyceride pool, different kinetic parameters can be calculated including the triglyceride production rate, fractional removal rate, maximal turnover rate of plasma triglycerides (V_{max}), and the triglyceride concentration at which the removal is operating at half-maximal velocity (K_m). Since the latter describes the relationship between the plasma triglyceride concentration and the triglyceride clearance, an increased K_m (high concentration/production ratio) indicates impaired removal which, if associated with a normal production rate, means that the hyperglyceridemia is caused solely by a removal defect. For instance, a drug-induced decrease in K_m without a decrease in the production would indicate that an improved removal is the triglyceride-lowering mechanism of this particular agent. It has been suggested (Nikkilä and Kekki, 1971a) that the low plasma triglyceride level in fertile-aged women is actually due to an efficient removal (low K_m). A similar efficient removal has also been found in hyperglyceridemic females (Olefsky et al., 1974). Oral contraceptives markedly enhance the production rate, but because the latter decrease the K_m the increase of the plasma triglyceride level is actually less than expected from the increased synthesis (Kekki and Nikkilä, 1971). The studies using the above mentioned techniques have indicated, in contrast to what was presented earlier, that in general, over-production, and not impaired removal, is the initiating event in most cases of type IV hypertriglyceridemia (Reaven et al., 1965; Nikkilä and Kekki, 1972a; Olefsky et al., 1974; Adams et al., 1974).

An intravenous fat tolerance test has also been used as a simple method for the estimation of the fractional removal rate of plasma triglycerides in man (Hallberg, 1965; Boberg et al., 1969; Carlson and Rössner, 1972). In this procedure a given amount of an artificial fat emulsion is injected intravenously and removal of the triglyceride emulsion is followed nephelometrically. The fractional removal rate (K_2, percentage/minute) is calculated from the semilogarithmic plot of the readings against time. This has been suggested to relay the removal of endogenous plasma triglycerides. Though this concept has been criticized (Adams et al., 1974), K_2' has been shown to correlate positively with the fractional turnover rate of both the secretion and the clearance methods (Rössner et al., 1974).

V. Mechanisms of Action of Lipid-Lowering Agents

The lipid-lowering drugs can be divided into unabsorbable and absorbable ones. The former group, including the clinically useful drugs, i.e., ion-exchange resins, neomycin, and plant sterols, functions within the intestinal lumen by interfering with the absorption of bile acids and cholesterol (cf. Miettinen, 1975a) and by thus enhancing cholesterol elimination into the stools. This action is then associated, despite enhanced cholesterol synthesis, with a serum cholesterol and LDL-lowering action, the exact mechanism of which is still unknown. These agents are suitable for the treatment of hypercholesterolemia. The most widely used absorbable drugs are clofibrate, nicotinic acid, and thyroid hormones, while some others are less frequently used (e.g., p-aminosalicylic acid) or are still under further clinical investigations (e.g., probucol and tibric acid). Current views on the effects of hypolipidemic agents on some parameters of lipid metabolism are summarized in Table II. Effects of

TABLE II

EFFECTS OF DIFFERENT HYPOLIPIDEMIC AGENTS ON VARIOUS PARAMETERS OF LIPID METABOLISM[a]

Hypolipidemic agent	Plasma lipids		Lipid synthesis		Fecal steroids			Intestinal absorption	
	Chol.	Trigl.	Trigl.	Chol.	BA	NS	Total	BA	Chol.
Unabsorbable agents									
Ion exchangers	–	=(+)	=+?	+	+	=	+	–	–=
Neomycin[b]	–	=	=?	+?	=+	+	+	=–	–
Plant sterols	–	=	=?	+?	=	+	+	=	–
Absorbable agents									
Clofibrate	–(=)	–	–	–(=+)	–=	+	+=	=	=
Nicotinic acid	–	–	–	+	=	+=	+=	=	=
Thyroid hormones[c]	–	–=	+	+	=	+	+	=	=
Probucol	–	=	?	–	+=	=–	+=	–=	–=

[a] Key to symbols: BA, bile acid; NS, neutral sterols; Chol., cholesterol; Trigl., triglycerides; +, increased; =, unchanged; –, decreased.

[b] Large doses of neomycin increase, while small doses (1–2 gm/day) have no effects on fecal bile acids.

[c] Judged from data on hypo- and hyperthyroid patients before and after therapy.

drugs on plasma triglyceride metabolism (Nikkilä, 1973), and on bile acid and cholesterol excretion (Miettinen, 1970b, 1973b, 1975a) have been reviewed.

A. Unabsorbable Agents

1. ION-EXCHANGE RESINS

Of the currently available ion-exchange resins, cholestyramine was the first to be introduced for the treatment of hypercholesterolemia. It was shown to reduce the serum cholesterol level in man by Bergen *et al.* (1959) and to enhance fecal bile acid excretion by Carey and Williams (1961) and later by Hashim and Van Itallie (1965). Subsequent studies have clearly shown that cholestyramine is an effective hypocholester-olemic agent (cf. Grundy, 1972; Levy *et al.*, 1973) which, in doses of 32 gm/day, increases in hypercholesterolemic patients the fecal cholesterol elimination as bile acids by approximately 1 gm/day (cf. Miettinen, 1970a, 1973b). Though the drug is ineffective on serum cholesterol in patients with the homozygous type II abnormality (Khachadurian, 1968), their fecal bile acid and total steroid excretion is increased by the resin (Miettinen, 1970c; Moutafis *et al.*, 1971; Grundy *et al.*, 1971), indicating that in those particular subjects the increased synthesis adequately balances the enhanced fecal elimination of cholesterol. In heterozygotes the increase in synthesis, although severalfold (cf. Table I), is apparently insufficient to fully compensate for fecal loss, resulting in a decrease of serum cholesterol. Factors influencing the magnitude of fecal bile acid increase and serum cholesterol decrease have been incompletely explored. The results obtained on a fairly large number of subjects (Miettinen, 1973b) have indicated that the increase in fecal bile acids correlates positively with body weight and negatively with the age of the patient. Despite a negative correlation of initial fecal bile acids with serum cholesterol and a positive one with cholestyramine-induced increase of fecal bile acids, the latter increase showed a positive correlation with the absolute and relative decreases of serum cholesterol. These findings suggest that the greater the cholestyramine-induced increase in the fecal elimination of cholesterol is, the greater is the decrease in serum cholesterol, and that the latter may be smaller in older subjects.

Colestipol (Parkinson *et al.*, 1970; Ryan and Jain, 1972) and PDX-chloride (DEAE-Sephadex cellulose; Howard and Hyams, 1971) are the two other bile acid sequestering agents used in the treatment of hypercholesterolemia. They are known to increase bile acid elimination into the stools in man by about the same quantities as cholestyramine does

(Miller *et al.*, 1973; Miettinen, 1973b, 1974a). The acetate-mevalonate test and serum methyl sterols indicate that cholesterol synthesis is enhanced compensatorily (Miettinen, 1974a), as by cholestyramine.

Cholestyramine administered to type IV hyperglyceridemic patients enhanced fecal bile acid excretion to a greater extent than in familial type IIA hypercholesterolemic subjects (1200 and 890 mg/day, respectively). Since, however, elimination as neutral sterols was decreased in the hyperglyceridemic subjects the overall increase in cholesterol elimination by cholestyramine only tended to be greater in type IV than type IIA subjects (1045 and 880 mg/day, respectively) yet serum cholesterol was decreased in type IIA but not in type IV patients (Miettinen, 1973a). In the type IIB abnormality the increase in cholestyramine-induced fecal bile acid and overall cholesterol elimination tended to be lower than in type IV but was associated with a significant serum cholesterol decrease.

Kinetic analysis of bile acid metabolism indicated that cholestyramine enhanced the total bile acid formation from 264 to 1116 mg/day in female type IIA patients and from 1044 to 1643 mg/day in male type IV subjects (Einarsson *et al.*, 1974). The former change was due primarily to enhanced production of cholic acid and the latter was caused exclusively by increased turnover of chenodeoxycholic acid. The pool size of cholic acid remained unchanged or was increased in both groups, while that of chenodeoxycholate was decreased or unchanged. Earlier kinetic studies of cholic acid in normolipidemic subjects revealed that cholestyramine markedly increased the turnover of cholic acid, had no effect on its pool size, and markedly reduced the relative amount of biliary deoxy- and chenodeoxycholate as if the increased synthesis of bile acids had proceeded primarily via the pathway of cholic acid (Garbutt and Kenney, 1972). A relative increase of biliary cholate by cholestyramine has also been reported by others (Juul and Van der Linden, 1969; Dam *et al.*, 1971; Wood *et al.*, 1972). In xanthomatotic type II subjects, Colestipol changed the biliary bile acid composition analogously to that of cholestyramine in normal persons (Miettinen, 1975, 1974a). Enhanced fecal bile acid excretion was due to an almost proportional increase of both cholic and chenodeoxycholic acids, indicating that despite the low biliary concentration of the latter its relative removal and synthesis were augmented almost to the same extent with cholate. The cholestyramine treatment of subjects with ileal dysfunction in whom chenodeoxycholate usually predominates in the bile decreased the biliary chenodeoxycholate markedly and increased its fecal output further (Miettinen, 1975b). Since the ion-exchange resins are known to more effectively bind di- than trihydroxy bile acids *in vitro* (Johns and Bates, 1969), the dissimilar changes in the metabolism of the two primary bile acids by these resins can be mostly ascribed to difference in their binding to the resin *in vivo*.

Bile acid sequestrants do not consistently increase fecal excretion of neutral sterols, suggesting that biliary secretion of cholesterol is decreased because absorption of exogenous and endogenous cholesterol is decreased (Hyun *et al.*, 1963; Splitter *et al.*, 1968). Cholestyramine administered to patients with a T-tube has actually been shown to reduce both biliary bile acid and cholesterol secretion, the decrement of cholic acid being less significant than that of chenodeoxycholic acid (Juul and Van der Linden, 1969; Sarles *et al.*, 1970). Administration of Colestipol with a fat meal simultaneously clearly disturbs micelle formation within the intestinal lumen so that marked amounts of fatty acids, sterols, phospholipids, and bile acids are precipitated, relatively little remaining in the micellar phase (Miettinen, 1974a). This transient disruption of the micellar phase of the intestinal contents during chronic bile acid sequestrant treatment presumably explains impaired cholesterol and fat absorption, particularly because enhanced bile acid synthesis keeps the total bile salt pool intact (Garbutt and Kenney, 1972; Einarsson *et al.*, 1974).

The marked increase of cholesterol elimination as bile acids by ion exchangers is associated with a decreased level of serum total and LDL cholesterol, a marked increase in cholesterol synthesis (cf. Table I), and an increased turnover of the LDL protein moiety (Langer *et al.*, 1969). This increases the fractional catabolic rate of LDL, suggesting that its synthesis and catabolism were increased. The exact mechanism of the latter phenomenon is unknown and becomes more interesting in view of the recent studies indicating that fibroblasts are able to degrade LDL molecules (Goldstein and Brown, 1973; Brown and Goldstein, 1974) and that catabolism of LDL *in vivo* probably does not take place in the liver but in extrahepatic and even extravisceral tissues (Steinberg, 1974).

Kinetic analysis of cholesterol metabolism "on" and "off" bile acid sequestrants have indicated that, although the agents markedly enhance the turnover rate of cholesterol in the rapidly miscible pool A, the size of only the plasma compartment of pool A is reduced significantly, while the size of the tissue compartment remains unchanged (Goodman and Noble, 1968; Goodman *et al.*, 1973a; Miller *et al.*, 1973) or is even expanded (Kudchodkar *et al.*, 1972b) even though hepatic cholesterol decreases (Miettinen, 1970c); there appears to be no change in fluxes between pools A and B by the ion exchangers. Although a long-term cholestyramine treatment is known to reduce in size cutaneous and tendon xanthomas as a sign of reduction of the slowly exchangeable pool of cholesterol, the kinetic data indicate that its pool size is not decreased. As a matter of fact, in the study by Goodman *et al.* (1973a) the maximum value of pool B determined with the two-pool model was increased from 52 to 74 gm by Colestipol. This finding suggests enhanced synthesis in that pool or methodological difficulties in estimating drug-induced changes in tissue cholesterol pools by the available kinetic methods. However, kinetic studies on patients before and after ileal bypass operation have shown that the pool sizes of both rapidly and slowly miscible cholesterol is reduced by this operation (Moore et al., 1969); it affects cholesterol metabolism principally in the same way as, though more strongly (Miettinen, 1973b) than, ion-exchange resins.

2. NEOMYCIN

Neomycin, a polybasic antibiotic, binds and precipitates bile acids, fatty acids phospholipids, cholesterol, and plant sterols both *in vitro* and *in vivo*, resulting in a disruption of lipid micelles within the intestinal lumen (Thompson et al., 1970, 1971; Miettinen, 1975a). This interference with micellar solubilization of intestinal lipids and sterols may thus account for impaired fat and cholesterol absorption, and ultimately for the well-known neomycin-induced decrease of serum cholesterol, demonstrated first by Samuel and Steiner (1959). However, neomycin is known to interfere with the absorption of a great variety of substances, indicating that it may induce malabsorption by different mechanisms including mucosal damage (cf. Faloon, 1970).

Earlier clinical studies employing nonspecific analytical methods and large neomycin doses showed a marked increase in fecal bile acids and neutral sterols (Goldsmith et al., 1960; Powell et al., 1962). N-Methylated neomycin, a derivative without antibacterial action, had a similar effect on fecal steroids in man (Van den Bosch and Claes, 1967). When small doses of neomycin (1–2 gm/day) are used serum cholesterol reduction is associated with an enhanced cholesterol elimination as fecal neutral sterols only (Miettinen, 1973b,c), as if cholesterol absorption, because of the disruption of lipid micelles, had been inhibited. Thus, in contrast to what was presented on the basis of the earlier clinical studies, neomycin actually is not a bile acid sequestrant. However, larger doses increase bile acid excretion (Table I), too, but reduce inconsistently the bile acid pool size and have little effect on cholic acid half-life (Hardison and Rosenberg, 1969). The effects of 6-gm doses on serum cholesterol and on fecal bile acids and neutral sterols can be seen even in fasting patients, in patients with liver cirrhosis, or with small intestinal bypass for treatment of massive obesity, regardless of whether the patients are normo- or hyperlipidemic (Rubulis et al., 1970; Schwob et al., 1972).

The initial fall in serum cholesterol induced by neomycin in Table I correlated positively with the increase in fecal steroids, suggesting that the greater the disturbance in cholesterol absorption the larger is the decrease in serum cholesterol. However, the larger the decrease in serum cholesterol the larger may also be the amount of cholesterol mobilized from tissues, which is subsequently exposed to biliary secretion and fecal elimination. Reexamination of the patients after 1 to 2 years on neomycin showed that the decreased serum cholesterol level was apparently associated with a permanently enhanced fecal steroid excretion, the increase in the latter still tending to correlate positively with the overall fall in serum cholesterol. Calculations revealed that on an average about 100 gm of additional cholesterol had been eliminated during the neomycin period by each patient. This amount of cholesterol can hardly be totally mobilized from tissue pools but is balanced, in part at least, by enhanced synthesis. Rat experiments have revealed that neomycin stimulates cholesterol synthesis from acetate in the intestinal mucosa but not in the liver; intestinal conversion of mevalonate to lipids, though not to cholesterol, is also enhanced and the intestinal cholesterol content is markedly reduced (Cayen, 1970). The acetate-mevalonate test actually indicated that neomycin stimulated cholesterol synthesis in man during the short-term experiments, even though this was associated with decreased or unchanged levels of serum methyl sterols (Table I). Despite the marked extra elimination of cholesterol during the long-term neomycin treatment, the acetate-mevalonate test and the serum methyl sterols suggested that cholesterol synthesis was decreased or unchanged. Though in patients with interrupted enterohepatic circulation of bile acids the enhanced cholesterol synthesis seems to be permanently reflected in the acetate-mevalonate test and serum methyl sterols (Miettinen, 1970a), it is possible that during neomycin treatment, particularly during a long-term therapy, the rate-limiting steps after mevalonate are enhanced proportionally to the enhancement of the step before mevalonate. Under these conditions increased carbon flow from acetate to cholesterol will not be detected by the two tests, particularly if methyl sterol absorption and intestinal metabolism is interfered by neomycin. In addition, it is interesting to note that the further increase in cholesterol elimination and decrease in serum cholesterol by neomycin in cholestyramine-treated subjects with markedly enhanced cholesterol production was associated, according to the two tests, with unchanged or even decreased synthesis (Table I). Comparison of the cholestyramine and neomycin effects in these experiments indicated that the increase in fecal elimination of cholesterol was much greater by cholestyramine than by neomycin, but because the former drug apparently caused a much greater and obviously qualitatively different compensatory increase in cholesterol synthesis than the latter one, the decrease in serum cholesterol effected by the two drugs was of the same magnitude. This suggests that, in contrast to what has been presented (cf. Dietschy and Wilson, 1970), compensatory increase in cholesterol synthesis is much smaller after the inhibition of cholesterol than of bile acid reabsorption and that the drugs which act by interfering with the cholesterol absorption are more effective serum cholesterol-lowering agents than those which interfere with the bile acid absorption.

Extra loss of body cholesterol by neomycin during long-term treatment could also be balanced by tissue mobilization of cholesterol. Kinetic analysis of cholesterol "off" and "on" neomycin revealed that in man the drug decreased the size of pool A by 33–44% (Samuel et al., 1968). The absolute decrease in pool size was much greater than the concomitant decrease in serum cholesterol, suggesting that the size of the tissue compartment was also reduced. Animal experiments indicate that intestinal cholesterol is decreased by the drug (Cayen, 1970). No definite information is available on the possible changes produced in pool B size by neomycin in man. More recent studies have confirmed the finding of neomycin-induced increase of fecal steroids as neutral sterols; absorption of dietary cholesterol was markedly reduced and the slope of plasma cholesterol specific activity-time curve was not changed, probably due to simultaneously enhanced cholesterol synthesis and mobilization of label tissue cholesterol (Sedaghat et al., 1975).

3. PLANT STEROLS

The mechanism by which plant sterols reduce serum cholesterol in man has been explored primarily by Grundy et al. (1969) with the sterol balance and isotopic techniques. Oral administration of large quantities (up to 11 gm/day) of plant sterols to hypercholesterolemic patients decreased the serum cholesterol by 20%. This was associated with a fall in the absorption percentage of dietary cholesterol (probably of endogenous, too) from 38–81% to 20–36% with no reduction in the delivery of endogenous cholesterol into the intestinal lumen. Thus, during a fairly long-term treatment several tens of grams of extra cholesterol from endogenous sources was eliminated into stools. That this was not balanced solely by mobilization of tissue cholesterol was indicated by an increased slope of the specific activity-time curve of serum cholesterol after the cholesterol concentration was decreased by plant sterols. This change in the slope, associated with the increased elimination of endogenous cholesterol, is most likely caused by enhanced cholesterol synthesis. No change was seen in the bile acid elimination, indicating that, as was the case with small doses of neomycin, enhanced cholesterol synthesis is not consistently associated with increased bile acid synthesis. Plant sterols can also decrease the slope of the specific activity-time curve (Sodhi *et al.*, 1973), indicating that flux of labeled tissue cholesterol together with reduced absorption of dietary cholesterol can counterbalance the increasing effect on the slope of enhanced synthesis.

The mechanism by which plant sterols inhibit cholesterol absorption is not known. It has been suggested that esterification and release of cholesterol into the lymph is inhibited or plant sterols block competitively the mucosal sites for cholesterol absorption (cf. Grundy *et al.*, 1969). Even though β-sitosterol, the major component of plant sterols is tranferred from the oil phase into the micellar phase of the intestinal contents less effectively than cholesterol (Miettinen and Siurala, 1971), large amounts of plant sterols may displace cholesterol from the mixed micelles during fat digestion, diminishing the absorption of cholesterol. As a matter of fact, addition of β-sitosterol to duodenal bile or intestinal contents obtained after a fat meal precipitates cholesterol and phospholipids but not bile acids from the mixed micelles of the intestinal contents in particular (Miettinen, 1975a).

B. Absorbable Agents

1. CLOFIBRATE

Clofibrate is the most widely used hypolipidemic agent which is more effective on serum triglycerides than on cholesterol. Although clofibrate inhibits adipose tissue lipolysis and reduces plasma free fatty acids, it also inhibits fatty acids, cholesterol, and triglyceride synthesis in the liver of experimental animals (cf. Nikkilä, 1973; Fallon *et al.*, 1972). These mechanisms may also play a decisive role in the hypolipidemic action of clofibrate in man, suggesting that inhibition of hepatic triglyceride production results subsequently in a decrease in VLDL production, especially in cases in which the overproduction of triglycerides is the primary cause of hyperglyceridemia.

Studies of plasma triglyceride kinetics off and on clofibrate (Kissebah *et al.*, 1974) and its derivative (Nikkilä and Kekki, 1972a) have actually suggested that in hyperglyceridemic patients the turnover of free fatty acids is markedly decreased and that the initially enhanced triglyceride turnover (usually in type IV subjects) is normalized while the clearance of endogenous triglycerides is unchanged but that of exogenous ones is improved. In hyperglyceridemic subjects with a normal triglyceride turnover (type V subjects) the impaired removal of both endogenous and exogenous triglycerides was significantly improved by clofibrate. Postheparin plasma lipolytic activity was also increased by the drug. These studies suggest that the major hypotriglyceridemic effect of clofibrate in man is to inhibit triglyceride and VLDL production, enhanced removal, probably based on lipoprotein lipase activation, being a contributing mechanism which may become important in cases with primarily impaired removal of triglycerides. This difference in action in different forms of hypertriglyceridemia may partly explain, in addition to different methodology, the earlier findings that activation of triglyceride removal, associated with an unchanged or decreased production, would have a basic role in the hypotriglyceridemic effect of clofibrate and its derivatives in man (Ryan and Schwartz, 1964, 1965; Spritz, 1965; Boberg et al., 1970; Bierman et al., 1970; Sodhi et al., 1971b). No information is available on the LDL turnover in clofibrate-treated type IV subjects in whom the turnover of VLDL is decreased; this would be particularly interesting in those subjects who respond to clofibrate with an increased LDL concentration (Strisower et al., 1968; Carlson et al., 1974). The turnover studies performed with labeled LDL in type II have indicated that absolute catabolism (and synthesis) of LDL is reduced proportionally to its diminished pool size (both extra- and intravascular), while the fractional catabolic rate is not changed (Walton et al., 1963; Scott and Hurley, 1969). These findings suggest that the decrease in the precursor (VLDL) production by clofibrate is followed by a decreased LDL synthesis.

The effect of clofibrate on cholesterol elimination, and cholesterol and bile acid kinetics have also been investigated. Since in hypertriglyceridemic subjects cholesterol synthesis and elimination are high (Miettinen, 1970a) and are frequently disproportionately increased in relation to body weight (Sodhi and Kudchodkar, 1973b), it can be expected that the clofibrate-induced decrease in the triglyceride level, and triglyceride (and VLDL) synthesis and turnover could be accompanied by a decrease in cholesterol elimination. In their extensive studies, Grundy et al. (1972) pointed out, however, that biliary cholesterol secretion was enhanced and that, irrespective of the type of hyperlipoproteinemia, clofibrate almost consistently enhanced the fecal elimination of cholesterol as neutral sterols while the fecal excretion of bile acids was frequently decreased. The net increase in total fecal steroid excretion, which was quite consistent in type IIA and V but less consistent in type IV, was not balanced by increased synthesis; on the contrary, evidence was obtained that the latter may have decreased and kinetic analysis of cholesterol indicated that tissue cholesterol was mobilized, resulting in a reduction of pool A and especially of pool B. The increased flux from the latter to pool A was frequently reflected in a decreased slope of the specific activity-time curve of plasma cholesterol, a finding recorded earlier also by others (Nestel *et al.*, 1965; Horlick *et al.*, 1971) and shown to correlate with the drop in the plasma cholesterol concentration or with the cholesterol calculated to be mobilized from the tissues (Sodhi *et al.*, 1973). Horlick and co-workers (1971), who with the acetate-mevalonate test showed that clofibrate inhibits cholesterol synthesis in man (Sodhi *et al.*, 1971a), reported also that this drug increased fecal elimination of cholesterol as bile acids and neutral sterols in type II subjects but was ineffective in type IV patients. Mitchell and Murchison (1972) found a decrease in bile acids but no change in the total elimination of cholesterol into feces by clofibrate.

Kinetic analysis of bile acids by Einarsson *et al.* (1973) revealed that clofibrate had no consistent effect on different parameters of bile acid metabolism in patients with type IIA and B abnormality. However, in patients with type IV the drug reduced the markedly elevated cholic acid production from 1076 to 333 mg/day, while the turnover of chenodeoxycholate was decreased insignificantly. The authors suggested that the secretion rate of VLDL is, in part, correlated with the formation of bile acids, and therefore clofibrate-induced reduction in VLDL production should decrease bile acid synthesis.

The basic reason for the quite consistent increase in fecal neutral sterol elimination by clofibrate is unknown. This may be secondary to the drug action on the mobilization of tissue cholesterol, removal of plasma lipoproteins or biliary secretion of cholesterol. Cholesterol absorption is not impaired, but increased biliary secretion of cholesterol has been repeatedly demonstrated (Thistle and Schoenfield, 1971; Grundy *et al.*, 1972; Pertsemlidis *et al.*, 1974). As shown above, mobilization of tissue cholesterol is enhanced, while the absolute turnovers of both LDL in type II subjects and its precursor VLDL in type IV subjects are decreased. Thus increased biliary secretion may be secondary to enhanced tissue mobilization of cholesterol, though the drug as such or the clofibrate-induced exposure of the liver to increased thyroid hormone action (Ruegamer *et al.*, 1969) may stimulate this secretion process.

Since clofibrate also inhibits cholesterol synthesis (cf. White, 1970), especially in the intestine (Grundy *et al.*, 1972), the drug in combination with bile acid sequestrants could be expected to cause a marked further decrease in the serum cholesterol level. Though the neutral sterol elimination is increased even under these conditions by clofibrate (Grundy *et al.*, 1972), kinetic analysis of cholesterol metabolism indicates that the drug is unable to block the increased rate of cholesterol synthesis resulting from bile acid sequestrant treatment (Goodman *et al.*, 1973a). The combination of clofibrate with a bile acid-binding resin does not seem consistently to further reduce the serum cholesterol level (Howard and Hyams, 1971; Grundy et al., 1972; Goodman et al., 1973a), but converts the bile markedly lithogenic (Grundy, 1974, personal communication).

2. NICOTINIC ACID

The effects of nicotinic acid on plasma lipids are comparable to those of clofibrate even though its free fatty acid- and cholesterol-lowering actions are more potent. Effects of nicotinic acid and its derivatives on different parameters of lipid metabolism have been extensively dealt with in a recent workshop (Gey and Carlson, 1971). Inhibition of lipolysis in adipose tissue by nicotinic acid reduces plasma free fatty acids and subsequently decreases hepatic influx of fatty acids. This in turn may result in a lowered synthesis of plasma triglycerides and of VLDL and later on its metabolic product LDL. Serial analyses of different parameters have actually indicated that the nicotinic acid-induced fall in plasma free fatty acids is followed by a fall in plasma triglycerides (VLDL) within 4 to 6 hours, while plasma cholesterol (LDL) is decreased after several days (Carlson et al., 1968). Kinetic analyses have finally indicated that in patients with type IV and V abnormality the turnover of plasma tirglycerides (Kissebah et al., 1974) and in patients with type II the turnover of labeled LDL (Langer and Levy, 1971) are clearly reduced by nicotinic acid. Furthermore, the hepatic esterification rate of plasma free fatty acids is reduced by nicotinic acid (Sailer and Bolzano, 1971) and in monkeys the drug decreases the incorporation of [^{14}C] threonine into plasma VLDL and LDL protein (Magide et al., 1973). Thus, although nicotinic acid enhances the removal of exogenous fat emulsion in man (Boberg et al., 1971), its major hypolipidemic mode of action appears to be the reduction of VLDL and LDL production.

The nicotinic acid-induced fall in lipoprotein synthesis and turnover can be expected to reduce cholesterol synthesis and also elimination. However, sterol balance studies have indicated that, analogously to clofibrate, nicotinic acid enhances cholesterol elimination in the majority of patients with type II abnormality, having no consistent effect on fecal bile acids (Miettinen, 1968b, 1971a; Sodhi et al., 1969). The increase in cholesterol elimination, which appears to be inconsistent in long-term use, correlates positively with the decrease in serum cholesterol. In addition, the decrease in the slope of the specific activity-time curve of plasma cholesterol (Miettinen, 1968b; Sodhi et al., 1969, 1973) indicates that the decrease in plasma cholesterol and the increase in fecal elimination of endogenous cholesterol as neutral sterols are associated with enhanced mobilization of tissue cholesterol, particularly because the

absorption of cholesterol is not impaired. Thus, mobilized tissue cholesterol enhances biliary cholesterol secretion and should result, as in the
case of clofibrate, in lithogenic bile and decreased tissue cholesterol. The
latter is seen in long-term treatment as a reduction in size of cholesterol
deposits and is also associated with a significant diminution in the liver
cholesterol (Miettinen, 1971a).

In some patients with heterozygous type II abnormality, nicotinic acid
has no increasing effect on, or even reduces, cholesterol elimination
(Miettinen, 1971a). Under these circumstances, the decrease in serum
cholesterol is primarily due to reduced lipoprotein and cholesterol synthesis, which also can flatten the specific activity-time curve of serum
cholesterol. The latter has been demonstrated in homozygous type II
patients in whom the effect of nicotinic acid was interpreted as reducing
serum cholesterol by inhibiting cholesterol synthesis, particularly because the fecal steroid elimination was unchanged (Moutafis and Myant,
1971). As a matter of fact, incorporation of [^{14}C]acetate to cholesterol
in a liver biopsy specimen was reduced during nicotinic acid treatment
in one of the homozygous children, a similar reduction to plasma cholesterol being also reported *in vivo* by Parsons (1961).

In some animal experiments, however, opposite results have been obtained (see Holmes, 1964). The acetate-mevalonate test and serum
methyl sterols suggested that, though in acute experiments the cholesterol
synthesis may have decreased when serum free fatty acids were lowered
by nicotinic acid, during chronic treatment after an overnight fast the
synthesis was increased (14 hours after the last dose of nicotinic acid)
according to both *in vivo* and *in vitro* experiments (Miettinen, 1971a).
Thus, the nicotinic acid effect on cholesterol synthesis may show a
diurnal variation during chronic treatment, when the relative lengths
of the periods of stimulation and inhibition during the day determine
the overall daily effect on synthesis.

It can be argued that clofibrate and nicotinic acid have no direct effect
on tissue mobilization of cholesterol but that this is solely secondary to
the decreased serum cholesterol level. As a matter of fact, the fall in
serum cholesterol correlated positively with the increase in fecal steroids
(Miettinen, 1971a) and also with the calculated amount of cholesterol
mobilized from the tissues (Sodhi *et al.*, 1973). Net mobilization may be
noted primarily in subjects with expanded tissue cholesterol pools.

3. THYROID HORMONES

The mode of hypolipidemic action of the thyroid hormones has been
studied primarily in hypothyroid patients, indicating that the hypocho-

lesterolemic effect is associated with improvement of cholesterol elimination by unknown mechanisms (cf. Myant, 1968; Miettinen, 1968a, 1970b). Dextrothyroxine, used in the treatment of hypercholesterolemic patients, can be expected to have a similar effect. Kinetic analysis of plasma triglycerides suggested that in hypothyroid patients the synthesis of triglycerides is normal while the fractional removal of both exogenous and endogenous triglycerides was reduced, accounting for the existing usually mild hyperglyceridemia (Nikkilä and Kekki, 1972b). In hyperthyroidism the production of triglycerides appeared to be enhanced, the fractional removal being normal and the efficiency of removal even improved. Furthermore, in hypothyroidism, the postheparin plasma lipolytic activity is low, because of the decreased hepatic lipase activity (Krauss et al., 1974), and correlated with the plasma triglyceride level and fractional endogenous triglyceride transport in thyroid imbalance (Porte et al., 1966; Kirkeby, 1968; Nikkilä and Kekki, 1972a). Since, in addition, turnover studies with labeled LDL have indicated that its removal is impaired in hypothyroidism (Walton et al., 1965) it seems logical to infere that, as has been generally accepted, the removal of the whole lipoprotein spectrum is impaired in hypothyroidism. Consequently, thyroid hormone treatment should enhance lipoproteins removal and may also stimulate their production.

Augmented lipoprotein metabolism by thyroid hormones should be associated with enhanced elimination of cholesterol, particularly because the fall in serum cholesterol may enhance mobilization of tissue cholesterol and because thyroid hormones are well known to stimulate cholesterol synthesis (cf. Myant, 1963). Sterol balance studies have actually shown that cholesterol synthesis is low in hypothyroid patients and that cholesterol elimination is impaired primarily in regard to neutral sterols and only to a small extent, if any, as bile acids (Miettinen, 1968a, 1970a, 1973a). The elimination as neutral sterols is clearly improved and, in contrast to the situation in the rat (Strand, 1963), only inconsistently as bile acids by thyroxine. In hyperthyroid patients the values are normal or tend to be supernormal. Furthermore, despite an apparently enhanced cholesterol synthesis the slope of the specific activity-time curve of plasma cholesterol is flattened by the fall in the serum cholesterol concentration, indicating mobilization of tissue cholesterol (Miettinen, 1968a). Thus, although, in contrast to the effect of clofibrate and nicotinic acid, the VLDL and LDL turnovers appear to be enhanced by thyroxine, cholesterol elimination undergoes almost the same pattern of change as by the two drugs, i.e., an increase of the fecal excretion of neutral sterols which is positively correlated with the decrease of the serum cholesterol level (Miettinen, 1970a).

4. PROBUCOL

Probucol is one of the most promising new hypocholesterolemic agents (Drake *et al.*, 1969; Kalams *et al.*, 1971; Danowski *et al.*, 1971; Miettinen, 1972a). It is well tolerated, reduces serum cholesterol by 9 to 27% according to different studies (cf. Miettinen and Toivonen, 1975), has an inconsistent effect on serum triglycerides, and is still under investigational use only. Animal experiments suggest that the drug reduces serum cholesterol by inhibiting cholesterol synthesis (Barnhart *et al.*, 1970). In patients with xanthomatotic type II hyperlipoproteinemia the 27% decrease in serum cholesterol by probucol was associated with a slight transient impairment of fat, water, bile acids, and cholesterol absorption but, in contrast to other absorbable hypocholesterolemic agents, with a reduced intestinal flux of endogenous cholesterol. Therefore, despite a slightly enhanced fecal bile acid excretion, the total cholesterol elimination was not increased (Miettinen, 1972a). Since, in addition, serum methyl sterols were initially decreased disproportionately to the decrease of serum cholesterol and the slope of the specific activity-time curve of serum cholesterol was decreased in most patients, it is apparent that cholesterol synthesis was reduced, particularly since mobilized tissue cholesterol contributed to the fecal steroids.

ACKNOWLEDGMENTS

Supported by grants from the Sigrid Jusélius Foundation and Finnish State Council for Medical Research, Finland.

REFERENCES

Adams, P. W., Kissebah, A. H., Harrigan, P., Stokes, T., and Wynn, V. (1974). *Eur. J. Clin. Invest.* **4,** 149.

Ahrens, E. H., Jr. (1970). *Advan. Metab. Disord.* **4,** 297.

Anonymous (1971). *Brit. Med. J.* **4,** 767, 775.

Barnhart, J. W., Sefranka, J. A., and McIntosh, D. D. (1970). *Amer. J. Clin. Nutr.* **23,** 1229.

Bergen, S. S., Van Itallie, T. B., Tennent, D. M., and Sebrell, W. H. (1959). *Proc. Soc. Exp. Biol. Med.* **102,** 676.

Bhattacharyya, A. K., Connor, W. E., and Spector, A. A. (1972). *J. Clin. Invest.* **51,** 2060.

Bierman, E. L., Brunzell, J. D., Bagdade, J. D., Lerner, R. L., Hazzard, W. R., and Porte, D., Jr. (1970). *Trans. Ass. Amer. Physicians* **83,** 211.

Boberg, J. (1971). *Acta Univ. Upsalaencis* **105,** 1.

Boberg, J., Carlson, L. A., and Hallberg, D. (1969). *J. Atheroscler. Res.* **9,** 159.

Boberg, J., Carlson, L. A. Fröberg, S. O., and Orö, L. (1970). *Atherosclerosis* **11,** 353.

Boberg, J., Carlson, L. A., Fröberg, S. O., Olsson, A., Orö, L., and Rössner, S. (1971). In "Metabolic Effects of Nicotinic Acid and Its Derivatives" (K. F. Gey and L. A. Carlson, eds.), p. 465. Hans Huber, Bern.

Boberg, J., Carlson, L. A., and Freyschuss, U. (1972a). *Eur. J. Clin. Invest.* **2**, 123.
Boberg, J., Carlson, L. A., Freyschuss, U., Lassers, B. W., and Wahlqvist, M. L. (1972b). *Eur. J. Clin. Invest.* **2**, 454.
Borgström, B. (1969). *J. Lipid Res.* **10**, 331.
Borkowski, A. J., Levin, S., Delcroix, C., Mahler, A., and Verhas, V. (1967). *J. Clin. Invest.* **46**, 797.
Bricker, L. A., Weis, H. J., and Siperstein, M. D. (1972). *J. Clin. Invest.* **51**, 197.
Brown, M. S., and Goldstein, J. L. (1974). *Science* **185**, 61.
Carey, J. B., Jr., and Williams, G. (1961). *J. Amer. Med. Ass.* **176**, 432.
Carlson, L. A., and Ekelund, L. G. (1963). *J. Clin. Invest.* **42**, 714.
Carlson, L. A., and Rössner, S. (1972). *Scand. J. Clin. Lab. Invest.* **29**, 271.
Carlson, L. A., Orö, L., and Östman, J. (1968). *J. Atheroscler. Res.* **8**, 667.
Carlson, L. A., Olsson, A., Orö, L., Rössner, S., and Waldius, G. (1974). *Proc. 3rd Int. Symp.*, p. 768.
Cayen, M. N. (1970). *Amer. J. Clin. Nutr.* **23**, 1234.
Chobanian, A. V., and Hollander, W. (1962). *J. Clin. Invest.* **41**, 1732.
Chobanian, A. V., Burrows, B. A., and Hollander, W. (1962). *J. Clin. Invest.* **41**, 1738.
Christakis, G., Rinzler, S. H., Archer, M., Winslow, G., Jampel, S., Stephenson, J., Friedman, G., Fein, H., Kraus, A., and James, G. (1966). *Amer. J. Pub. Health* **56**, 299.
Connor, W. E., and Lin, D. S. (1974). *J. Clin. Invest.* **53**, 1062.
Crouse, J. R., Grundy, S. M., and Ahrens, E. H., Jr. (1972). *J. Clin. Invest.* **51**, 1292.
Dam, H., Prange, I., Krogh Jensen, M., Kallehauge, H. E., and Fenger, H. J. (1971). *Z. Ernaehrungswiss.* **10**, 188.
Danielsson, H., and Einarsson, K. (1969). In "The Biological Basis of Medicine" (E. E. Bittar and N. Bittar, eds.), Vol. 5, p. 279. Academic Press, New York.
Danowski, T. S., Vester, J. W., Sunder, J. H., Gonzales, A. R., Khurana, R. C., and Jung, Y. (1971). *Clin. Pharmacol. Ther.* **12**, 929.
Davignon, J., Simmonds, W. J., and Ahrens, E. H., Jr. (1968). *J. Clin. Invest.* **47**, 127.
Dayton, S., Pearce, M. L., Hashimoto, S., Dixon, W. J., and Tomiyasu, U. (1969). *Circulation Suppl.* **40**, II, 1.
Denbesten, L., Connor, W. E., Kent, T. H., and Lin, D. (1970). *J. Lipid Res.* **11**, 341.
Dietschy, J. M. (1968). *J. Clin. Invest.* **47**, 286.
Dietschy, J. M., and Gamel, W. G. (1971). *J. Clin. Invest.* **50**, 872.
Dietschy, J. M., and Brown, M. S. (1974). *J. Lipid Res.* **15**, 508.
Dietschy, J. M., and Wilson, J. D. (1970). *New Engl. J. Med.* **282**, 1128, 1179, 1241.
Dorr, A. E., Martin, W. B., and Freyburger, W. A. (1974). *Int. Symp. on Drugs Affecting Lipid Metabolism 5th Milan, 1974 Abst.* p. 46.
Drake, J. W., Bradford, R. H., McDearmon, M., and Furman, R. H. (1969). *Metabolism* **18**, 916.
Eaton, R. P. (1971). *J. Lipid Res.* **12**, 491.
Eaton, R. P., Berman, M., and Steinberg, D. (1969). *J. Clin. Invest.* **48**, 1560.
Ehnholm, C., Huttunen, J. K., Kinnunen, P. J., Miettinen, T. A., and Nikkilä, E. A. (1975). *New Engl. J. Med.* **292**, 1314.
Einarsson, K., Hellström, K., and Kallner, M. (1973). *Eur. J. Clin. Invest.* **3**, 345.
Einarsson, K., Hellström, K., and Kallner, M. (1974). *Eur. J. Clin. Invest.* **4**, 405.
Fallon, H. J., Adams, L. L., and Lamb, R. G. (1972). *Lipids* **7**, 106.

Faloon, W. W. (1970). *Amer. J. Clin. Nutr.* 23, 645.

Farguhar, J. W., Gross, R. C., Wagner, R. M., and Reaven, G. M. (1965). *J. Lipid Res.* 6, 119.

Fujiwara, T., Hirono, H., and Arakawa, T. (1965). *Tohoku J. Exp. Med.* 87, 155.

Garbutt, J. T., and Kenney, T. J. (1972). *J. Clin. Invest.* 51, 2781.

Gey, K. F., and Carlson, L. A. (1971). "Metabolic Effects of Nicotinic Acid and Its Derivatives." Hans Huber, Bern.

Gitlin, D., Cornwell, D. G., Nakasato, D., Oncley, J. L., Hughes, W. L., Jr., and Janeway, C. A. (1958). *J. Clin. Invest.* 37, 172.

Glueck, C. J. (1971). *Metabolism* 20, 691.

Glueck, C. J., Ford, S., Jr., Steiner, P., and Fallat, R. (1973). *Metabolism* 22, 807.

Goldsmith, G. A., Hamilton, J. G., and Miller, O. N. (1960). *Arch. Intern. Med.* 105, 512.

Goldstein, J. L., and Brown, M. S. (1973). *Proc. Nat. Acad. Sci. U.S.* 70, 2804.

Goodman, D. S. (1964). *J. Clin. Invest.* 43, 1480.

Goodman, D. S., and Noble, R. P. (1968). *J. Clin. Invest.* 47, 231.

Goodman, D. S., Noble, R. P., and Dell, R. B. (1973a). *J. Clin. Invest.* 52, 2646.

Goodman, D. S., Noble, R. P., and Dell, R. B. (1973b). *J. Lipid Res.* 14, 178.

Gould, R. G. (1959). *In* "Hormones and Atherosclerosis" (G. Pincus, ed.), p. 75. Academic Press, New York.

Gould, R. G., and Swyryd, E. A. (1966). *J. Lipid Res.* 7, 698.

Grundy, S. M. (1972). *Arch. Intern. Med.* 130, 638.

Grundy, S. M., and Ahrens, E. H., Jr. (1966). *J. Clin. Invest.* 45, 1503.

Grundy, S. M., and Ahrens, E. H., Jr. (1969). *J. Lipid Res.* 10, 91.

Grundy, S. M., and Metzger, A. L. (1972). *Gastroenterology* 62, 1200.

Grundy, S. M., Ahrens, E. H., Jr., and Miettinen, T. A. (1965). *J. Lipid Res.* 6, 397.

Grundy, S. M., Ahrens, E. H., and Salen, G. (1968). *J. Lipid Res.* 9, 374.

Grundy, S. M., Ahrens, E. H., Jr., and Davignon, J. (1969). *J. Lipid Res.* 9, 304.

Grundy, S. M., Ahrens, E. H., Jr., and Salen, G. (1971). *J. Lab. Clin. Med.* 78, 94.

Grundy, S. M., Ahrens, E. H., Jr., Salen, G., Schreibman, P. H., and Nestel, P. J. (1972). *J. Lipid Res.* 13, 531.

Hallberg, D. (1965). *Acta Physiol. Scand.* 64, 306.

Hamprecht, B., Nüssler, C., and Lynen, F. (1969). *FEBS Lett.* 4, 117.

Hardison, W. G. M., and Rosenberg, I. H. (1969). *J. Lab. Clin. Med.* 74, 564.

Hashim, S. A., and Van Itallie, T. B. (1965). *J. Amer. Med. Ass.* 192, 289.

Havel, R. J., Kane, J. P., Balasse, E. O., Segel, N., and Bosso, L. V. (1970). *J. Clin. Invest.* 49, 2017.

Holmes, W. L. (1964). *In* "Lipid Pharmacology" (R. Paoletti, ed.), p. 132. Academic Press, New York.

Horlick, L., Kudchodkar, B. J., and Sodhi, H. S. (1971). *Circulation* 43, 299.

Howard, A. N., and Hyams, D. E. (1971). *Brit. Med. J.* 3, 25.

Hyun, S. A., Vahouny, G. V., and Treadwell, C. R. (1963). *Proc. Soc. Exp. Biol. Med.* 112, 496.

Johns, W. H., and Bates, T. (1969). *J. Pharm. Sci.* 58, 179.

Juul, A. H., and Van der Linden, W. (1969). *Acta Physiol. Pharmacol. Neerl.* 15, 469.

Kalams, Z., Dacquisto, M., and Kornett, G. S. (1971). *Curr. Ther. Res.* 13, 692.

Kekki, M., and Nikkilä, E. A. (1971). *Metabolism* 20, 878.

Kekki, M., and Nikkilä, E. A. (1975). *Scand. J. Clin. Lab. Invest.* 35, 171.

Kekki, M., Miettinen, T. A., and Wahlström, B. (1973). *In* "Regulation and Control in Physiological Systems" (A. S. Iberall and A. C. Guyton, eds.), pp. 71–73. International Federation of Automatic Control.

Khachadurian, A. K. (1968). *J. Atheroscler. Res.* **8**, 177.

Kirkeby, K. (1968). *Acta Endocrinol.* **59**, 555.

Kissebah, A. H., Adams, P. W., Harrigan, P., and Wynn, V. (1974). *Eur. J. Clin. Invest.* **4**, 163.

Kottke, B. A., and Subbiah, M. T. R. (1972). *J. Lab. Clin. Med.* **80**, 530.

Krasno, L. R., and Kidera, G. J. (1972). *J. Amer. Med. Ass.* **219**, 845.

Krauss, R. M., Levy, R. I., and Fredrickson, D. S. (1974). *J. Clin. Invest.* **54**, 1107.

Kudchodkar, B. J., Sodhi, H. S., and Horlick, L. (1972a). *Metabolism* **21**, 343.

Kudchodkar, B. J., Sodhi, H. S., and Horlick, L. (1972b). *Clin. Res.* **20**, 944.

Kudchodkar, B. J., Sodhi, H. S., and Horlick, L. (1973). *Metabolism* **22**, 155.

Langer, T., and Levy, R. J. (1971). *In* "Metabolic Effects of Nicotinic Acid and Its Derivatives" (K. F. Gey and L. A. Carlson, eds.), p. 641. Hans Huber, Bern.

Langer, T., Levy, R. I., and Fredrickson, D. S. (1969). *Circulation Suppl.* II, **40**, 14.

Langer, T., Strober, W., and Levy, R. I. (1972). *J. Clin. Invest.* **51**, 1528.

Leren, P. (1966). *Acta Med. Scand. Suppl.*, p. 466.

Levy, R. I., Fredrickson, D. S., Stone, N. J., Bilheimer, D. W., Brown, W. V., Glueck, C. J., Gotto, A. M., Herbert, P. N., Kwiterovich, P. O., Langer, T., LaRosa, J., Lux, S. E., Rider, A. K., Shulman, R. S., and Sloan, H. R. (1973). *Ann. Intern. Med.* **79**, 51.

Lindstedt, S. (1957). *Acta Physiol. Scand.* **40**, 1.

Lindstedt, S. (1970). *Proc. 2nd Int. Symp.* p. 262.

Liu, G. C. K., Schreibman, P. H., Samuel, P., and Ahrens, E. H., Jr. (1974). *J. Clin. Invest.* **53**, 47a.

London, I. M., and Rittenberg, D. (1950). *J. Biol. Chem.* **184**, 687.

Loud, A. V. (1955). Thesis, M.I.T., Cambridge, Massachusetts.

Magide, A. A., Reichl, D., and Myant, N. B. (1973). *Clin. Sci.* **44**, 20P.

Miettinen, T. A. (1966). *Proc. IIIrd Meet. FEBS, Warzawa, 1966,* p. 104.

Miettinen, T. A. (1968a). *J. Lab. Clin. Med.* **71**, 537.

Miettinen, T. A. (1968b). *Clin. Chim. Acta* **20**, 43.

Miettinen, T. A. (1968c). *Ann. Med. Exp. Biol. Fenn.* **46**, 172.

Miettinen, T. A. (1969). *Life Sci.* **8**, 713.

Miettinen, T. A. (1970a). *Ann. Clin. Res.* **2**, 300.

Miettinen, T. A. (1970b). *Proc. 2nd Int. Symp.*, p. 508. Springer-Verlag, Berlin.

Miettinen, T. A. (1970c). *Proc. 2nd Int. Symp.*, p. 558. Springer-Verlag, Berlin.

Miettinen, T. A. (1971a). *In* "Metabolic Effects of Nicotinic Acid and Its Derivatives" (K. F. Gey and L. A. Carlson, eds.), pp. 649, 677. Hans Huber, Bern.

Miettinen, T. A. (1971b). *Ann. Clin. Res.* **3**, 264.

Miettinen, T. A. (1972a). *Atherosclerosis* **15**, 163.

Miettinen, T. A. (1972b). *Gut* **13**, 682.

Miettinen, T. A. (1973a). *In* "The Bile Acids" (P. Nair and D. Kritchewsky, eds.), Vol. 2, p. 191. Plenum, New York.

Miettinen, T. A., (1973b). *Excerpta Med. Int. Congr. Ser.* No. **283**, p. 77.

Miettinen, T. A. (1973c). *Eur. J. Clin. Invest.* **3**, 256.

Miettinen, T. A. (1974a). *Eur. J. Clin. Invest.* **4**, 365.

Miettinen, T. A. (1974b). *Thromb. Res.* Suppl. 1, **4**, 41.

Miettinen, T. A. (1975). *Proc. 9th Int. Meet. Pharmacol., 1975 Helsinki,* in press.

Miettinen, T. A. (1976). *In* "Lipid Absorption: Biochemical and Clinical Aspects," (R. Bihmer, ed.,), p. 151. MTP Publ., London.

Miettinen, T. A., and Siurala, M. (1971). Z. Klin. Chem. Klin. Biochem. 9, 47.
Miettinen, T. A., and Toivonen, I. (1975). Postgrad. Med. J. Suppl. 8, 51, 71.
Miettinen, T. A., Ahrens, E. H., Jr., and Grundy, S. M. (1965). J. Lipid Res. 6, 411.
Miettinen, M., Turpeinen, O., Karvonen, M. J., Elosuo, R., and Paavilainen, E. (1972). Lancet ii, 835.
Miller, N. E., Clifton-Bligh, P., and Nestel, P. J. (1973). J. Lab. Clin. Med. 82, 876.
Mitchell, W. D., and Murchison, L. E. (1972). Clin. Chim. Acta 36, 153.
Moir, N. J., Gaylor, J. L., and Yanni, J. B. (1970). Arch. Biochem. Biophys. 141, 465.
Moore, R. B., Frantz, I. D., Jr., and Buchwald, H. (1969). Surgery 65, 98.
Moore, R. B., Frantz, I. D., Jr., Varco, R. L., and Buchwald, H. (1970). Proc. 2nd Symp., p. 295.
Moutafis, C. D., and Myant, N. B. (1969). Clin. Sci. 37, 443.
Moutafis, C. D., and Myant, N. B. (1971). In "Metabolic Effects of Nicotinic Acid and Its Derivatives" (K. F. Gey and L. A. Carlson, eds.), p. 659. Hans Huber, Bern.
Moutafis, C. D., Myant, N. B., Mancini, M., and Oriente, P. (1971). Atherosclerosis 14, 247.
Myant, N. B. (1968). In "The Biological Basis of Medicine" (E. E. Bittar and N. Bittar, eds.), Vol. 2, p. 193. Academic Press, New York.
Nestel, P. J., Hirsch, E. Z., and Couzens, E. A. (1965). J. Clin. Invest. 44, 891.
Nestel, P. J., Whyte, H. M., and Goodman, D. S. (1969). J. Clin. Invest. 48, 982.
Nikkilä, E. A. (1972). In "Pharmacological Control of Lipid Metabolism" (W. L. Holmes, R. Paoletti, and D. Kritchewsky, eds.), p. 113. Plenum, New York.
Nikkilä, E. A., and Kekki, M. (1971a). Acta Med. Scand. 190, 49.
Nikkilä, E. A., and Kekki, M. (1971b). Scand. J. Clin. Lab. Invest. 27, 97.
Nikkilä, E. A., and Kekki, M. (1972a). Eur. J. Clin. Invest. 2, 231.
Nikkilä, E. A., and Kekki, M. (1972b). J. Clin. Invest. 51, 2103.
Olefsky, J., Farquhar, J. W., and Reaven, G. M. (1974). Eur. J. Clin. Invest. 4, 121.
Parsons, W. B., Jr. (1961). Circulation 24, 1099.
Parkinson, T. M., Gundersen, K., and Nelson, N. A. (1970). Atherosclerosis 11, 531.
Pawliger, D. F., and Shipp, J. C. (1965). J. Clin. Invest. 44, 1084.
Perl, W., and Samuel, P. (1969). Circ. Res. 25, 191.
Pertsemlidis, D., Panveliwalla, D., and Ahrens, E. H., Jr. (1974). Gastroenterology 66, 565.
Porte, D., Jr., and Bierman, E. L. (1969). J. Lab. Clin. Med. 73, 631.
Porte, D., Jr., O'Hara, D. D., and Williams, R. H. (1966). Metabolism 15, 107.
Powell, R. C., Nunes, W. T., Harding, R. S., and Vacca, J. B. (1962). Amer. J. Clin. Nutr. 11, 156.
Quarfordt, S. H., and Greenfield, M. F. (1973). J. Clin. Invest. 52, 1937.
Quarfordt, S. H., Frank, A., Shames, D. M., Berman, M., and Steinberg, D. (1970). J. Clin. Invest. 49, 2281.
Quintao, E., Grundy, S. M., and Ahrens, E. H., Jr. (1971a). J. Lipid Res. 12, 221.
Quintao, E., Grundy, S. M., and Ahrens, E. H., Jr. (1971b). J. Lipid Res. 12, 233.
Reaven, G. M., Hill, D. B., Gross, R. C., and Farguhar, J. W. (1965). J. Clin. Invest. 44, 1826.
Rössner, S., Boberg, J., Carlson, L. A., Freyschuss, U., and Lassers, B. W. (1974). Eur. J. Clin. Invest. 4, 109.
Rubulis, A., Rubert, M., and Faloon, W. W. (1970). Amer. J. Clin. Nutr. 23, 1251.
Ruegamer, W. R., Ryan, N. T., Richert, D. A., and Westerfeld, W. W. (1969). Biochem. Pharmacol. 18, 613.

Ryan, J. R., and Jain, A. (1972). *J. Clin. Pharmacol.* **12**, 268.
Ryan, W. G., and Schwartz, T. B. (1964). *J. Lab. Clin. Med.* **64**, 1001.
Ryan, W. G., and Schwartz, T. B. (1965). *Metabolism* **14**, 1243.
Sailer, S., and Bolzano, K. (1971). In "Metabolic Effects of Nicotinic Acid and Its Derivatives" (K. F. Gey and L. A. Carlson, eds.), p. 479. Hans Huber, Bern.
Sailer, S., Sandhofer, F., and Braunsteiner, H. (1966). *Klin. Wochenschr.* **44**, 1032.
Samuel, P., and Lieberman, S. (1973). *J. Lipid Res.* **14**, 189.
Samuel, P., and Perl, W. (1970). *J. Clin. Invest.* **49**, 346.
Samuel, P., and Steiner, A. (1959). *Proc. Exp. Biol. Med.* **100**, 193.
Samuel, P., Holtzman, C. M., Meilman, E., and Perl, W. (1968). *J. Clin. Invest.* **47**, 1806.
Sarles, H., Crotte, C., Gerolami, A., Mule, A., Domingo, N., and Hauton, J. (1970). *Scand. J. Gastroenterol.* **5**, 603.
Schwob, D., Rubulis, A., Lim, E. C., Sherman, C. D., and Faloon, W. W. (1972). *Amer. J. Clin. Nutr.* **25**, 987.
Scott, P. J., and Hurley, P. J. (1969). *J. Atheroscler. Res.* **9**, 25.
Sedaghat, A., Samuel, P., Crouse, J. R., and Ahrens, E. H., Jr. (1975). *J. Clin. Invest.* **55**, 12.
Sodhi, H. S., and Kudchodkar, B. J. (1973a). *Metabolism* **22**, 895.
Sodhi, H. S., and Kudchodkar, B. J. (1973b). *Lancet* i, 513.
Sodhi, H. S., Horlick, L., and Kudchodkar, B. (1969). *Clin. Res.* **17**, 395 and 663.
Sodhi, H. S., Kudchodkar, B. J., Horlick, L., and Weder, C. H. (1971a). *Metabolism* **20**, 348.
Sodhi, H. S., Kudchodkar, B. J., and Horlick, L. (1971b). *Metabolism* **20**, 309.
Sodhi, H. S., Kudchodkar, B. J., and Horlick, L. (1973). *Atherosclerosis* **17**, 1.
Sodhi, H. S., Kudchodkar, B. J., Varughese, P., and Duncan, D. (1974). *Proc. Soc. Exp. Biol. Med.* **145**, 107.
Splitter, S. D., Michaels, G. D., Schlierf, G., Wood, P.D.S., and Kinsell, L. W. (1968). *Metabolism* **17**, 1129.
Spritz, N. (1965). *Circulation* Suppl. II, **32**, 201.
Steinberg, D. (1974). *Proc. 3rd Int. Symp.*, p. 658. Springer-Verlag, Berlin-Heidelberg-New York.
Strand, O. (1963). *J. Lipid Res.* **4**, 305.
Strisower, E. H., Adamson, G., and Strisower, B. (1968). *Amer. J. Med.* **45**, 488.
Taylor, C. B., Mikkelson, B., Anderson, J. A., and Forman, D. T. (1966). *Arch. Pathol.* **81**, 213.
Thistle, J. L., and Schoenfield, L. J. (1971). *Gastroenterology* **61**, 488.
Thompson, G. R., MacMahon, M., and Claes, P. (1970). *Eur. J. Clin. Invest.* **1**, 40.
Thompson, G. R., Barrowman, J., Gutierrez, L., and Dowling, R. H. (1971). *J. Clin. Invest.* **50**, 319.
Walton, K. W., Scott, P. J., Jones, J. V., Fletcher, R. F., and Whitehead, T. (1963). *J. Atheroscler. Res.* **3**, 396.
Walton, K. W., Scott, P. J., Dykes, P. W., and Davies, J. W. L. (1965). *Clin. Sci.* **29**, 217.
Van Berge Henegouwen, G. P. (1974). Thesis, University of Nijmegen. The Netherlands.
Van den Bosch, J. F., and Claes, P. J. (1967). *Progr. Biochem. Pharmacol.* **2**, 97.
White, L. W. (1970). *Proc. 2nd Int. Symp.*, p. 545. Springer-Verlag, New York.
Wood, P. D., Shioda, R., Estrich, D. L., and Splitter, S. (1972). *Metabolism* **21**, 107.
Zilversmit, D. B. (1972). *Proc. Soc. Exp. Biol. Med.* **140**, 862.
Zilversmit, D. B., and Hughes, L. B. (1974). *J. Lipid Res.* **15**, 465.

Chapter 4

DIETARY AND DRUG

REGULATION OF

CHOLESTEROL METABOLISM

IN MAN

SCOTT M. GRUNDY

Department of Medicine, Division of Metabolism,
The University of California, San Diego, California
and
The Veterans Administration Hospital
San Diego, California

I. Introduction

The observation that hypercholesterolemia is associated with an increased risk for atherosclerosis has been a strong stimulus for development of therapeutic methods for lowering of plasma cholesterol. However, it is now well recognized that cholesterol does not circulate freely in plasma but is transported in association with other lipids and specific proteins (apoproteins). Since alterations in plasma cholesterol levels are necessarily accompanied by changes in concentrations of whole lipoprotein complexes, the interrelationships between the metabolism of cholesterol and that of other lipids and apoproteins has become a subject of considerable interest. In this chapter, attention will be directed toward means for regulation of plasma lipoprotein concentrations through the use of diets and drugs that alter the metabolism of lipids, especially cholesterol. The discussion will be introduced by a brief summary of current concepts of cholesterol metabolism in man. The potential adverse effects of cholesterol lowering will also be examined, and special attention will be given to the possibility that decreases in plasma lipoproteins and cholesterol may sometimes be associated with formation of cholesterol gallstones.

II. Cholesterol Metabolism in Man

The liver represents a key organ in the metabolism of cholesterol in man. It plays an important role in the regulation of cholesterol concentrations in body compartments, especially the plasma, and it is also the major site of cholesterol degradation and excretion. Hepatic cholesterol is derived from three sources: (a) from that synthesized locally within the liver; (b) from cholesterol absorbed by the intestine; and (c) from that produced in other tissues. The numerous biochemical steps in hepatic synthesis of cholesterol from acetate have been the subject of intensive investigation, and most of the steps have been successfully elucidated (Bloch, 1965). The rate of cholesterol synthesis in the liver is regulated through feedback inhibition by cholesterol derived from other sources, especially the gut (Gould, 1951; Langdon and Bloch, 1953; Frantz *et al.*, 1954; Siperstein and Guest, 1960).

Cholesterol in the liver can also have several fates: (a) it can be secreted into plasma in association with lipoproteins; (b) it can be partially converted into the primary bile acids (cholic and chenodeoxycholic acids); and (c) it can be secreted into bile as cholesterol itself. Each of these processes is important for regulation of cholesterol concentrations in plasma and bile.

Cholesterol is secreted into plasma in conjunction with triglyceride-rich lipoproteins called "very low-density lipoproteins" or VLDL. As triglycerides are removed from VLDL through the action of lipolytic enzymes, a "remnant" particle is produced (Havel et al., 1973; Havel and Kane, 1973). This "remnant" can be either removed by the liver or transformed into another lipoprotein, "low-density lipoprotein" (LDL). The detailed steps involved in the conversion of VLDL into LDL are not known, but one step is apparently the esterification of cholesterol through transfer of a fatty acid molecule from lecithin (Norum et al., 1971; Forte et al., 1971). This transesterification is catalyzed in plasma by the enzyme, lecithin-cholesterol acyltransferase (LCAT) (Glomset, 1968). The fate of the cholesterol ester in LDL and sites of ester hydrolysis are unknown. Recent studies by Sniderman et al. (1974) indicate that LDL is not removed from plasma primarily by hepatic uptake, but other tissues (such as connective tissue or the reticuloendothelial system) appear to be the major sites of LDL disposal. Following uptake and degradation of LDL complexes by peripheral tissues, cholesterol must be eventually returned to the liver for excretion or conversion to bile acids; however, mechanisms for transport of cholesterol from periphery to the liver have not been determined.

In a normal person, about 750 to 1000 mg of cholesterol are synthesized daily, and about one-third of each day's production is converted into bile acids (200–350 mg/day) (Grundy and Ahrens, 1969). Bile acid formation occurs entirely within the liver, and newly synthesized bile acids become part of the body pool; this pool is located largely within the enterohepatic space including liver, gallbladder, and intestine. During the day when food is ingested, bile acids are constantly recycling between the intestine and liver. A normal pool of 2 to 3 gm recycles 5–6 times per day (Borgström et al., 1963). Bile acids are reabsorbed in the lower one-third of the small intestine (Baker and Searle, 1960; Borgström et al., 1963; Lack and Wiener, 1963), and at each turn of the enterohepatic circulation (EHC), 97–98% of bile acids are reabsorbed (Grundy et al., 1972a). Unabsorbed bile acids pass into the large intestine and are subsequently excreted in feces (Grundy et al., 1965). In the steady state, excretion of bile acids should equal approximately the conversion of cholesterol into bile acids. The rate of formation of bile acids depends on several factors. One is the ingestion of food; bile acid synthesis is enhanced during feeding and suppressed in fasting (Grundy, 1972b, 1974). Another factor is feedback inhibition by bile acids; as bile acids return to the liver they inhibit their own production (Bergström and Danielsson, 1958; Shefer et al., 1969). As discussed below, interruption of the EHC of bile acids results in a marked increase in bile acid synthesis.

A third fate of hepatic cholesterol is the direct secretion into bile. Normally, the liver secretes 1000–1500 mg of cholesterol into bile each day (Grundy and Metzger, 1972). In its passage through the intestinal tract, about 50% of cholesterol is absorbed, the remainder passing into feces. In the large intestine, cholesterol is transformed by bacteria into two other neutral steroids—coprostanol and coprostanone (Miettinen et al., 1965). When the diet contains cholesterol, fecal neutral steroids are derived from both endogenous and exogenous sources.

Fecal neutral steroids and bile acids (acidic steroids) can be quantitatively measured through the use of gas–liquid chromatography (Miettinen et al., 1965; Grundy et al., 1965). Radioactive cholesterol can be used to label the endogenous pool of cholesterol, and thus fecal endogenous neutral steroids and unabsorbed dietary cholesterol can be distinguished. Absorption of dietary cholesterol can be calculated as the difference between cholesterol intake and unabsorbed dietary cholesterol (Grundy and Ahrens, 1969). Also, in the steady state, the synthesis of cholesterol should be equal to the difference between cholesterol intake and excretion of cholesterol and all its products (cholesterol balance). Isotope kinetic techniques using radioactive cholesterol can also be used to independently estimate cholesterol production and pool sizes (Goodman and Noble, 1968; Goodman et al., 1973; Perl and Samuel, 1969; Samuel and Lieberman, 1973). These techniques gives values for cholesterol production that are comparable to those obtained by the cholesterol balance method (Grundy and Ahrens, 1969).

Since cholesterol is completely insoluble in aqueous solutions, special mechanisms are required for its transport and excretion. In plasma and intracellular compartments, transport of cholesterol occurs through the action of solubilizing proteins (lipoproteins). However, solubilization in bile occurs by another mechanism, namely, by formation of mixed micelles with bile acids and lecithin (Hofmann and Small, 1967). Up to 10 molecules of cholesterol can be dissolved by the combination of 100 molecules of these two solubilizing lipids. However, at higher percentages of cholesterol, micellar solutions become supersaturated and there is a tendency for precipitation of cholesterol (Admirand and Small, 1968). Bile that is supersaturated with cholesterol has been found to be associated with an increased prevalence of cholesterol gallstones, and hence supersaturated bile is frequently called "lithogenic bile."

In the discussion to follow, the effects of various dietary regimens on plasma lipids will be examined; these will include: (a) low cholesterol diets, (b) low fat diets. (c) low calorie diets (weight reduction), and (d) diets rich in polyunsaturated fats. Thereafter the following chemotherapeutic agents will be·discussed: (a) plant sterols, (b) bile acid sequestering agents, (c) clofibrate, and (d) other hypolipidemic agents including nicotinic acid, thyroid hormone, and neomycin. Finally there will be a short discussion on use of agents in combination for treatment of hyperlipidemia.

III. Low Cholesterol Diets

The early observation in many animal species that excess dietary cholesterol causes a marked hypercholesterolemia and atherosclerosis naturally led to speculation that dietary cholesterol might also be important for human atherosclerosis. For this reason, several studies have been carried out to determine the effects of dietary cholesterol on plasma cholesterol levels in man. These studies have uniformly shown that excess cholesterol in the diet does not produce hypercholesterolemia of the magnitude found in some animals, but there are conflicting data on the extent to which plasma cholesterol levels are affected by dietary cholesterol. While some workers (Keys et al., 1956; Ahrens, 1957) have reported that dietary cholesterol has almost no influence on plasma cholesterol, later studies have claimed that significant increases in plasma concentrations can be induced by adding cholesterol to the diet (Beveridge et al., 1959; Brown and Page, 1965; Connor et al., 1961a,b). In these latter reports, addition of appreciable quantities of cholesterol to "cholesterol-free" diets usually caused a 10–15% increase in plasma cholesterol.

Since intake of large amounts of cholesterol does not induce marked hypercholesterolemia in man, as it does in many animal species, we might speculate about the mechanisms that protect man from high cholesterol levels. These mechanisms could include: (a) limited absorption of cholesterol; (b) feedback inhibition on cholesterol synthesis; (c) increased reexcretion of cholesterol or its metabolites; or (d) storage of cholesterol in body pools outside the plasma compartment. Each of these possibilities will be discussed in turn.

Dietary cholesterol is incompletely absorbed from the intestinal tract. Most reports indicate that only 25–50% of ingested cholesterol is absorbed by the intestine (Kaplan et al., 1963; Wilson and Lindsey, 1965; Borgström, 1969; Quintão et al., 1971); thus when intakes are in the range of 750 mg/day, only about 200–400 mg are actually absorbed. The limitation of cholesterol absorption is apparently related to the amount that can be solubilized in mixed micelles containing bile acids and digestive products of dietary fats. However, if dietary intakes of cholesterol are markedly increased, absorption can also be enhanced. Borgström (1969) reported that absorption of cholesterol is generally proportional to intake, and Quintão et al. (1971) showed that up to 1 gm of exogenous cholesterol can be absorbed each day when intake is increased to 3 gm. Nevertheless, regardless of intake, less than one-half of ingested cholesterol is absorbed, and this limitation may be significant for preventing hypercholesterolemia. In this regard, it has also been shown that familial hypercholesterolemia in man is not due to excessive absorption of cholesterol (Connor and Lin, 1974).

In several animal species, which are not prone to hypercholesterolemia, ingestion of large quantities of cholesterol causes feedback inhibition of cholesterol synthesis, and by this mechanism accumulation of cholesterol in plasma and tissue compartments is prevented. Inhibition of cholesterol synthesis in the liver has been demonstrated for the rat (Tomkins *et al.*, 1953) and dog (Gould *et al.*, 1953; Pertsemlidis *et al.*, 1973a), and for primates (Wilson, 1972). However, there has been a controversy as to whether cholesterol feeding in man causes a significant inhibition of cholesterol synthesis. Using isotope dilution techniques with constant oral feeding of radioactive cholesterol, Taylor *et al.* (1960) obtained evidence that cholesterol synthesis in man is not significantly reduced by dietary cholesterol. Since feedback inhibition of cholesterol synthesis is known to occur in the liver of animals, these authors postulated that most of human cholesterol synthesis must occur outside the liver. Subsequently, human studies have indicated that feeding of large amounts of cholesterol does in fact inhibit hepatic cholesterogenesis (Bhattathiry and Siperstein, 1963), but it remains unknown whether extrahepatic synthesis of cholesterol represents an appreciable portion of the total production in human subjects or whether excess dietary cholesterol affects peripheral production of cholesterol.

Recently, Quintão *et al.* (1971) have studied the effects of large amounts of dietary cholesterol on production of cholesterol in the whole body. These workers found that many patients show significant feedback inhibition of whole body production of cholesterol when dietary cholesterol is fed in excessive amounts, and in some patients, synthesis of cholesterol is almost totally inhibited. On the other hand, some patients in their study did not demonstrate feedback inhibition despite absorption of significant quantities of cholesterol, and these patients tended to accumulate cholesterol in body pools. Therefore, the feedback regulatory mechanism seems to be variable, and the particular response of an individual may be an important factor in determining the cholesterol content of plasma and other body pools.

If newly absorbed cholesterol should exceed the decrement in cholesterol synthesis due to feedback inhibition, accumulation of cholesterol in tissue pools would occur unless other mechanisms are available to dispose of this excess cholesterol. One such mechanism is the reexcretion of cholesterol from the body. Reexcretion can occur in either of two ways: (a) as cholesterol itself, or (b) as bile acids. Wilson (1964) showed that cholesterol feeding in rats results in an increased conversion of cholesterol into bile acids; thus, in this species, excess accumulation of cholesterol is prevented by removal as bile acids. A similar mechanism has been shown for the dog by Pertsemlidis *et al.* (1973a). In contrast, an increased conversion of cholesterol into bile acids has not been found in man; rather, reexcretion occurs as cholesterol itself. Grundy *et al.* (1969) and Quintão *et al.* (1971) have shown that the feeding of large amounts of cholesterol produces an increased excretion of endogenous neutral steroids but not of acidic steroids. This increase in endogenous neutral steroids was found to be derived from an increment in biliary cholesterol. Thus, in addition to inhibition of cholesterol synthesis, reexcretion of cholesterol is another mechanism for prevention of hypercholesterolemia during feeding of high cholesterol diets.

Despite the three mechanisms listed above (limited cholesterol absorption, feedback inhibition, and reexcretion) excess dietary cholesterol can still lead to an accumulation of cholesterol in tissue pools. As mentioned above, accumulation is especially likely to occur in patients who fail to demonstrate feedback inhibition (Grundy *et al.*, 1969; Quintão *et al.*, 1971). In fact, increments in total body cholesterol up to 20 gm have been found in some patients of these studies. Nevertheless, despite considerable accumulation in tissue pools, plasma cholesterol levels do not always increase appreciably. Thus, another barrier for development of hypercholesterolemia may be mechanisms that prevent movement of cholesterol from tissue stores into the plasma compartment. For example, feeding of cholesterol to animals has been reported to increase liver concentrations without causing significant increases in plasma cholesterol (Ho and Taylor, 1968). The liver thus appears to act as a buffer between cholesterol in the diet and that in the plasma compartment. On the other hand, accumulations of excess dietary cholesterol might also occur outside the liver; if the arterial wall is one site of accumulation, concentrations of tissue cholesterol could be even more important for determining rates of atherogenesis than plasma cholesterol concentrations. The possibility that chronic ingestion of large quantities of cholesterol could accelerate atherogenesis by local deposition in tissues has not been adequately explored.

Despite several mechanisms for prevention of hypercholesterolemia by dietary cholesterol, an important factor regulating plasma cholesterol concentrations must be the total amount of cholesterol absorbed from the intestine. When absorption of both endogenous and exogenous cholesterol is inhibited by feeding of plant sterols (see below), concentrations of cholesterol in plasma are decreased by approximately 15% (Farquhar *et al.*, 1956; Grundy *et al.*, 1969). Likewise, when absorption of cholesterol is increased to a maximum by feeding large quantities of cholesterol, levels in the plasma are frequently increased by 15% (Connor *et al.*, 1964). Therefore, within the full range of cholesterol absorption from zero to maximal, plasma cholesterol concentrations can be made to vary by approximately 30%.

In addition to affecting plasma cholesterol, the quantity of cholesterol in the diet may also influence biliary lipid composition. As mentioned above, one mechanism for prevention of hypercholesterolemia is the re-excretion of cholesterol in bile. Although careful studies have not been carried out on the effects of dietary cholesterol on bile lipid composition in man, studies in the prairie dog and monkey have shown that an excess of dietary cholesterol can induce a lithogenic bile and cholesterol gall-stones (Brenneman *et al.*, 1972; Osuga and Portman, 1971). When these species are fed large quantities of cholesterol in the diet, they presumably do not have the capacity to convert the excess of newly absorbed cho-lesterol into bile acids; as noted, a similar lack of increase in bile acid formation following ingestion of increased cholesterol has been found in man (Grundy *et al.*, 1969; Quintão *et al.*, 1971). It is of interest that cholesterol feeding does not lead to gallstone formation in the dog, an animal which has the capacity to transform excess dietary cholesterol into bile acids (Pertsemlidis *et al.*, 1973a,b). While it seems unlikely that the quantity of cholesterol ingested in the usual human diet has a profound influence on bile lipid composition, an excess of dietary cholesterol could nevertheless tip the balance toward gallstone formation in subjects whose bile composition is marginally lithogenic.

To summarize, several mechanisms exist in man that prevent excessive accumulation of cholesterol in tissue and plasma compartments, and these undoubtedly retard development of severe atherosclerosis, as occurs in certain animals fed high cholesterol diets. Nevertheless, intakes of dietary cholesterol in the range of 500 to 700 mg/day probably cause a modest increase in plasma cholesterol concentrations (5–15%). In addition, pro-longed intakes of cholesterol at this level may lead to increases in tissue concentrations of cholesterol that are not reflected in the plasma compart-ment. For these reasons, the use of low cholesterol diets must be con-sidered as one approach to reduction of plasma cholesterol and even possibly for retardation of atherogenesis.

IV. Low Fat Diets

Several epidemiological studies have reported that the incidence of atherosclerotic disease is relatively low in the geographical areas where the intake of total fat is also low (Keys, 1953; Keys and Anderson, 1954; Keys and Grande, 1957). Thus, diets containing a low proportion of total calories as fat might reduce plasma cholesterol and thereby prevent de-velopment of atherosclerosis. However, a low intake of fat is not the only dietary consideration in areas of the world where the incidence of atherosclerosis is low. First, intakes of total calories and even protein are often reduced if not marginal, and varying degrees of malnutrition could contribute to low plasma lipids. Second, diets are commonly devoid of significant amounts of animal fats so that intakes of cholesterol are also low. Third, common malabsorption syndromes, such as tropical sprue, may affect lipid levels; and finally, genetic or racial differences between population groups may influence plasma cholesterol and triglycerides (Sievers, 1968; Ho et al., 1971, 1974; Feldman, 1972). Therefore, it is presently not possible to conclude on the basis of epidemiological studies that low fat diets per se reduce plasma lipids or that they retard development of atherosclerosis.

As long as calories are maintained at weight maintenance levels, a low fat diet is synonymous with a high carbohydrate diet. Since the proportion of total calories derived from proteins can vary only a few percentage points (Ahrens, 1957), body weights can be maintained on low fat diets only by isocaloric substitution with carbohydrates. Although it has been widely assumed that substitution of carbohydrate for fat will cause a decrease in plasma cholesterol (Connor, 1968), this assumption has not been adequately proved. On the other hand, high carbohydrate diets clearly have the opposite effect on plasma triglycerides. The carbohydrate "induction" or accentuation of hypertriglyceridemia is a well recognized phenomenon; ingestion of high carbohydrate intakes can produce a marked increase in triglycerides, especially in those patients in whom triglyceride levels are initially elevated (Ahrens et al., 1961; Knittle and Ahrens, 1964). In some patients with hyperlipidemia, plasma cholesterol levels may even be increased on high carbohydrate diets because of greater VLDL concentrations (Schreibman, personal communication, 1974). The mechanism for hypertriglyceridemia resulting from intake of high increased carbohydrate has not been elucidated; however, increased hepatic synthesis of triglycerides from glucose may cause a greater secretion of triglycerides into plasma as VLDL (Knittle and Ahrens, 1964).

Effects of isocaloric exchange of fat and carbohydrates on various parameters of cholesterol metabolism have not been extensively studied. The few cholesterol balance studies carried out during such an exchange have provided inconclusive results. Whyte et al. (1973) reported that high carbohydrate diets fed to normal subjects frequently cause an increased excretion of bile acids, but neutral steroids are often decreased. Studies at the Rockefeller University have shown that during carbohydrate "induction" of hyperlipidemia, fecal excretion of total steroids is usually reduced (Schreibman, personal communication, 1974); however, in those patients in whom concentrations of cholesterol did not rise, fecal steroid excretion was generally unchanged. In neither of these studies was sufficient evidence obtained for the concept that high carbohydrate diets produce a decrease in cholesterol synthesis, as compared with high fat diets.

An increasing incidence of gallstones in several countries (e.g., Japan) has been attributed to the incorporation of greater amounts of fat into the diet. Therefore, subjects on low fat diets could possibly have a less lithogenic bile than those on diets rich in fat. This possibility has thus far not been examined adequately. However, in one study, Grundy and Metzger (1972) could find no consistent difference in hepatic secretion of biliary lipids following exchange of low fat for high fat diets, and in other studies, these workers (Grundy et al., 1972a; Bennion and Grundy, 1975) have suggested that excessive caloric intake and obesity may be more significant in production of lithogenic bile than the type of calories ingested (see below).

V. Low Calorie Diets (Weight Reduction)

The relation between obesity, the hypercaloric state, and plasma lipids has been a subject of considerable controversy. Some workers claim that there is a positive correlation between these parameters, but others suggest that relationships are nonexistent or only modest. Clearly, many obese subjects do not have hyperlipidemia, and extreme obesity is often associated with surprisingly low concentrations of plasma lipids. Nevertheless, the possibility must be considered that excess caloric intake with obesity may contribute (a) to higher levels of plasma lipids in otherwise normal subjects and (b) to accentuation of hyperlipidemia in patients with an underlying metabolic defect. By the same token, weight reduction may lead to a significant reduction in plasma lipids in both groups.

Several epidemiological studies suggest that plasma lipids are higher and atherosclerosis is increased in areas where intakes of total calories are greatest (Montenegro and Solberg, 1968; Scrimshaw and Guzman, 1968). In fact, increased caloric intakes and obesity may be as much responsible for increased plasma lipids in these areas as the distribution of calories between fats and carbohydrates (Smith and Levine, 1964; Albrink and Meigs, 1965). Direct evidence that total body weight can have an effect on levels of triglyceride and cholesterol has been presented in a recent study by Olefsky et al. (1974). These workers examined effects of weight reduction on plasma lipid concentrations in a group of 36 patients who had varying degrees of obesity. Lipid levels were measured during maintenance of weight at a steady state before and after weight reduction. In this group, the mean weight loss was about

11 kg. Weight reduction produced an average decrease in plasma triglycerides of 44% and a decrease in cholesterol of 21%. In other studies, triglyceride reductions have been commonly observed with decreasing weight, but cholesterol concentrations have not always been found to decline. For example, Wilson and Lees (1972) noted that patients with increased VLDL concentrations consistently lowered triglycerides with caloric restriction but failed to decrease cholesterol; with decreasing weight, the VLDL fraction fell, but LDL concentration rose slightly without a change in total plasma cholesterol. However, these latter findings may represent a more short-term effect of weight loss, while the results of Olefsky et al. (1974) may be indicative of long-term changes. Clearly, more studies are needed to determine the correlation between body weight and plasma lipid levels within a homogeneous population group.

Although the influence of obesity on plasma lipids in otherwise normal subjects has not been thoroughly elucidated, there is ample evidence that excess body weight can accentuate hyperlipidemia in patients with an underlying metabolic disorder. Patients with hyperlipidemia, especially those with hypertriglyceridemia, are very commonly obese; Fredrickson et al. (1967) have noted that patients with type V lipoprotein patterns are especially likely to be obese. Also, obese diabetic patients commonly have hyperlipidemia (Albrink, 1962). Finally, as recently suggested by Brunzell et al. (1974), obesity may affect the manner in which hyperlipidemia is manifested. In patients who have combined hyperlipidemia (i.e., those with a tendency for increases in both VLDL and LDL), the degree of obesity usually determines which lipoprotein predominates; obese patients have increases mainly in triglycerides (VLDL) while the nonobese show increases in cholesterol (LDL).

The mechanisms by which obesity affects plasma lipid concentrations remain to be fully determined. Olefsky et al. (1974) propose that increased body weight enhances triglyceride concentrations by accelerating hepatic production of VLDL. They suggest that this effect is due to hyperinsulinemia, secondary to resistance to insulin-mediated glucose uptake at the cellular level; presumably, excess insulin might stimulate triglyceride or VLDL production by the liver. However, increased hepatic synthesis of triglyceride in obesity and the hypercaloric state could also be the result of an excess flux of fatty acids or glucose into the liver. It seems more likely that obesity causes an overproduction of plasma triglycerides than a defect in their removal. Although many patients with increased triglycerides appear to have diminished removal of VLDL by peripheral tissues, there is no evidence that obesity interferes with the removal process. Nevertheless, in the face of a removal defect, an overproduction of triglycerides by obesity could produce a marked increase in plasma VLDL.

Another example of an increase in lipid synthesis in obesity is found with cholesterol. Several studies have revealed an overproduction of cholesterol in the obese state (Nestel *et al.*, 1969, 1973; Miettinen, 1971a). The obese patient must, therefore, cope with a greater load of endogenous cholesterol. As patients on high cholesterol diets, obese subjects must dispose of a large daily influx of new cholesterol. In obesity, negative feedback does not prevent increased production of new cholesterol, and in a manner analogous to high cholesterol diets, greater synthesis may increase concentrations of plasma cholesterol.

Not only may increased production of cholesterol in obesity enhance plasma cholesterol, it can also increase biliary cholesterol. Studies carried out in our laboratory have shown that biliary secretion of cholesterol is roughly proportional to total body weight, or to cholesterol synthesis (Grundy *et al.*, 1974). A greater hepatic secretion of cholesterol can be a significant factor in production of lithogenic bile, as shown in American Indian women with cholesterol gallstones (Grundy *et al.*, 1972a). An excess in biliary cholesterol, combined with decreases in bile acid pool sizes, causes a highly lithogenic bile and gallstone formation in these patients. While mechanisms for the bile acid deficiency have not been entirely defined, excess biliary cholesterol seems to be largely due to obesity. In fact, obesity per se, without a reduction in bile acid pools, can lead to lithogenic bile. Studies recently carried out in our laboratory have shown that markedly obese men have highly lithogenic bile despite increases in production and pool sizes of bile acids (Bennion and Grundy, 1975); this greater lithogenicity can be attributed entirely to increased outputs of biliary cholesterol. This finding is in accord with recent epidemiological studies which have confirmed the clinical impression that obese subjects have an abnormally high incidence of cholesterol stones (Friedman *et al.*, 1966; Marinovic *et al.*, 1972).

Just as weight reduction can cause a decrease in plasma cholesterol in many subjects, a reduction in weight can also decrease lithogenicity of bile. Recently, Bennion and Grundy (1975) have shown that weight loss in both American Indians and Caucasian subjects decreases lithogenicity of bile. However, the reduction in lithogenicity occurs only after reestablishment of weight maintenance at a lower level. During the period of caloric restriction, lithogenicity of bile can actually increase, as shown by Schreibman *et al.* (1974) and in our own laboratory (Bennion and Grundy, 1975). This increased lithogenicity during active weight loss may be due to two factors: (a) reduction in synthesis of solubilizing lipids (bile acids and phospholipids), and (b) mobilization of cholesterol from adipose tissue stores.

The studies described above strongly suggest that obesity and the hypercaloric state are associated with increased production of lipids by the body. The site of this excess production has not been fully elucidated, but the liver is a prime candidate. If so, hepatic cholesterol and triglycerides may "spill" into the plasma compartment to enhance circulating lipids. Likewise, all biliary lipids are increased by obesity, but in many cases cholesterol in bile is disproportionately enhanced; in such cases, bile becomes highly lithogenic. If significant weight reduction can be accomplished, lipids in both plasma and bile will be reduced, and hence major risk factors for atherosclerosis and gallstones should be decreased.

VI. Polyunsaturated Fats

Another dietary approach for reduction of plasma cholesterol is through the use of polyunsaturated fats. It has now been established that diets rich in polyunsaturated fats can lower plasma cholesterol independently of the sterol content of the diet (Spritz et al., 1965). The extent to which cholesterol levels can be reduced for any given patient depends on a variety of factors, but reductions in the range of 20 to 30% can usually be achieved if the dietary fat is entirely of vegetable origin (e.g., corn oil or safflower oil).

Because of the cholesterol-lowering action of polyunsaturated fats, there has been widespread interest in the use of these fats in the possible prevention of atherosclerosis. In fact, several clinical trials have been carried out, and some actually suggest that polyunsaturated fats can retard or prevent clinical manifestations of atherosclerosis (Leren, 1966; Miettinen et al., 1972). If future trials can confirm the preliminary findings, the use of polyunsaturated fats will undoubtedly greatly increase.

While the cholesterol-lowering effect of polyunsaturated fats has been consistently demonstrated, the mechanisms for this effect have not been elucidated. The question of how polyunsaturated fats cause a reduction in plasma cholesterol may be related to their safety as well as beneficial effects, and therefore a large number of studies have been carried out in both animals and man in the attempt to determine the way in which cholesterol lowering occurs. A variety of different actions have been reported, but which actions are related to reduction of plasma cholesterol are unknown. Theoretically, a decrease in plasma cholesterol could be due to alterations in cholesterol metabolism or to changes in the metabolism of other constituents of lipoproteins. A point of considerable controversy has been whether the lowering of plasma cholesterol is associated with a mobilization of cholesterol from the body.

In an early study, Hellman *et al.* (1957) reported that feeding of polyunsaturated fats to one patient caused an increased excretion of cholesterol in feces simultaneously with the fall in plasma cholesterol. Additional workers subsequently claimed that an increased excretion of cholesterol (as neutral steroids) or bile acids occurred following institution of these fats (Gordon *et al.*, 1957; Antonis and Bersohn, 1962; Wood *et al.*, 1966; Moore *et al.*, 1968a; Connor *et al.*, 1969; Nestel *et al.*, 1973). However, others presented conflicting results. Spritz *et al.* (1965) reported that exchange of polyunsaturated fats for saturated fats failed to induce an increase in fecal steroids in five patients, and similar findings were published in 5 of 6 patients by Avigan and Steinberg (1965). These conflicting results left open the question of whether cholesterol that disappears from the plasma compartment is removed from the body. If not, the decrement in plasma cholesterol might enter directly into tissue pools and, hence, might accelerate rather than retard atherosclerosis.

The patients who were employed in the latter two studies (Spritz *et al.*, 1965; Avigan and Steinberg, 1965) differed considerably from one another in concentrations of plasma lipids, and their types of hyperlipoproteinemia were not well defined. In a more recent study, Grundy and Ahrens (1970) found no consistent change in fecal steroid excretion in eight patients with familial hypercholesterolemia during exchange of saturated and polyunsaturated fats. Although some patients had slight increases in acidic steroids on polyunsaturated fats, it was rarely possible to account for the decrement in plasma cholesterol by an increase in fecal steroids. Since most of the hypercholesterolemic patients failed to show an increase in excretion of cholesterol or its products, it was concluded that polyunsaturated fats need not cause a greater excretion of steroids to produce a reduction in plasma cholesterol and that they probably cause a redistribution of cholesterol between plasma and tissue pools.

These findings in patients with familial hypercholesterolemia have not been confirmed in normal subjects and in patients with other forms of hyperlipidemia. At least four studies in normal subjects (Wood *et al.*, 1966; Moore *et al.*, 1968a; Connor *et al.*, 1969; Nestel *et al.*, 1973) have shown that polyunsaturated fats cause a small but consistent increase in fecal steroids, either neutral steroids or bile acids. An even greater increase in fecal steroids has been observed in patients with hypertriglyceridemia. In two hypertriglyceridemic patients studied at The Rockefeller University by Grundy and Ahrens (1966, 1970) and in several additional patients with increased triglycerides who were studied in San Diego by Grundy (1975), polyunsaturated fats were found to significantly, and in some patients markedly, increase steroid excretion. Increases frequently occurred in both neutral and acidic steroid fractions. Thus, except in familial hypercholesterolemia, polyunsaturated fats appear to cause a mobilization of cholesterol or its products from the body simultaneously with the reduction in plasma cholesterol. In most patients with hypertriglyceridemia, increases in fecal steroids were found to exceed the decrement in plasma cholesterol. This finding suggests that cholesterol is also mobilized from tissue pools; however, direct measurements of whole body cholesterol must be carried out to prove this latter possibility.

There are several mechanisms by which polyunsaturated fats might enhance fecal excretion of steroids. First, they might inhibit reabsorption of cholesterol or bile acids; second, they might increase the rate of conversion of cholesterol into bile acids; and third, polyunsaturated fats could promote hepatic secretion of cholesterol. Finally, the possibility must also be considered that increased steroid excretion is associated with enhanced synthesis of cholesterol in some patients. Studies carried out in our laboratory (Grundy, 1975) suggest that cholesterol synthesis may in fact be increased in some patients with hypertriglyceridemia. Whether a greater synthesis of cholesterol is a primary action of these fats or whether it is a compensatory response to mobilization of tissue cholesterol remains to be proved. As yet, various possible mechanisms for the action of polyunsaturated fats on the metabolism of cholesterol and bile acids in the enterohepatic circulation have not been dissected, and additional studies are required.

Not only is the question of mobilization of cholesterol on polyunsaturated fats important from the viewpoint of tissue cholesterol, but it is also related to another potential problem, namely, cholesterol gallstones. Sturdevant et al. (1973) have reported that the incidence of cholesterol stones is enhanced in patients who ingest polyunsaturated fats for long periods. Obviously, additional studies are necessary to prove that these fats truly promote gallstone formation, but if they can cause a flux of excess cholesterol through the biliary tree, the potential for lithogenic bile must exist. Studies in our laboratory indicate that lithogenic bile develops in some patients but not in all who are fed polyunsaturated fats (Grundy, 1975). When increased lithogenicity occurs it appears to be related to secretion of greater quantities of cholesterol in bile and not to a reduction in pool size of bile acids.

Since patients with familial hypercholesterolemia show a reduction in plasma cholesterol without a consistent change in fecal steroids, the reciprocal change in plasma cholesterol and fecal steroids in other categories of patients may not be casually related. Thus, at least two questions must be raised about the actions of polyunsaturated fats: (a) whether an increased excretion of steroids contributes to lowering of plasma cholesterol at all, and (b) whether mechanisms unrelated to the metabolism of cholesterol play a major role in the decrease of plasma cholesterol. First, in those subjects who show an increased steroid excretion, the greater output may not be the only factor responsible for the decreased plasma cholesterol. The possibility must, nevertheless, be considered that an increased excretion could at least contribute to plasma cholesterol lowering. By continuous removal of excessive amounts of cholesterol from the body, the availability of cholesterol for transfer into the plasma compartment is decreased. Therefore, even if mobilization is not the primary mechanism for cholesterol reduction, it may be contributory by allowing this lowering to occur to a maximum extent.

If polyunsaturated fats have more than one action leading to a decrease in plasma cholesterol, they might also affect the metabolism of constituents of lipoproteins other than cholesterol. One possibility is that these fats alter the metabolism of plasma triglycerides. It is of interest to note that despite extensive clinical studies on polyunsaturated fats, their effects on concentrations of plasma triglycerides have never been fully characterized. Generally, in normal subjects and in patients with hypercholesterolemia, these fats seem to have little or no effect on triglyceride concentrations (Spritz and Mishkel, 1969), but recent studies in our laboratory show that polyunsaturated fats almost always cause a lowering of triglycerides in patients with hypertriglyceridemia (Grundy, 1975). In fact, when triglyceride levels are significantly elevated, the degree of cholesterol reductions is closely related to changes in triglyceride concentrations. Since plasma triglycerides are carried almost exclusively in very low-density lipoproteins (VLDL), polyunsaturated fats must have an influence on VLDL concentrations as well as on levels of low-density lipoproteins (LDL). VLDL concentrations could be reduced either by decreasing their rate of secretion into plasma or by enhancement of their removal. Both mechanisms could be involved, as suggested by different workers (Bagdade et al., 1970; Nestel and Barter, 1971).

While cholesterol reduction in hypertriglyceridemic patients seems to be related to decreases in VLDL, changes in normal subjects and hypercholesterolemic patients occur predominantly in the LDL fraction (Spritz and Mishkel, 1969). In both groups of subjects, cholesterol levels can be significantly reduced without changes in triglyceride concentrations. Again, LDL concentrations could be affected in at least two ways; rates of LDL formation could be reduced or removal could be increased. Another interesting possibility has also been suggested by Spritz and Mishkel (1969); these workers postulate that polyunsaturated fats, as compared to saturated fats, alter the steric configuration of lipid molecules (cholesterol esters and phospholipids) and this change may alter the structure of LDL in such a way as to reduce its capacity to transport as much cholesterol.

In view of the many different and variable effects of polyunsaturated fats on sterol and lipoprotein metabolism, it is reasonable that these fats should produce multiple metabolic alterations, as compared to saturated fats. Ingestion of polyunsaturated fats causes changes in the fatty acid composition of many lipids (cholesterol esters, triglycerides, and phospholipids). Since the physical properties of tissue and plasma lipids are undoubtedly altered by these changes, it would not be surprising if rates of metabolic processes involving these lipids were also altered. Thus, the plasma cholesterol-lowering action of polyunsaturated fats is probably the result of multiple actions, and the magnitude of the decrease may be dependent on changes in the metabolism of cholesterol, triglycerides, and phospholipids as well as upon their steric arrangement in lipoproteins.

VII. Plant Sterols

In addition to polyunsaturated fats, plant oils contain another group of lipids that are capable of lowering plasma cholesterol; these are the plant sterols. The major component of the phytosterols is β-sitosterol, but other sterols include campesterol and stigmasterol. The average diet provides a daily intake of plant sterols in the range of 250 to 300 mg/day, and the effects of these intakes on plasma cholesterol are probably negligible. However, when plant sterols are fed at the level of 10 to 15 gm/day, a distinct lowering of cholesterol occurs (Farquhar et al., 1956; Grundy et al., 1969); reductions in the range of 10 to 20% have usually been reported. A recent study has shown that even smaller intakes of plant sterols may produce a significant decrease in plasma cholesterol (Grundy et al., 1971).

Plant sterols cause reduction in plasma cholesterol by interfering with intestinal absorption of both endogenous and exogenous cholesterol. This inhibition of cholesterol absorption leads to a marked increase in fecal excretion of neutral steroids. It should be emphasized that these sterols have no effect on reabsorption or excretion of bile acids (Grundy et al., 1969). The fall in plasma cholesterol during plant sterol feeding is probably due to an increased drain on tissue pools of cholesterol, especially the liver pool. This mechanism illustrates that lowering of plasma lipoproteins can occur following alteration in the metabolism of only one constituent of the lipoprotein, namely cholesterol. Although plant sterol feeding causes a marked drain on body cholesterol by interrupting its intestinal reabsorption, the degree of cholesterol lowering is rather limited in many patients. This limitation is due to a release in feedback inhibition on cholesterol synthesis. After a period of a few weeks of plant sterol feeding, a new steady state is reached in which plasma cholesterol concentrations are reduced but, at the same time, rates of cholesterol synthesis are enhanced.

Although the extent of lowering of plasma cholesterol with plant sterols is often rather modest, it has been generally thought that this approach has the advantage of being completely safe. Since absorption of plant sterols by the intestine is very low, it was assumed than only small quantities of plant sterols enter body pools. Most studies have indicated that absorption of β-sitosterol is less than 5% of a given dose (Gould, 1955; Gould et al., 1969; Salen et al., 1970). Earlier studies using the sterol balance technique have suggested that up to 50% of β-sitosterol is absorbed (Ivy et al., 1955; Swell et al., 1956), but these studies failed to take into account the possible degradation of the sterol molecule in its passage through the intestinal tract (Grundy et al., 1968). A detailed study of the metabolism of β-sitosterol in man has been carried out more recently by Salen et al. (1970). Using isotope kinetic techniques, these workers demonstrated that less than 5% of β-sitosterol is absorbed, and at normal intakes (250–450 mg/day), absorption is less than 10 mg/day.

The reasons for the low absorption of β-sitosterol, as compared to cholesterol, are not fully understood. In the rat, Sylvén and Borgström (1969) found that the ratio of β-sitosterol to cholesterol in intestinal lymphatics is identical to that in the intestinal mucosa. They therefore suggested that the difference in absorption was not due to differences in the incorporation of these sterols into chylomicrons but perhaps to a lesser micellar solubilization or a slower transport into the mucosal cell. Nevertheless, other work suggests that the rate of esterification of β-sitosterol may be one factor limiting its absorption; Kuksis and Huang (1962) reported that the small quantities of β-sitosterol in the thoracic duct of dogs are largely unesterified while most of the cholesterol in chyle is esterified.

Although esterification of β-sitosterol by the intestinal mucosa may be a rate-limiting step in its absorption, esterification of plasma β-sitosterol seems to proceed at approximately the same rate as that of cholesterol (Salen et al., 1970). Also, β-sitosterol is converted into bile acids similarly to cholesterol. On the other hand, β-sitosterol appears to be secreted into bile more rapidly than cholesterol, and this increased secretion, coupled with reduced absorption, tends to keep total body pools of β-sitosterol at low levels. Salen et al. (1970) have reported that total body pools of β-sitosterol in normal subjects range from 75 to 200 mg.

The importance of maintaining small body pools of β-sitosterol is illustrated by the recent report of Bhattacharyya and Connor (1975); these workers have carried out studies on two sisters who apparently have an inborn error in the metabolism of β-sitosterol. These patients showed an excessive absorption of β-sitosterol (about 25% of dietary intake) which leads to hyper-β-sitosterolemia and xanthomatosis. Plasma levels of β-sitosterol were about 20 mg/100 ml (normal < 1 mg/100 ml), and plant sterols constituted about 15% of total sterols. It is of interest that despite the absence of hypercholesterolemia, increases in plasma concentrations of β-sitosterol lead to extensive deposition of cholesterol as well as β-sitosterol in tissues. This finding suggests that appreciable quantities of β-sitosterol in plasma sterols leads to an instability of lipoproteins so that both cholesterol and β-sitosterol are deposited in tissues. An analogous finding occurs in another rare disease, cerebrotendinous xanthomatosis; in this disease, abnormally high plasma levels of another sterol, cholestanol, are observed (Menkes et al., 1968). Similarly, an increase in cholestanol in plasma leads to severe xanthomatosis without hypercholesterolemia in which cholesterol is again the major sterol deposited in tissues (Salen, 1971; Salen and Grundy, 1973).

The findings that relatively low percentages of abnormal sterols in plasma can cause tissue deposition of cholesterol raises the question of whether feeding of large amounts of plant sterols is in fact completely safe. However, in their study of five patients on low and very high intakes of plant sterols, Salen et al. (1970) showed that plasma levels of β-sitosterol were slightly enhanced during the high intakes, but in no case did plasma β-sitosterol exceed 2.0 mg/100 ml. In all cases, β-sitosterol represented less than 1.0% of total sterols. Despite daily intakes of several grams of β-sitosterol, marked accumulation of plant sterols in the body is apparently prevented in normal and hyperlipidemic adults. However, a recent report indicates that infants and children are more susceptible to hyper-β-sitosterolemia than adults when the diet contains significant quantities of plant sterols (Mellies et al., 1974).

VIII. Bile Acid Sequestering Agents

Not only can plasma cholesterol be reduced by inhibition of cholesterol absorption, but significant reductions can be achieved by interruption of the enterohepatic circulation (EHC) of bile acids. This can be accomplished medically by the use of resins that bind bile acids in the intestinal tract (e.g., cholestyramine and Colestipol·). A surgical approach for interrupting the EHC of bile acids has also been introduced by Buchwald (1964); this is the ileal exclusion operation in which the lower one-third of the small intestine is excluded from the fecal stream. This procedure prevents the reabsorption of bile acids since intestinal contents bypass the region where conjugated bile acids are mainly reabsorbed (Borgström et al., 1963; Lack and Weiner, 1963; Holt, 1964). Both medical and surgical approaches to interrupting the EHC of bile acids lead to a decrease in plasma cholesterol; the mechanisms for the cholesterol lowering are not entirely understood, but they are related to marked alterations in the metabolism of both bile acids and cholesterol.

First, interruption of the EHC causes an increase in fecal excretion of bile acids; daily excretion rates are increased 3- to 15-fold (Hashim and Van Itallie, 1965; Moore et al., 1968b; 1969; Miettinen, 1969; Grundy et al., 1971). This increase inevitably leads to a greater conversion of cholesterol into bile acids, and it apparently leads to a drain on liver pools of cholesterol.

Another factor causing a reduced hepatic cholesterol may be a decrease in cholesterol absorption (Buchwald, 1964, 1965; Grundy et al., 1971). Since bile acids are required for cholesterol absorption, a reduction of bile acids within the EHC may lead to a decrease in absorption of cholesterol; without adequate bile acids, cholesterol cannot be solubilized for absorption. Some patients clearly show a decreased cholesterol absorption, as reflected by enhanced excretion of neutral steroids. However, in other patients a greater synthesis of bile acids may be sufficient to prevent depletion of bile acids, and a reduced absorption of cholesterol is not consistently found (Grundy et al., 1971).

Since interruption of the EHC of bile acids causes a drain on body cholesterol by increasing the conversion of cholesterol into bile acids and by decreasing absorption of cholesterol, body pools of cholesterol would rapidly become depleted if a compensatory increase in synthesis did not occur. Evidence from several sources indicates that a marked increase in cholesterol synthesis does occur which ultimately limits the degree to which body pools of cholesterol can be reduced. Increases in cholesterol synthesis probably occur both in the liver and in the intestinal tract (Moore et al., 1968b, 1969; Moutafis and Myant, 1968; Dietschy, 1968; Grundy et al., 1971). Mechanisms for enhanced cholesterol synthesis have not been fully elucidated, but several factors are probably involved. First, decreased absorption of cholesterol causes increased synthesis in the liver by release of feedback inhibition. Second, bile acids regulate cholesterol synthesis in the intestine, and decreasing the availability of bile acids within the intestinal lumen allows for an increased mucosal production (Grundy et al., 1971). Finally, bile acids may have a direct inhibitory effect on cholesterol synthesis in the liver (Fimognari and Rodwell, 1965; Back et al., 1969), and a decreased return of bile acids to the liver may allow for an increased hepatic production. While several lines of evidence support the existence of such direct feedback, the biochemical mechanisms for inhibition of cholesterol synthesis by bile acids are unknown. Despite increased synthesis of cholesterol, both ileal exclusion and bile acid sequestering agents significantly lower plasma cholesterol in most patients. In patients with hypercholesterolemia, cholestyramine in doses of about 16 gm/day produce average reductions of plasma cholesterol in the range of 20 to 30%; in about one-third of such patients, reductions are even greater (Grundy, 1972a). Although fewer patients have been studied with ileal exclusion, reductions in plasma cholesterol have been reported to be marked. Buchwald and associates (Buchwald et al., 1968; Buchwald and Varco, 1967) claim that the operation causes decreases in plasma cholesterol of greater than 30% in about two-thirds of patients—in their patients, an average decrease of 35 to 40% was found. Similar results have been reported in a smaller number of patients by Miettinen and Lempinen (1970). The finding of significant decreases in plasma cholesterol in patients with familial hypercholesterolemia does not extend to those with the homozygous form of the disease; the limited number of patients with homozygous familial hypercholesterolemia who have been treated with cholestyramine or ileal exclusion have generally failed to respond (Johnston et al., 1967; Grundy et al., 1971).

On the whole, the ileal exclusion operation would seem to produce a greater lowering of plasma cholesterol than drugs such as cholestyramine (Grundy, 1972). This greater lowering is probably due to a more severe inhibition of bile acid reabsorption. However, mechanisms for cholesterol reduction are essentially the same for the medical and surgical approaches, and it would thus seem reasonable to evaluate the response to a binding resin in a given patient before carrying out the operation. If the patient should fail to respond to large doses of the resin, it is unlikely that he would show a marked reduction in plasma control following the operation.

In the treatment of hyperlipidemia, bile acid-binding resins or the ileal exclusion operation would seem to be indicated primarily for those in whom the major increase is in the cholesterol fraction (or LDL), i.e., those patients with the type II pattern. Not only does increased bile acid excretion produce a drain on body cholesterol, it also seems to enhance the removal of the protein moiety of LDL (Levy and Langer,

1971). This important observation suggests that the metabolic fates of both cholesterol and apoproteins of plasma lipoproteins are interdependent.

When resins have been given to patients with other forms of hyperlipidemia (types III, IV, and V), variable responses have been reported (Fallon and Woods, 1968; Bressler et al., 1966). In general, these resins will cause a decrease in plasma cholesterol, but levels of triglycerides (and VLDL) are often increased (Grundy et al., 1971).

Since prolonged interruption of the EHC of bile acids produces a significant reduction in plasma cholesterol concentrations, the question can also be asked as to whether tissue pools of cholesterol outside the plasma compartment are also reduced. One indication of a decrease in tissue cholesterol is the observation that the size of tendon xanthoma can be reduced in patients with hypercholesterolemia who are treated with either cholestyramine or ileal exclusion (Hashim and Van Itallie, 1965; Khachadurian 1968; Buchwald, 1969); a decrease in xanthomas has been noted even in patients whose plasma concentrations are not significantly reduced (Khachadurian, 1968). Moore et al. (1969), using an isotope kinetic technique, reported that total body cholesterol is reduced following ileal exclusion, but Goodman and Noble (1968) found no corresponding change in hyperlipidemic patients treated with cholestyramine. Likewise, Pertsemlidis et al. (1973b) observed no significant decrease in tissue pools of cholesterol after prolonged interruption of the EHC in dogs; however, this finding does not rule out the possibility that the same approach would cause a mobilization of tissue cholesterol in hyperlipidemic patients who have large accumulations of cholesterol, such as those with tendon xanthomatosis.

Thus far, serious side effects have generally not been identified following interruption of the EHC of bile acids. A variety of minor gastrointestinal discomforts including constipation have been associated with bile acid-binding resins, and when the dose exceeds 25 gm/day, patients may develop significant steatorrhea. While constipation is the main gastrointestinal effect of the resins, the major complication of ileal exclusion is diarrhea. Most patients have frequent watery stools in the first few months after the operation, but many of them make a gradual adjustment over several months to the point where they have only a moderate increase in stool frequency. However, some patients have a persistent problem with diarrhea after ileal exclusion.

Another theoretical side effect of interruption of the EHC of bile acids is the formation of cholesterol gallstones. A deficiency of bile acids in the EHC has been reported to be associated with an increased frequency of gallstones in several studies (Vlahcevic et al., 1970; Danzinger et al.,

1971; Grundy et al., 1972a). Also, patients with ileal resection for lower intestinal disease apparently have a greater prevalence of cholesterol gallstones than normal subjects (Heaton and Read, 1969). Despite the possibility of gallstone formation, an increased incidence of gallstones has not yet been reported in patients treated with binding resins or ileal resection. One protective mechanism may be a marked compensatory increase in bile acid synthesis that replenishes the pool of bile acids; Garbutt and Kenney (1972) found no reduction in total bile acid pools in normal subjects treated with cholestyramine. Likewise, bile has not been reported to become highly supersaturated with cholesterol in patients treated with cholestyramine. In our laboratory, we have noted that treatment of hyperlipidemic patients with Colestipol generally does not greatly enhance the lithogenicity of bile, but in one patient, lithogenicity increased markedly on Colestipol therapy, and gallstone formation occurred within a few weeks after starting the drug (Grundy, unpublished data). Therefore, further studies are clearly indicated to determine whether treatment with resins or ileal exclusion will, in fact, increase the likelihood of gallstone formation.

IX. Clofibrate

During the past decade, the drug clofibrate has become recognized as an effective and relatively nontoxic agent for lowering plasma cholesterol and triglycerides. The drug has been used increasingly for treatment of hyperlipidemias in which triglycerides are primarily elevated (e.g., types III, IV, and V). While clofibrate can frequently produce a profound lowering of triglycerides, its potential for reduction of cholesterol in the plasma compartment seems to be more modest, that is, clofibrate has a greater effect on plasma concentrations of VLDL than on LDL. Mechanisms of VLDL reduction have not been thoroughly elucidated. Some workers have suggested that the production of VLDL by the liver may be reduced (Cenedella et al., 1968; Steinberg, 1970; Bierman et al., 1970), but others indicate that degradation and removal of lipoproteins are increased (Spritz, 1965; Ryan and Schwartz, 1965; Wolfe et al., 1973). Both mechanisms may actually be involved (Bierman et al., 1970). Although clofibrate appears to predominately affect plasma triglyceride and VLDL concentrations, many patients also have some decrease in cholesterol levels (Grundy et al., 1972b). In addition, the drug can have a profound effect on cholesterol metabolism that is not reflected in the plasma compartment, as shown by Grundy et al. (1972b). In this latter report, results were presented from several kinds of long-term studies in

29 patients; these included plasma lipid analyses, cholesterol balance data, isotope kinetic studies, and measurement of hepatic secretion of biliary lipids. The study showed that clofibrate causes a lowering of plasma lipids in all types of patients with hyperlipidemia except those with the fat-induced variety (type I); the most striking changes were seen in patients with a primary elevation in VLDL (types III, IV, and V), but significant changes were usually noted in LDL concentrations in patients with type II lipoprotein patterns.

In most hyperlipidemic patients of this study, clofibrate caused an increased fecal excretion of cholesterol that was consistent and often of considerable magnitude. Other workers had previously shown similar but not always identical results (Horlick et al., 1971). The increased excretion of cholesterol was due to an increased secretion of cholesterol into bile and not to a decreased reabsorption. Also, in a significant number of patients, clofibrate caused a reduction in fecal excretion of bile acids; thus, the drug appears to partially inhibit the synthesis of bile acids at the same time it causes the liver to secrete more cholesterol into bile. In most patients, the increased excretion of cholesterol exceeds the decrease in bile acid excretion, and. thus a greater cholesterol output cannot be explained simply by a decreased conversion of cholesterol into bile acids.

The increase in output of total steroids on clofibrate could be due either to a greater synthesis of cholesterol or to a mobilization of cholesterol from preexisting tissue pools. Evidence from isotope kinetic studies tends to rule out the first possibility (Nestel et al., 1965; Grundy et al., 1972b). With introduction of clofibrate, most patients show a significant reduction in the rate of decay of specific radioactivity-time curves of plasma cholesterol—this finding suggests that cholesterol synthesis is actually decreased by the drug, as indicated by studies in animals (Cayen and Dvornik, 1970). In addition, direct measurements of total body pools of cholesterol by isotope kinetic techniques imply that pools are significantly reduced after treatment with the drug for 2 to 3 years (Grundy et al., 1972b). Using the sterol balance technique in eight patients, these workers estimated that from 6 to 47 gm (average 22 gm) of cholesterol were mobilized over a period of 1 to 3 months in these patients, the reduction in total body pools of cholesterol far exceeded the decrement in plasma cholesterol. This mobilization was noted for patients with familial hypercholesterolemia (type II) as well as for those with hyper-triglyceridemia. Finally, another indication that clofibrate can cause a mobilization of cholesterol from tissue pools is the rapid resolution of xanthomatosis in patients with severe hyperlipidemia.

As indicated above, flux of cholesterol from tissue pools is associated with an increased secretion of cholesterol into bile (Grundy et al., 1972b). This excess secretion of biliary cholesterol results in an enhanced lithogenicity of bile at least during the period of cholesterol mobilization (Pertsemlidis et al., 1974). In this period, patients treated with clofibrate may be at greater risk for cholesterol gallstone formation than normal subjects. In fact, gallstone formation following the introduction of clofibrate therapy has been reported (Ahrens, 1973). The lithogenicity of bile during treatment may be further enhanced by decreased formation of bile acids (Grundy et al., 1973; Einarsson et al., 1973). Since many patients with hypertriglyceridemia are probably at increased risk for gallstone formation because of obesity, introduction of clofibrate could even further increase lithogenicity. Therefore, before treatment of obese subjects with clofibrate it may be prudent to recommend weight reduction; since lithogenicity of bile is decreased after weight loss, clofibrate should be less likely to induce cholesterol gallstones.

X. Other Hypolipidemic Drugs

Several additional drugs have been found to decrease plasma lipids, and some of these are currently being examined as potential agents for treatment of hyperlipidemia. Three of these agents will be briefly reviewed.

A. Nicotinic Acid

Nicotinic acid causes a lowering of both plasma cholesterol and triglycerides in patients with differing types of hyperlipidemia. In familial hypercholesterolemia (type II), cholesterol and LDL levels are predominately reduced, and lowering of triglycerides (and VLDL) occurs in hypertriglyceridemia (types IV and V). Recent studies suggest that triglyceride lowering with nicotinic acid may be even more pronounced than cholesterol reduction (Carlson and Walldius, 1972). Whether decreases in plasma VLDL and LDL are the result of reduced lipoprotein secretion or enhanced removal remains to be determined. Langer and Levy (1971) propose that the drug decreases the influx of VLDL into plasma, and since VLDL may be a precursor to LDL, concentrations of the latter would also be decreased. On the other hand, Boberg et al. (1971) and Fröberg et al. (1971) favor the concept that nicotinic acid improves removal of triglycerides. Further studies are thus required to elucidate mechanisms by which plasma lipoproteins are decreased.

The effects of nicotinic acid on cholesterol metabolism have been examined by Miettinen (1968a, 1971b, 1972c). In these studies, cholesterol balance measurements were carried out in a large number of patients with hypercholesterolemia before and during treatment with nicotinic acid. Lowering of plasma cholesterol was associated with a transitory increase in excretion of endogenous cholesterol. Changes in bile acid excretion were less marked. At the same time, isotope kinetic studies suggested that cholesterol production in the body as a whole may actually have been decreased (Miettinen, 1968a, 1971c). These findings are thus similar to those reported for clofibrate (Grundy et al., 1972b), and again they suggest that mobilization of tissue cholesterol occurs along with plasma cholesterol reduction. The marked reduction in plasma VLDL by nicotinic acid together with an increased fecal excretion of cholesterol suggest that clofibrate and nicotinic acid may act by similar mechanisms. However, their actions are probably not identical; although both drugs cause a lowering of VLDL, nicotinic acid seems to have a greater effect on LDL concentrations.

Unfortunately, the effectiveness of nicotinic acid for reduction of plasma lipids is somewhat offset by its side effects. Cutaneous flushing and gastrointestinal irritation are a distinct disadvantage of this agent. Also, the possibility has not been completely excluded that long-term administration on nicotinic acid may have a deleterious effect on hepatic function.

B. Thyroid Hormone

The hypocholesterolemic action of thyroid hormone has been long recognized, and attempts have been made to treat elevated levels of plasma cholesterol with this hormone. D-Thyroxine has been the form of the hormone used most recently for reduction of plasma cholesterol. Decreases in plasma cholesterol could be due to effects on either plasma lipoproteins or cholesterol metabolism. Evidence has been presented by Walton et al. (1965) that excess circulating thyroid hormone (hyperthyroidism) is associated with enhanced removal of LDL from plasma, and this mechanism may account for the decrease in LDL levels following treatment with D-thyroxine. About the same time, Miettinen (1968b) found that thyroid hormone induced a marked increase in fecal excretion of cholesterol (but not of bile acids). Whether enhanced excretion of cholesterol was due to increased removal of plasma cholesterol or to enhanced synthesis of cholesterol was not resolved in this study.

The beneficial effects that might be achieved by cholesterol lowering with D-thyroxine have been largely offset by the potential danger that results from the use of this drug in patients with atherosclerotic disease; it may precipitate cardiac arrhythmias or angina pectoris in such patients.

C. Neomycin

Another agent with hypercholesterolemic properties is neomycin; oral administration of neomycin sulfate (2 gm daily) significantly lowers plasma cholesterol concentrations (Samuel and Steiner, 1959; Samuel and Waithe, 1961; Samuel et al., 1967, 1968). The mechanisms by which neomycin causes a decrease in plasma cholesterol have not been fully determined; however, the drug probably affects plasma cholesterol through actions within the intestinal tract. Increases in excretions of both cholesterol and bile acids have been reported (Hamilton et al., 1964; Hardison and Rosenberg, 1969); these changes are probably related to decreases in sterol absorption. Samuel et al. (1973) have also suggested that plasma cholesterol lowering by neomycin may be related to changes in bacterial flora, possibly due to prevention of formation of deoxycholic acid; in other words, these authors postulate that the molecular specificity of bile acids within the EHC may influence levels of plasma cholesterol. They present evidence that an excess of deoxycholic acid is frequently associated with hypercholesterolemia, and a decrease in this bile acid might thus reduce cholesterol concentrations. While such a mechanism could play a role in cholesterol lowering, the more obvious mechanism for neomycin is the inhibition in sterol reabsorption that has also been shown to occur during treatment with the drug.

XI. Combined Therapy for Hyperlipidemia

In the therapeutic approach to many diseases, there has been an increasing interest in the use of combined therapy. Typical examples are found in the treatment of hypertension and chemotherapy of cancer. Likewise, it has become apparent that the use of agents in combination may be advantageous in treatment of many patients with hyperlipidemia. Combinations can include both diets and drugs. Since agents employed for lowering of plasma lipids usually produce their effects by different mechanisms, their actions may often be additive or even synergistic.

The first step in lowering plasma lipids in hyperlipidemic patients should probably be dietary change. If a person is overweight, caloric restriction and weight reduction should be the first line of therapy. The addition of large quantities of polyunsaturated fats to the diet should not be prescribed in obese subjects since these fats may only enhance obesity and increase plasma lipids. If hyperlipidemia persists after weight reduction, the use of polyunsaturated fats will usually produce the greatest lowering of plasma lipids. Also, removal of cholesterol from the diet of patients treated with polyunsaturated fats should cause a further decrease in cholesterol levels.

In many patients with hyperlipidemia, weight reduction, polyunsaturated fats, and low cholesterol intakes are still not sufficient to bring plasma lipids to normal levels. In these patients, additional lipid lowering can be achieved through the use of drugs. If concentrations of triglycerides and VLDL are increased, clofibrate or nicotinic acid in addition to dietary change may be effective in establishment of normal lipid levels. Likewise, the use of bile acid sequestering agents, plant sterols, or neomycin can produce distinct lowering of LDL levels beyond that which can be achieved by diet. Therefore, these latter drugs could be used alone or in combination in patients with type II lipoprotein patterns.

Finally, the combination of bile acid sequestering agents and drugs such as clofibrate or nicotinic acid have several attractive features. In patients with type II hyperlipoproteinemia both clofibrate and nicotinic acid appear to cause a mobilization of cholesterol from tissue pools. Therefore, the addition of such an agent to the sequestering agents should have an additive effect on the movement of cholesterol from body pools. In addition, such a combination should produce a maximum lowering of both VLDL and LDL concentrations. Thus, through the use of a combination of diets and multiple drugs, it should be possible to bring plasma lipid concentrations to near normal levels in most patients with hyperlipidemia. It will, therefore, be important for future studies to determine whether these combinations are safe as well as effective.

REFERENCES

Admirand, W. H., and Small, D. M. (1968). *J. Clin. Invest.* 49, 1043–1052.

Ahrens, E. H., Jr., (1957). *Amer. J. Med.* 23, 928–952.

Ahrens, E. H. (1973). *New Engl. J .Med.* 288, 620–626.

Ahrens, E. H., Jr., Hirsch, J., Oette, K., Farquhar, J. W.; and Stein, Y. (1961). *Trans. Ass. Amer. Physicians* 74, 134.

Albrink, M. J. (1962). *Arch. Intern. Med.* 109, 345–359.

Albrink, M., and Meigs, J. (1965). *Ann. N.Y. Acad. Sci.* 131, 673–683.

Antonis, A., and Bersohn, I. (1962). *Amer. J. Clin. Nutr.* 11, 142–155.

Avigan, J., and Steinberg, D. (1965). *J. Clin. Invest.* 44, 1845–1856.

Back, P., Hamprecht, B., and Lynen, F. (1969). *Arch. Biochem. Biophys.* 133, 11–21.

Bagdade, J. D., Hazzard, W. R., and Carlin, J. (1970). *Metabolism* 19, 1020–1024.

Baker, R. D., and Searle, G. W. (1960). *Proc. Soc. Exp. Biol. Med.* 105, 521–523.

Bennion, L. J., and Grundy, S. M. (1975). *J. Clin. Invest.* 56, 996–1011.

Bergström, S., and Danielsson, H. (1958). *Acta Physiol. Scand.* **43**, 1–7.
Beveridge, J. M. R., Connell, W. F., Haust, H. L., and Mayer, G. A. (1959). *Can. J. Biochem. Physiol.* **37**, 575–582.
Bhattacharyya, A. K., and Connor, W. A. (1975). *J. Clin. Invest.* **53**, 1033–1043.
Bhattathiry, E. P. M., and Siperstein, M. D. (1963). *J. Clin. Invest.* **42**, 1613–1618.
Bierman, E. L., Brunzell, J. D., Bagdade, J. D., Lerner, R. L., Hazzard, W. R., and Porte, D., Jr. (1970). *Trans. Ass. Amer. Physicians* **83**, 211–224.
Bloch, K. (1965). *Science* **150**, 19–28.
Boberg, J., Carlson, L. A., Fröberg, S., Olsson, A., Orö, L., and Rossner, S. (1971). *In* "Metabolic Effects of Nicotinic Acids and Its Derivatives" (K. F. Gey and L. A. Carlson, eds.), pp. 465–470. Hans Huber, Bern.
Borgström, B. (1969). *J. Lipid Res.* **10**, 331–337.
Borgström, B., Lundh, G., and Hofmann, A. (1963). *Gastroenterology* **45**, 229–238.
Brenneman, D. E., Connor, W. E., Forker, E. L., and DenBesten, L. (1972). *J. Clin. Invest.* **51**, 1495–1503.
Bressler, R., Nowlin, J., and Bogdonoff, M. D. (1966). *Southern Med. J.* **59**, 1097–1103.
Brown, H. B., and Page, I. H. (1965). *J. Amer. Diet. Ass.* **46**, 189–192.
Brunzell, J. D., Hazzard, W. R., Motulsky, A. G., and Bierman, E. L. (1974). *Clin. Res.* **22**, 462a.
Buchwald, H. (1964). *Circulation* **29**, 713–720.
Buchwald, H. (1965). *Surgery* **58**, 22–36.
Buchwald, H. (1969). *J. Atheroscler. Res.* **10**, 1–4.
Buchwald, H., and Varco, R. L. (1967). *Surgery* **124**, 1231–1238.
Buchwald, H., Moore, R. B., Lee, G. B., Frantz, I., and Varco, R. L. (1968). *Arch. Surg. (Chicago)* **98**, 275–282.
Carlson, L. A., and Walldius, G. (1972). *In* "Pharmacological Control of Lipid Metabolism" (W. Holmes, R. Paoletti, and D. Kritchevsky, eds.), pp. 165–178. Plenum Press, New York.
Cayen, M. N., and Dvornik, D. (1970). *Can. J. Biochem.* **48**, 1022–1023.
Cenedella, R. J., Jarrell, J. J., and Saxe, L. H. (1968). *J. Atheroscler. Res.* **8**, 903–911.
Connor, W. E. (1968). *Med. Clin. N. Amer.* **52**, 1249–1260.
Connor, W. E., and Lin, D. S. (1974). *J. Clin. Invest.* **3**, 1062–1070.
Connor, W. E., Hodges, R. E., and Bleiler, R. E. (1961a). *J. Lab. Clin. Med.* **57**, 331–342.
Connor, W. E., Hodges, R. E., and Bleiler, R. E. (1961b). *J. Clin. Invest.* **40**, 894–901.
Connor, W. E., Stone, D. B., and Hodges, R. E. (1964). *J. Clin. Invest.* **43**, 1691–1696.
Connor, W. E., Witiak, D. T., Stone, D. B., and Armstrong, M. L. (1969). *J. Clin. Invest.* **48**, 1363–1375.
Danzinger, R. G., Hofmann, A. F., Schoenfeld, L. J., and Thistle, J. L. (1971). *J. Clin. Invest.* **50**, 24a.
Dietschy, J. M. (1968). *J. Clin. Invest.* **47**, 286–300.
Einarsson, K., Hellström, K., and Kallner, M. (1973). *Eur. J. Clin. Invest.* **3**, 345–351.
Fallon, H. J., and Woods, J. W. (1968). *J. Amer. Med. Ass.* **204**, 1161–1164.
Farquhar, J. W., Smith, R. E., and Dempsey, M. E. (1956). *Circulation* **14**, 77–82.
Feldman, S. A. (1972). *Arch. Pathol.* **94**, 42–58.

Fimognari, G. M., and Rodwell, V. M. (1965). *Science* **147**, 1038.

Forte, T., Norum, K. R., Glomset, J. A., and Nichols, A. V. (1971). *J. Clin. Invest.* **50**, 1141–1148.

Frantz, I. D., Jr., Schneider, H. S., and Henkelman, B. T. (1954). *J. Biol. Chem.* **206**, 465–469.

Fredrickson, D. S., Levy, R. I., and Lees, R. S. (1967). *New Engl. J. Med.* **276**, 34–44, 94–103, 148–156, 215–226, 273–281.

Friedman, G. D., Kannel, W. B., and Dauber, T. R. (1966). *J. Chron. Dis.* **19**, 273–274.

Fröberg, S., Boberg, J., Carlson, L. A., and Eriksson, M. (1971). In "Metabolic Effects of Nicotinic Acid and Its Derivatives" (K. F. Gey and L. A. Carlson, eds.), pp. 167–181. Hans Huber, Bern.

Garbutt, J. T., and Kenney, T. J. (1972). *J. Clin. Invest.* **51**, 2781–2788.

Glomset, J. A. (1968). *J. Lipid Res.* **9**, 155–167.

Goodman, D. S., and Noble, R. P. (1968). *J. Clin. Invest.* **47**, 231–241.

Goodman, D. S., Noble, R. P., and Dell, R. B. (1973). *J. Lipid Res.* **14**, 178–188.

Gordon, H., Lewis, B., Eales, L., and Brock, J. F. (1957). *Lancet* ii, 1299–1306.

Gould, R. G. (1951). *Amer. J. Med.* **11**, 209–227.

Gould, R. G. (1955). *Trans. N.Y. Acad. Sci.* **18**, 129–134.

Gould, R. G., Taylor, C. B., Hagerman, J. S., Warner, I., and Campbell, D. J. (1953). *J. Biol. Chem.* **201**, 519–528.

Gould, R. G., Jones, R. J., LeRoy, G. V., Wissler, R. W., and Taylor, C. B. (1969). *Metabolism* **18**, 652–662.

Grundy, S. M. (1972a). *Arch. Intern. Med.* **130**, 638–648.

Grundy, S. M. (1972b). *Gastroenterology* **63**, 201–203.

Grundy, S. M. (1974). *J. Clin. Invest.* **53**, 115a.

Grundy, S. M. (1975). *J. Clin. Invest.* **55**, 269–282.

Grundy, S. M., and Ahrens, E. H., Jr., (1966). *J. Clin. Invest.* **45**, 1503–1515.

Grundy, S. M., and Ahrens, E. H., Jr. (1969). *J. Lipid Res.* **10**, 91–107.

Grundy, S. M., and Ahrens, E. H., Jr. (1970). *J. Clin. Invest.* **49**, 1135–1152.

Grundy, S. M., and Metzger, A. L. (1972). *Gastroenterology* **62**, 1200–1217.

Grundy, S. M., Ahrens, E. H., Jr., and Miettinen, T. A. (1965). *J. Lipid Res.* **6**, 397–410.

Grundy, S. M., Ahrens, E. H., Jr., and Salen, G. (1968). *J. Lipid Res.* **9**, 374–387.

Grundy, S. M., Ahrens, E. H., Jr., and Davignon, J. (1969). *J. Lipid Res.* **10**, 304–315.

Grundy, S. M., Ahrens, E. H., J., and Salen, G. (1971). *J. Lab. Clin. Med.* **78**, 94–121.

Grundy, S. M., Metzger, A. L., and Adler, R. D. (1972a). *J. Clin. Invest.* **51**, 3026–3043.

Grundy, S. M., Ahrens, E. H., Jr., Salen, G., Schreibman, P. H., and Nestel, P. J. (1972b). *J. Lipid Res.* **13**, 531–551.

Grundy, S. M., Duane, W. C., Adler, R. D., Aron, J. M., and Metzger, A. L. (1973). *Metabolism* **23**, 67–73.

Grundy, S. M., Duane, W. C., Adler, R. D., Aron, J. M., and Metzger, A. L. (1974). *Metabolism* **23**, 67–73.

Hamilton, J. G., Muldrey, J. E., McCracker, B. H., Goldsmith, G. A., and Miller, O. N. (1964). *J. Amer. Oil Chem. Soc.* **41**, 760–762.

Hardison, W. G. M., and Rosenberg, I. H. (1969). *J. Lab. Clin. Med.* **74**, 564–573.

Hashim, S. A., and Van Itallie, T. B. (1965). *J. Amer. Med. Ass.* **192**, 289–293.

Havel, R. J., and Kane, J. P. (1973). *Proc. Nat. Acad. Sci. U.S.* **70**, 2015–2019.

Havel, R. J., Kane, J. P., and Kosyap, M. W. (1973). *J. Clin. Invest.* **52**, 32–38.
Heaton, K. W., and Read, A. E. (1969). *Brit. Med. J.* **3**, 494–496.
Hellman, L., Rosenfeld, R. S., Insull, W., Jr., and Ahrens, E. H., Jr. (1957). *J. Clin. Invest.* **36**, 898 (Abstr.).
Ho, K. J., and Taylor, C. B. (1968). *Arch. Pathol.* **86**, 585–596.
Ho, K., Biss, K., Mikkelson, B., Lewis, L. A., and Taylor, C. B. (1971). *Arch. Pathol.* **91**, 387–410.
Ho, K., Biss, K., and Taylor, C. B. (1974). *Arch. Pathol.* **97**, 307–315.
Hofmann, A. F., and Small, D. M. (1967). *Annu. Rev. Med.* **18**, 333–376.
Holt, P. R. (1964). *Amer. J. Physiol.* **207**, 1–7.
Horlick, L., Kudchodkar, B. J., and Sodhi, H. S. (1971). *Circulation* **43**, 299–309.
Ivy, A. C., Lin, T., and Karvinen, E. (1955). *Amer. J. Physiol.* **183**, 79–85.
Johnston, D. A., Davis, J. A., and Myant, N. B. (1967). *Proc. Roy. Soc. Med.* **60**, 746–748.
Kaplan, J. A., Cox, G. E., and Taylor, C. (1963). *Arch. Pathol.* **76**, 359–368.
Keys, A. (1953). *J. Mount Sinai Hosp. N.Y.* **20**, 118–139.
Keys, A., and Anderson, J. T. (1954). *Symp. Atheroscler. Publ. 338*, p. 181. Nat. Acad. Sci., Nat. Res. Council, Washington, D.C.
Keys, A., and Grande, F. (1957). *Amer. J. Pub. Health*, **47**, 1520–1541.
Keys, A., Anderson, J. T., Mickelsen, O., Adelson, S. F., and Fidanza, F. (1956). *J. Nutr.* **59**, 39–56.
Khachadurian, A. K. (1968). *J. Atheroscler. Res.* **8**, 177–188.
Knittle, J. L. and Ahrens, E. H., Jr. (1964). *J. Clin. Invest.* **43**, 485–495.
Kuksis, A., and Huang, T. C. (1962). *Can. J. Biochem.* **40**, 1493–1504.
Lack, L., and Wiener, I. M. (1963). *Fed. Proc. Fed. Amer. Soc. Exp. Biol.* **22**, 1334–1338.
Langdon, R. G., and Bloch, K. (1953). *J. Biol. Chem.* **202**, 77–81.
Langer, T., and Levy, R. I. (1971). In "Metabolic Effects of Nicotinic Acid and Its Derivatives" (K. F. Gey and L. A. Carlson, eds.), p. 641. Hans Huber, Bern.
Leren, P. (1966). *Acta Med. Scand. Suppl.* **466**, 3–92.
Levy, R. I., and Langer, T. (1971). *Proc. 4th Int. Symp. Drugs Affecting Lipid Metabol.* pp. 155–163.
Marinovic, I., Guerra, C., and Larach, G. (1972). *Rev. Med. Chile* **100**, 1320–1327.
Mellies, M., Glueck, C. J., Sweeney, C., and Ishikawa, T. (1974). *Circulation* (Suppl. 3) **48** (Abstr.).
Menkes, J. H., Schimschock, J. R., and Swanson, P. D. (1968). *Arch. Neurol. (Chicago)* **19**, 47–53.
Miettinen, T. A. (1968a). *Clin. Chim. Acta* **20**, 43–51.
Miettinen, T. A. (1968b). *J. Lab. Clin. Med.* **71**, 537–547.
Miettinen, T. A. (1969). *Scand. J. lCin. Lab. Invest.* **23** (Suppl. 108), 56.
Miettinen, T. A. (1971a). *Circulation* **44**, 842–850.
Miettinen, T. A. (1971b). In "Metabolic Effects of Nicotinic Acid and Its Derivatives" (K. F. Gey and L. A. Carlson, eds.), pp. 677–686. Hans Huber, Bern.
Miettinen, T. A. (1971c). In "Metabolic Effects of Nicotinic Acid and Its Derivatives" (K. F. Gey and L. A. Carlson, eds.), pp. 649–658. Hans Huber, Bern.
Miettinen, T. A., and Lempinen, M. (1970). *Scand. J. Clin. Lab. Invest.* **25** (Suppl. 113), 55.
Miettinen, T. A., Ahrens, E. H., Jr., and Grundy, S. M. (1965). *J. Lipid Res.* **6**, 411–424.
Miettinen, M., Karvonen, M. J., Turpeinen, O., Elosuo, R., and Paavilainen, E. (1972). *Lancet* ii, 835–838.

Miller, N. E., Clifton-Bliger, P., and Nestel, P. J. (1973). *J. Lab. Clin. Med.* **82,** 876–890.

Montenegro, M. R., and Solberg, L. A. (1968). *Lab. Invest.* **18,** 594–603.

Moore, R. B., Anderson, J. T., Taylor, H. L., Keys, A., and Frantz, I. D., Jr. (1968a). *J. Clin. Invest.* **47,** 1517–1534.

Moore, R. B., Crane, C. A., and Frantz, I. D. (1968b). *J. Clin. Invest.* **47,** 1664–1671.

Moore, R. B., Frantz, I. D., Jr., and Buchwald, H. (1969). *Surgery* **65,** 98–108.

Moutafis, C. D., and Myant, N. B. (1968). *Clin. Sci.* **34,** 541–548.

Nestel, P. J., and Barter, P. (1971). *Clin. Sci.* **40,** 345–350.

Nestel, P. J., Hirsch, E. Z., and Couzens, E. A. (1965). *J. Clin. Invest.* **44,** 891–896.

Nestel, P. J., Whyte, H. M., and Goodman, D. S. (1969). *J. Clin. Invest.* **48,** 982–991.

Nestel, P. J., Havenstein, N., Whyte, H. M., Scott, L. J., and Cook, L. J. (1973). *New Engl. J. Med.* **288,** 379–382.

Norum, K. R., Glomset, J. A., Nichols, A. V., and Forte, T. (1971). *J. Clin. Invest.* **50,** 1131–1140.

Olefsky, J., Reaven, G. M., and Farquhar, J. W. (1974). *J. Clin. Invest.* **53,** 64–76.

Osuga, T., and Portman, O. W. (1971). *Proc. Soc. Exp. Biol. Med.* **136,** 722–726.

Perl, W., and Samuel, P. (1969). *Circ. Res.* **25,** 191–199.

Pertsemlidis, D., Kirchman, E. H., and Ahrens, E. H., Jr. (1973a). *J. Clin. Invest.* **52,** 2353–2367.

Pertsemlidis, D., Kirchman, E. H., and Ahrens, E. H., Jr. (1973b). *J. Clin. Invest.* **52,** 2368–2378.

Pertsemlidis, D., Panueliwalla, D., and Ahrens, E. H., Jr. (1974). *Gastroenterology* **66,** 565–573.

Quintão, E., Grundy, S. M., and Ahrens, E. H., Jr. (1971). *J. Lipid Res.* **12,** 221–232.

Ryan, W. G., and Schwartz, T. B. (1965). *Metabolism* **14,** 1243–1254.

Salen, G. (1971). *Ann. Intern. Med.* **75,** 843–851.

Salen, G., and Grundy, S. (1973). *J. Clin. Invest.* **52,** 2822–2835.

Salen, G., Ahrens, E. H., Jr., and Grundy, S. M. (1970). *J. Clin. Invest.* **49,** 952–967.

Samuel, P., and Leiberman, S. (1973). *J. Lipid Res.* **14,** 189–196.

Samuel, P., and Steiner, P. (1959). *Proc. Soc. Exp. Biol. Med.* **100,** 193–195.

Samuel, P., and Lieberman, S. (1973). *J. Lipid Res.* **14,** 189–196.

Samuel, P., Holtzman, C. M., and Goldstein, J. (1967). *Circulation* **35,** 932–945.

Samuel, P., Holtzman, C. M., Meilman, E., and Perl, W. (1968). *J. Clin. Invest.* **47,** 1806–1818.

Samuel, P. H., Holtzman, C. M. and Meilman, E. (1973). *Circ. Res.* **33,** 393–402.

Schreibman, P. H., Pertsemlidis, D., Liu, G. C. K., and Ahrens, E. H., Jr. (1974). *J. Clin. Invest.* **53,** 72a (Abstr.).

Scrimshaw, N. S., and Guzman, M. A. (1968). *Lab. Invest.* **18,** 623–628.

Shefer, S., Hauser, S., Bekersky, I., and Mosbach, E. (1969). *J. Lipid Res.* **10,** 646–655.

Sievers, M. L. (1968). *J. Chronic Dis.* **21,** 107–115.

Siperstein, M. D., and Guest, M. J. (1960). *J. Clin. Invest.* **39,** 642–652.

Smith, M., and Levine, R. (1964). *Med. Clin. N. Amer.* **48,** 1387–1397.

Sniderman, A. D., Carew, T. E., Chandler, J. G., and Steinberg, D. (1974). *Science* **183,** 526–528.

Spritz, N. (1965). *Circulation* **31/32** (Suppl. II), 201.

Spritz, N., and Mishkel, M. A. (1969). *J. Clin. Invest.* **48,** 78–86.

Spritz, N., Ahrens, E. H., Jr., and Grundy, S. M. (1965). *J. Clin. Invest.* **44,** 1482–1493.

Steinberg, D. (1970). *Proc. 2nd Int. Symp. Atherosclerosis,* p. 500. Springer-Verlag,

Sturdevant, R. A. L., Pearce, M. L., and Dayton, S. (1973). *Neew Engl. J. Med.* **288,** 24–27.

Swell, L., Boiter, T. A., Field, H., Jr., and Treadwell, C. R. (1956), *J. Nutr.* **58,** 385–398.

Sylvén, C., and Borgström, B. (1969). *J. Lipid Res.* **10,** 179–182.

Taylor, C. B., Patton, D., Yogi, N., and Cox, G. E. (1960). *Proc. Soc. Exp. Biol. Med.* **103,** 768.

Tomkins, G. M., Sheppard, H., and Chaikoff, I. L. (1953). *J. Biol. Chem.* **201,** 137–141.

Vlahcevic, Z. R., Bell, C. C., Buhoc, I., Farrar, J. T., and Swell, L. (1970). *Gastroenterology* **59,** 165–173.

Walton, K. W., Scott, P. J., Dykes, P. W., and Davies, J. W. L. (1965). *Clin. Sci.* **29,** 217–238.

Whyte, H. M., Nestel, P. J., and Pryke, E. S. (1973). *J. Lab. Clin. Med.* **81,** 818–828.

Wilson, D., and Lees, R. (1972). *J. Clin. Invest.* **51,** 1051–1057.

Wilson, J. D. (1964). *J. Lipid Res.* **5,** 409–417.

Wilson, J. D. (1972). *J. Clin. Invest.* **52,** 1450–1458.

Wilson, J. D., and Lindsey, C. A., Jr. (1965). *J. Clin. Invest.* **44,** 1805–1814.

Wolfe, B. M., Kane, J. P., Havel, R. J., and Brewster, H. P. (1973). *J. Clin. Invest.* **52,** 2146–2159.

Wood, P. D. S., Shioda, R., and Kinsell, L. W. (1966). *Lancet* **ii,** 604–607.

Chapter 5

DIET AND DRUGS

IN OBESITY CONTROL

DANIEL L. AZARNOFF AND DON W. SHOEMAN

Clinical Pharmacology-Toxicology Center,
Departments of Medicine and Pharmacology,
University of Kansas Medical Center,
Kansas City, Kansas

I. Introduction

In this discussion we do not intend to undertake an extensive review of the cumulated knowledge of the control of food intake, etiology, risks, metabolic changes, etc., associated with obesity. However, in order to discuss the rational use of diet and drugs in the treatment of obesity, it will be necessary to outline for the reader a few of the pertinent observations about this disorder.

II. Definition of Obesity

Obesity is a disturbance of energy regulation which can occur by a variety of mechanisms. This disturbance in regulation leads to an enlargement of the adipose tissue fat depots. Whatever the etiology, obesity is due to the ingestion of more calories than are being utilized.

$$\text{Caloric intake} \rightarrow \text{caloric stores} \rightarrow \text{energy expenditure}$$

It is obvious from this representation that caloric stores can be decreased by reducing caloric intake or by increasing energy expenditure.

Adipose tissue is in a dynamic metabolic state, storing excess calories as triglycerides and releasing them as free fatty acids when energy is required. Obesity is not an inborn error of metabolism (Vallence-Owen, 1964), and obese individuals actually have an increased free fatty acid turnover (Issekutz et al., 1967). It is essential to discriminate between obesity and overweight. Obesity is significant when the fat content exceeds 25% of body weight in males and 30% in females. It has been estimated that as many as 30% of males and 20% of females are significantly overweight and that the percentage increases to more than 50% in both sexes after age 40 (Montegriffo, 1968). The United States Public Health Service (1966) reports that 25 to 45% of American adults over 30 years of age are more than 20% overweight. As pointed out by Gastineau (1972), unless weight is gradually lost as one grows older, the fat depot is increased since muscle mass is gradually lost. Since body composition cannot be measured routinely, anthropometric measurements may be used instead. In a study of 67,739 obese individuals, Bernstein et al. (1973) found the best relationship to obesity between height and weight was W/H^2 in the male and W/H in the female. However, body adiposity probably correlates best with skin fold measurement (Allen et al., 1956; Crook et al., 1966).

III. Etiology and Classification

In experimental animals, obesity can be classified on the basis of the mechanisms which produce it and include hypothalamic lesions, endocrine manipulations, dietary manipulations, reduction of physical activity, and by genetic transmission (Bray and York, 1971). This classification has been extended to humans by Bray (1973). As in diabetes, there appears to be a juvenile onset and an adult onset form of the disorder with markedly different intensities. In the adult onset form there is an increase primarily in the size of the existing number of adipocytes, whereas in the juvenile onset form there is also a marked increase in the number of adipocytes as well as an increase in cell size (Hirsch and Knittle, 1970; Bray, 1970). The volume of adipocytes may increase as much as three- to fourfold in these individuals. Weight reduction does not decrease the number, but rather only the size of the cells.

IV. Metabolic Derangements

The metabolic changes found in obese individuals are relatively constant and may be modified by weight reduction and diet (Rabinowitz, 1970). Fasting insulin levels as well as the response to intravenous and oral glucose loads are elevated (Bagdade et al., 1967). Fasting insulin levels correlate directly with body weight (Bagdade et al., 1967) and size of adipocytes (Stern et al., 1972). The elevated fasting insulin levels in obese individuals are associated with peripheral resistance to this hormone (Rabinowitz and Zierler, 1962) and increased levels of circulating amino acids (Felig et al., 1969). Whether the hyperinsulinemia is secondary to the resistance is not definitely established. Obese individuals also have signficant elevations in fasting plasma FFA levels (Kalkhoff et al., 1971), a situation associated with inhibition of glucose oxidation (Randle et al., 1963). After weight reduction, insulin responses to glucose, tolbutamide, and glucagon return to normal as do the low growth hormone levels (Kalkhoff et al., 1971). Grey and Kipnis (1971) have demonstrated that carbohydrate restriction reduces basal hyperinsulinemia in the obese; indeed, diabetes mellitus can be controlled in 80% of obese diabetic patients by carbohydrate restriction without weight reduction (Wall et al., 1973). It is·of interest to note that one "fad" diet recommends eating avocados in significant amounts. This fruit contains large quantities of d-manoheptulose, a known inhibitor of insulin secretion (Simon and Kraicer, 1966). Further evaluation may be warranted.

Following studies with the genetically obese sand rat, Robertson *et al.* (1973) suggested that hyperinsulinemia may be responsible for the elevated triglyceride levels. Olefsky *et al.* (1974) report data that support the concept that insulin resistance in obese humans leads to hyperinsulinemia which, in turn, stimulates accelerated hepatic triglyceride production and hypertriglyceridemia. Weight reduction in turn decreases VLDL-triglyceride production rates by 40%. Weight loss lowers each of the above factors in sequence, an observation which strengthens the hypothesis of the authors. Although plasma and liver triglyceride levels are elevated in obese individuals, there is little if any increase in plasma cholesterol levels although production and fecal elimination are twice those of the controls (Miettinen, 1971).

V. Risks

Obesity has only relatively recently been recognized as a disease state. The first evidence appeared in 1901 when an increased mortality in overweight individuals was seen in data compiled from life insurance holders (Rogers, 1901). This shortened life expectancy was found mainly in males and primarily associated with cardiovascular disorders. In 1959, these studies were extended to five million individuals and the association between obesity and coronary heart disease in males reconfirmed (Soc. Actuaries, 1959). Little difference in mortality was found in females. There are no other studies of this magnitude although smaller studies both confirm and deny these observations (Kannel *et al.*, 1967; Dunn *et al.*, 1970; Carlson and Bottiger, 1972; Keys *et al.*, 1972). In the study by Keys *et al.* (1972), for example, the association did not exist after other risk factors, such as elevated serum cholesterols, were factored out. In addition to the large number of subjects, the insurance companies' statistics have the advantage of long-term follow-up.

It is becoming increasingly obvious that the incidence of coronary heart disease in patients with hyperlipoproteinemia should not be compared to that in those with statistically "normal" lipid levels, but rather to those with the best prognosis (cholesterol < 200 mg/dl). The same is true for comparisons with obesity. In any particular comparison, the results obtained may well depend upon the definition of obesity used for that particular study. For example, Kannel *et al.* (1967) found that obesity was only a weak risk factor of coronary heart disease in the Framingham study until body weight exceeded 125% of the mean. In one study, males 15–19 years old on entry were followed for 16 to 19 years. Underweight individuals had 72% and average weight individuals 90% of the expected mortality, whereas the slight, moderate, and markedly over-weight individuals had 118, 141, and 221%, respectively (Gubner, 1957).

Irrespective of etiology, the treatment of obesity is directed toward a reduction of the excess adipose tissue and is accomplished by decreasing caloric intake and/or increasing energy expenditure. Oxidation of one pound of fat liberates about 3500 kcal. We can readily calculate that if we decrease intake or increase expenditure as little as 200–500 kcal per day, a weight loss of 20 to 50 pounds per year will be achieved.

VI. Dietary Control of Obesity

A. Food Intake

Environmental cues such as perceived time, lighting, palatability and availability of food as well as metabolic cues control food intake (Hoebel, 1971). Obese individuals respond quite differently than lean individuals to environmental cues (Schachter, 1968). Hirsch (1972) has collated the available data into an intriguing systems analysis of the control of food intake. The newer aspects of control of food intake have been reviewed by Lepkovsky (1973).

B. Fasting

Fasting is not the ideal manner in which to lose weight. Weight loss following long-term fasting is derived approximately 65% from fat-free rather than adipose tissue (Benoit et al., 1965; Ball et al., 1970). Fasting produces many adverse metabolic changes such as hyperuricemia, protein and water loss, anemia, malabsorption, electrolyte disorders, and even death. Long-term fasting should be carried out only in a hospital under a physician's care and is now rarely prescribed since its hazards are now better understood. However, Duncan et al. (1963) recommend repeated 1- to 2-day periods of fasting as useful in a weight reduction program.

C. Frequency of Meals

There is little doubt that weight loss follows sufficient caloric restriction, but what do we know about the effect of the composition of the diet or level of calories at each meal on weight loss? In a series of epidemiological studies in several hundred elderly men, it was observed that excess weight, hypercholesterolemia, impaired glucose tolerance, and ischemic heart disease are negatively correlated with the frequency of meals per day. Those who ate only three to four meals per day were not only significantly more overweight than those who ate five to six meals per day, but their skin fold thickness was also greater. Those eating four to five meals per day were intermediate (Hejda and Fabry, 1964; Fabry *et al.*, 1964). In another study, Fabry and his colleagues (1966) studied teenage children in three boarding schools serving three, five, or seven meals per day, respectively. In the 11–16 year olds, there was a significant increase in the percentage of weight compared to height as well as increase in skin fold thickness in the children in the school which served just three meals per day compared to the others. Bray (1972) has reported that lipogenesis was greater in obese individuals who gorged (one 4-hour feeding) 5000 kcal/day than in those that nibbled (ten feedings at 2-hour intervals). Lipogenesis was estimated as [^{14}C]pyruvate incorporation into glyceride glycerol and fatty acids by adipose tissue biopsies. Differences in time factors that affect nutrient availability, uptake, and metabolism as well as levels of enzyme activity have been stated by Sassoon (1973) to account for the effect of differences in meal frequency. He recommends that daily energy requirements be supplied primarily as fats and slowly available starches to minimize factors contributing to obesity. Weight loss, on the other hand, is not affected by the frequency of meals (Bortz *et al.*, 1966; Finkelstein and Fryer, 1971).

D. High Protein Diets

Obese persons use carbohydrates, fat, and protein normally during rest and exercise and are not unusually efficient in the use of these energy substrates (Nelson *et al.*, 1973). They also mobilize fat stores for energy as readily as lean individuals. Yet a variety of manipulations of the composition of dietary nutrients has been suggested as the best way to lose weight. These include high protein, high fat, and low carbohydrate modifications. The first is predicated on the concept that the specific dynamic action required for protein digestion and assimilation will aid in weight loss. The concept is untenable, however, since Bradford and Jourdan (1973) demonstrated that specific dynamic action, measured as increased oxygen consumption for several hours at bed rest following a test meal compared to a basal period, is not significantly different between isocaloric diets containing either 40 gm protein or no protein. There was also no difference between isoprotein dosages of varying caloric content. Therefore, there is no advantage in treating obesity with a high protein diet, which is also expensive.

E. High Fat Diets

The effect of the high fat diet is related, at least in part, to the decreased palatability of this type of diet. Normal individuals can absorb and digest more than 600 gm of fat per day (Kasper, 1970). Kasper *et al.* (1973) studied five volunteers who ate increasing amounts of fat in a formula diet maintained constant in carbohydrate and protein. The volunteers gained weight at a rate much slower than expected from their caloric intake, particularly if the fat was given as corn oil which contains 45% linoleic acid. This discrepancy was less obvious if olive oil, which contains only 7% linoleic acid, was used. Hirsch and Van Italie (1973) have vigorously criticized the design and authors' interpretations of this study.

The mechanism of the weight loss associated with the low carbohydrate diet even in the face of a high caloric intake has been explained by the loss of ketone bodies in urine. The oxidation of ketones yields about 4.5 kcal/gm so that at most 100 kcal/day are lost in this manner even during continuous fasting (Grande, 1968; Duell and Gulick, 1932). Indeed, if the caloric intake is sufficient, humans can maintain body weight on protein and fat alone (Tolstoi, 1929; McClelland and DuBois, 1930).

F. Low Carbohydrate Diets

Pennington (1951) suggested that a high fat–low carbohydrate diet would result in weight loss regardless of caloric intake. He later tried to explain this phenomenon on the basis of differences in the metabolism of pyruvate in obese individuals (Pennington, 1955). Further support was supposedly provided by Kekwick and Pawan (1956) when they reported that obese patients lost more weight over an 8- to 10-day period on a high fat (90% of calories) diet than when carbohydrate was replaced isocalorically for the fat. However, when a similar study was repeated by Pilkington *et al.* (1960), the difference in the rate of weight loss was observed again at 8 to 10 days, but by 18 to 24 days the rate was the same for both diets. These authors concluded that the temporary difference in rate of weight loss was due to changes in water balance. Other investigators have also concluded that the temporary changes in rate of weight loss from low carbohydrate diets is secondary to transient changes in water balance (Olesen and Quaade, 1960, Werner, 1955; Worthington and Taylor, 1974). In two long-term studies of several months duration with diet modification at a minimum of

3- to 4-week intervals, weight loss was determined primarily by caloric content, changes in the composition of the diet having little or no effect on the rate of weight loss (Kinsell et al., 1964; Pilkington et al., 1960). Bell et al. (1969) have observed that the sodium loss secondary to starvation is arrested by administration of carbohydrate, but not fat and protein. Also, carbohydrate administration following a fast increases salt and water retention (Hood et al., 1970). In this study four obese females received 1000 kcal/day with varying proportions of carbohydrate. Each individual received 3, 6, 12, 25, and 50% of the calories as sucrose in a random order for 8-day periods. There was no difference in the 1.2 kg/week rate of weight loss in any period. The authors suggest that the previously reported changes in salt and water balance may have been acute responses secondary to rapid changes in the composition of the diet. In this study where the changes in composition were made gradually, markedly different responses were not seen.

In another study Kasper et al. (1973) studied 25 obese individuals with varying degrees of caloric restriction and fat to carbohydrate ratios. The weight loss of 0.3 kg/day on the high fat–low carbohydrate, 1707-kcal diet persisted for 45 days and the authors concluded that water balance was not involved since the patients were studied for 45 days. In fact, individual treatment periods were only 6–14 days. It is of interest to note that if fat was exchanged isocalorically for glucose, weight loss ceased and that when carbohydrate restriction was continued on an outpatient basis, the body weight continued to decrease despite an increase in calories to 2000–2300 per day. Increased energy output emitted in the form of heat was suggested as the mechanism of the observed effect. Studies of this type are suspect since the effect of short-term low carbohydrate diets is known to be influenced by water balance. Further careful studies are needed. Young et al. (1971) have demonstrated that moderately obese college males on an 1800-kcal, 155-gm protein diet containing 104, 60, or 30 gm carbohydrate lost 11.9, 12.8, and 16.1 kg, respectively, in 9 weeks. Importantly, density measurements were used to demonstrate that 75, 84, and 95%, respectively, of the weight loss was fat. There were no changes in sodium balance and the three diets were effective in controlling hunger. An excellent review of the history and clinical evaluation of the low carbohydrate diet has been published by the American Medical Association Council on Foods and Nutrition (1973).

It is useful to know what response is observed following manipulation of the composition of isocaloric diets in the metabolic ward, but to treat patients we must also know what happens in the outside world since the satiety effect of various nutrients in the diet can markedly alter compliance with caloric restrictions. Yudkin and Carey (1960) observed that when allowed a free choice diet specified only as low carbohydrate (50–60 gm/day), obese patients actually did not alter their protein and fat intake significantly. Later, Stock and Yudkin (1970) repeated and corroborated the study, reporting that if 10–20 ounces of whole milk are included in the diet, there is not a decrease and perhaps even an improved intake of essential nutrients.

VII. Drug Control of Obesity

A. Dinitrophenol and Thyroxine

In the past, prevailing opinion as to the metabolic cause of obesity has guided the selection of drugs for its treatment. Thus, thyroid hormones and dinitrophenol (DNP) were introduced quite early in the hope of increasing the utilization of excess caloric intake. The recognition of the high toxicity of DNP led to its abandonment. Thyroid hormones, however, continue to attract attention despite a lack of clear-cut evidence of effectiveness (Penick and Stunkard, 1972). A group of 12 obese outpatients lost more weight on a 1000-kcal diet during 1 month of daily 225 μg triiodothyronine (T_3) administration than while receiving a placebo. The administration of 150 μg T_3 daily increases oxygen uptake and fat and protein catabolism, but the weight loss that ensues is nearly 80% lean body mass (Bray et al., 1973). Thyroid hormones are held to be effective in conjunction with restriction of caloric intake (Gershberg, 1972). Hollingsworth et al. (1970), however, in a study which compared the combination of an 800 kcal/day diet with a placebo or 75 μg T_3 t.i.d. found that the weight loss with T_3 was significantly greater only during the first 8 weeks of their 24-week study, and that no difference existed at the 48-week follow-up. Lamki et al. (1973) also failed to demonstrate a significant effect of L-thyroxine (0.9 or 0.3 mg/day) over a 1200- or 600-kcal diet on weight loss in four massively obese males. Thyroid hormones also appear to increase the very moderate short-term weight loss achieved with amphetamine in the absence of dietary restrictions (Kaplan and Jose, 1970).

B. Chorionic Gonadotropin

The use of another hormone, human chorionic gonadotropin (HCG) in the treatment of obesity has attracted a small number of very enthusiastic proponents. Simeons (1969) recommends a defined 550-kcal diet in conjunction with 125 U HCG daily, six times a week for a total of 40 injections. Many investigators have failed to demonstrate the efficiency of HCG in weight reduction programs (Frank, 1974; Craig et al., 1963). This discrepancy in effectiveness has been attributed by some authors to noncompliance with the recommended dosage schedule and dietary restrictions (Gusman, 1969). Despite the doubtful efficacy of the Simeons program (Rivlin, 1975), commercial clinics offering this treatment have proliferated. The mechanism of action of HCG as an adjunct in the treatment of obesity remains unknown, but it has been suggested that it makes the severe caloric restriction more tolerable, possibly through fat mobilization. Asher and Harper (1973) compared 125 U HCG with placebo injections in 40 female patients on a 500-kcal diet and found that the HCG group lost more weight than did the placebo group, although all patients did not complete the 6-week study. Hirsch and Van Italie (1973) reviewed this study and found that significantly fewer injections were received by the placebo group and that within the placebo group the number of injections correlated positively with the percentage of weight loss. Thus, not the drug, but injections per se appear to influence the rate of weight loss. Studies comparing the effects of drugs in addition to such extreme caloric restrictions have little meaning since few "free living" patients will readily adhere to diets when receiving drugs. Similarly, few but the most masochistic patients will tolerate the discomfort and few can afford the expense of daily injections.

C. Biguanides

Adult onset diabetics lose weight when treated with biguanides, an observation which has prompted their use in the treatment of nondiabetic, obese patients. When care is taken to exclude diabetics (usually on the basis of glucose tolerance) from the population tested, these agents are without effect on weight loss. Hart and Cohen (1970) investigated the effect of phenformin in conjunction with a 1200-kcal diet using a double blind cross-over trial. They found no difference in weight loss between placebo and phenformin groups. Lawson et al. (1970) compared 2 gm fenfluramine, placebo, and 1.5 or 3 gm metformin and also found no difference between either dose of metformin and placebo. The metabolic effects of biguanides depend upon the initial metabolic state of the individual in whom they are measured. Vermeulen and Rottiers (1972) found no effect of metformin on fasting blood glucose or on intravenous glucose tolerance in obese, nondiabetic women. Similarly, Schatz et al. (1972) found effects only in those obese patients whose glucose tolerances were abnormal. Mace *et al.* (1972), however, found that phenformin treatment led to decreases in plasma glycerol and triglycerides and increases in ketones and cholesterol in normal (nonobese) subjects. In spite of these actions, it does not appear that biguanides have any weight-lowering effect in nondiabetic patients, and are potentially harmful in maturity onset diabetes (Sapeika, 1974).

D. Amphetamine

The intake of calories may be more amenable to drug treatment than the subsequent utilization of these calories. Amphetamine is perhaps the best known anorexic agent and remains the standard by which other agents are measured. Equally well known are the major drawbacks to the chronic use of amphetamine and amphetaminelike drugs: central nervous system (CNS) stimulation (Tinklenberg, 1971), potential for abuse, and the tolerance which develops to their effect on rate of loss of weight (Edison, 1971). The abuse potential and tolerance were suggested as early as 1937, although according to Fineberg (1972) tolerance to the anorexic effects has never been adequately demonstrated in man. Anderson (1974) considers the danger of abuse sufficient to proscribe the use of amphetamine in weight reduction programs. Laboratory animals become tolerant to the weight loss produced by amphetamine either incorporated in the diet (Panksepp and Booth, 1973) or injected (Tormey and Lasagna, 1960). The undesired CNS stimulation produced by amphetamine when used for its anorexic effect is not effectively eliminated by fixed combination with barbituates. Combination of amobarbital with phenmetrazine had little effect on side effects except insomnia; in addition, amobarbital reduced the effectiveness of phenmetrazine in producing weight loss (Hadler, 1969). Meprobamate, however, in combination with *d*-amphetamine lowered the incidence of stimulatory CNS side effects without decreasing the weight loss produced in females on a 1500- and males on a 1200-kcal diet (Noble, 1972).

Amphetamine enters the brain via a saturable transport process in addition to simple diffusion (Pardridge and Connor, 1973), where it exerts a number of effects on neurotransmitters therein. The whole brain content of norepinephrine is lowered by amphetamine, whereas serotonin (5-HT) is unaffected and dopamine increased (Holtzman and Jewett, 1971). Hypothalamic norepinephrine is also depleted by amphetamine. Injection of norepinephrine into the lateral hypothalamus causes hyperphagia, whereas injection of amphetamine into this area reduces food intake (Blundel *et al.*, 1973). Injection of α- and β-adrenergic agents into the hypothalamus of rats has revealed: (1)

that the medial hypothalamus is sensitive to α- whereas the lateral
hypothalamus is sensitive to β-adrenergic agonists and (2) that α
stimulation results in eating, while β stimulation suppresses eating behavior (Leibowitz, 1970). Further evidence for norepinephrine involvement is provided by the effect of α-methyl-*p*-tyrosine (MT) which
blocks norepinephrine synthesis. MT blocks the anorexic effect of amphetamine at a dose which has no effect by itself (Abdallah and White,
1970). At higher doses, it produced an anorexic effect which was additive to that of amphetamine.

E. Amphetaminelike Drugs

Diethylpropion, phenmetrazine, phendimetrazine, phentermine, and
chlorphentermine seem to resemble amphetamine closely in action and
hazard, if not in structure. Diethylpropion, like amphetamine, does not
enhance the rate of weight loss achieved by dietary restriction (Silverman and Okun, 1971). Previous treatment with amphetamine or phenmetrazine may render patients more susceptible to the schizophrenic-
like psychosis seen after treatment with diethylpropion (Editorial, 1970)
and phenmetrazine (Ananth, 1971).

Chlorphentermine possesses a longer plasma half-life (41 hours) in
man than does phentermine (Beckett and Brookes, 1971; Jun and Triggs,
1970) and, in the rat, accumulates in tissue to a much greater extent
(H. Lullman *et al.*, 1973a). This accumulation is accompanied by ultrastructural changes interpreted to reflect drug-induced lipidoses (H.
Lullman *et al.*, 1973b) which extend to the fetus in pregnant animals
(R. Lullman-Rauch, 1973). Coupling of phentermine to resin prolongs
its effect for 10 hours (Silverstone, 1972). Phentermine resin is more
effective than placebo in combination with a diet containing 60% of the
calories necessary to maintain ideal body weight (Truant *et al.*, 1972).

F. Fenfluramine

The seriousness of the side effects of amphetamine led to a search for
drugs possessing a more specific anorexic action. The first product of this
search was fenfluramine, an amphetamine derivative which actually
produces sedation in addition to exerting an anorexic effect. The studies
of Tang and Kirch (1971) in the monkey are an example of many which
demonstrate these actions. Fink and Shapiro (1969) and Fink *et al.*
(1970) using EEG patterns to classify the clinical activity of a number
of drugs, found that the effects of 40 mg fenfluramine and 50 mg amobarbital were similar in human subjects. Whereas fenfluramine is a

sedative in most species, it is a stimulant in rabbits. Funderburk and Ward (1970) suggested that cortical stimulation in rabbits may have been due to an unusual metabolite. However, later work by Funderburk *et al.* (1971a) showed that the metabolic pattern in the rabbit was actually similar to that in the cat. Thus, the mechanism of stimulation in the rabbit is unknown, but might be related to the degree to which 5-hydroxytryptamine (5-HT) is lowered, an effect to which the rabbit appears resistant. The anorexic effect of fenfluramine appears to be independent of its effects on 5-HT as β-methoxy derivatives which do not effect brain 5-HT concentrations retain anorexic activity (Cattabeni *et al.*, 1972). Fenfluramine antagonizes the CNS stimulant effect of amphetamine in rats and mice (Berger *et al.*, 1973). The combination of fenfluramine with amphetamine is additive with respect to effect on food consumption but synergistic with respect to toxicity (Yelnosky and Lawlor, 1970). While all anorexics seem to stimulate central gamma receptors (Hoyer and Van Zwieten, 1972) and interact with noradrenergic neurons (Gropetti *et al.*, 1972), the unique activity of fenfluramine may be related to its additional 5-HT-like properties. Methisergide (Southgate *et al.*, 1971) and AHR -3009 (Funderburk *et al.*, 1971b), two 5-HT antagonists, block the reduction of food intake produced by fenfluramine but not that produced by amphetamine, chlorphentermine, or diethylpropion. Fenfluramine also differs from amphetamine in the effect produced after intrahypothalamic injection (Blundell *et al.*, 1973) in a way consistent with the hypothesis that different neurohumors are involved. Lesions of the midbrain raphe which resulted in selective degeneration of serotonergic neurons were without effect on food intake and blocked the anorexic effect of fenfluramine while that of amphetamine remained unaffected (Samanin *et al.*, 1972).

Fenfluramine influences the electrical activity in the hypothalamic feeding centers (Chhina *et al.*, 1971) although this may be secondary to increased peripheral glucose utilization. The "glycolytic" effect of fenfluramine refers to increased glucose uptake in muscle (Turtle and Burgess, 1973) in addition to other peripheral effects (Sapeika, 1972), although these actions probably do not contribute to the therapeutic effect of this drug (Garrow *et al.*, 1972). Durnin and Womersley (1973) found no effect of fenfluramine on oxygen consumption during exercise, or was any more fat lost than in a similar group who lost weight through dietary restriction alone.

Fenfluramine treatment leads to modest weight loss in the absence of dietary restriction (Hadler, 1971; Elliott, 1970). Its effectiveness decreased with time, however, even in the face of increasing doses (Goldrick *et al.*, 1973). Stunkard *et al.* (1973) compared 20 mg fenfluramine,

5 mg amphetamine, and placebo administered t.i.d. Similar amounts of weight (greater than placebo) were lost, although the rate of weight loss decreased during treatment with both agents. The placebo group contained the largest number of dropouts in this study. Blood pressure often falls during drug treatment (Seedat and Reddy, 1974). Although the fall in blood pressure may reflect the degree of weight loss (Aikman et al., 1971), evidence exists that fenfluramine may possess intrinsic hypotensive effects. Waal-Manning and Simpson (1969) studied fenfluramine in obese hypertensives receiving antihypertensive drugs. While the rate of weight loss decreased during the fenfluramine phase of their cross-over study, the fall in blood pressure was increased. Thus, the antihypertensive effect was dissociated from weight loss. The authors suggested that the hypotensive effect of fenfluramine could be due to direct central effects, catecholamine or serotonin depletion, and/or peripheral vasodilatation. Mroczek et al. (1974) found no significant effect on cardiac output or total peripheral vascular resistance in a group of hypertensive obese patients who, in the absence of dietary control, also failed to lose weight after the first week of treatment.

Not all investigators have found fenfluramine to be of additional benefit when combined with effective dietary control. Kaufman and Blondheim (1973) and Durnin and Womersley (1973) using adults and Court (1972) using children did not demonstrate any difference between fenfluramine combined with diet and with diet alone. Datey et al. (1973) found that a number of patients failed to respond to increasing doses of fenfluramine. Two reached a dose of 240 mg/day without effect. Fenfluramine improves the glucose tolerance of obese diabetics only in the absence of dietary restriction (Dykes, 1973); however, it produced a further decrease in insulin secretion over diet alone in another group of patients in the same study. Chronic amphetamine or mazindol treatment leads to hyperinsulinemia (Sirtori et al., 1971).

Although fenfluramine is not considered a stimulant, depression results from its withdrawal (Oswald et al., 1971). Chronic fenfluramine treatment leads to altered sleep patterns (Lewis et al., 1971), but does not reduce REM sleep (Oswald, 1969). Harding (1972) reported fenfluramine dependence in four depressed patients, successfully treated with amitriptyline, who became depressed upon fenfluramine withdrawal. Successful withdrawal in one patient required much longer than is normally the case (Harding, 1971). Golding (1970) also reported agitated depression on withdrawal of fenfluramine in a patient with previous illness. Steel and Briggs (1972) found a pronounced decline in mood after stopping fenfluramine in obese (nondepressed) patients as well. Depression is common during fenfluramine treatment (Imlah, 1970) and a number of unusual side effects have been described: vivid dreams (Hooper, 1971), nightmares (Alvi, 1969), extreme sleepiness (Ellis, 1960), collapse or paranoia during administration with chlordiazepoxide (Brandon, 1969), and one case of hemolytic anemia (Nussey, 1973).

The β-benzyloxyethyl derivative of fenfluramine (S992) is four times as potent as fenfluramine in antagonizing the effect of amphetamine in the rat tail artery (Jespersen and Bonaccursi, 1969). Like fenfluramine (Pawan, 1969), it causes weight loss in patients not restricted in diet (Pawan et al., 1971a,b). Fenpororex (N^2-cyanoethyl amphetamine) inhibits the pressure response to amphetamine in cats (Berry et al., 1971); however, this compound is converted to amphetamine in man, 30% of a 2.4 mg dose of the compound being excreted as amphetamine in 24 hours (Tognoni et al., 1972).

G. Aminorex

Aminorex (2-amino-5-phenyl-2-oxazoline) in the absence of dietary restriction has a significant effect on weight loss although tolerance to its effect did appear during a 6-month study (Sandoval et al., 1971). One patient developed pulmonary hypertension. Kew (1970) found no evidence for it in a study which examined cardiovascular parameters in addition to weight loss. However, the preponderance of evidence is that pulmonary hypertension is a severe and not infrequent adverse effect of this drug (Wirz and Arbenz, 1970; Gurtner et al., 1968). Hadler (1970) found no difference in the weight loss produced by 30, 75, or 105 mg aminorex per day although he concluded, on the basis of a dose response curve, that the drug was active.

H. Mazindol

Mazindol is another compound totally unrelated structurally to amphetamine. Hadler (1972) and Elmaleh and Miller (1974) found it to be more effective than placebo in conjunction with a 1000-kcal diet. Mazindol is also more effective than amphetamine and little tolerance develops to its anorexic effect (DeFelice et al., 1973). Goldrick et al. (1974) found no difference between fenfluramine and mazindol in conjunction with a 1000-kcal diet. Sharma et al. (1973) demonstrated a significant effect over placebo combined with a 600-kcal diet in children. Thus, it would appear that mazindol, in contrast to amphetamine, is effective in combination with caloric restriction. In conjunction with a 1200-kcal diet, however, Thorpe et al. (1975) demonstrated a significant effect over placebo only during the first 2 weeks of an 8-week study.

The glucose intolerance and hyperinsulinemia produced by this drug leaves room for concern regarding toxicity (Sirtori *et al.*, 1971).

VIII. Conclusions and Recommendations

Obese individuals accept reducing regimens because they believe the weight loss will decrease morbidity and mortality as well as for psychic and cosmetic reasons. Whatever the reasons, the long-term results of attempts to reduce the weight of obese individuals have been generally disappointing (Young *et al.*, 1955; Stunkard and McLaren-Hume, 1959; Glennon, 1966; Sohar and Sneh, 1973). Even so, attempts should be made to reduce adipose tissue stores, especially in the markedly obese. The ideal approach is both a reduction in caloric intake and an increase in energy expenditure by increasing physical activity. Diets containing fewer than 1200 kcal are not well tolerated. Reeducation is essential in developing new, life-long eating habits. To be effective the diet must, in addition to reducing caloric intake, be palatable, provide the essential nutrients, and satisfy hunger, and the foodstuffs should be readily accessible and economically feasible.

The education process should include methods for determining the size of portions of food. Weighing each portion is possible, but unnecessary. Plastic models of different size portions or other types of visual aids are more meaningful than indicating an "average" or "small" serving of this or that. Aside from caloric content, the reducing diet need not vary from the composition of any well-balanced diet although some consideration may be given to a reduction in the percentage of carbohydrates, especially simple sugars.

The *sine qua non* to satisfactory, long-term weight reducing regimens is the determination and will power of the patient. Drugs may be used as temporary crutches to tide the patient over the early phases of caloric restriction, but at present they have no role in long-term therapy. Fenfluramine and mazindol appear reasonably effective for this purpose. Drug therapy must be prescribed as an adjunct to, not instead of, a reduction in caloric intake. Group programs such as Weight Watchers or TOPS (Take Off Pounds Sensibly) and the newer behavior modification techniques appear to offer the best chance for success.

As a last resort in the markedly obese patient, surgical approaches such as small bowel and gastric bypass have been tried with varying degrees of success. These procedures have distinct drawbacks in addition to the operative morbidity and mortality (Editorial, 1967; Drenick *et al.*, 1970).

REFERENCES

Abdallah, A., and White, H. D. (1970). Arch. Int. Pharmacodyn. Ther. 188, 271.
Aikman, P. M., Lamell, M., Lawrence, N., Levine, D. A., Levy, E., Lister, H. K. N., Ludwig, Z. J. (1971). Practitioner 207, 101.
Allen, T. H., Peng, M. T., Chen, K. P., Huang, T. F., Chang, C., and Fang, H. S. (1956). Metabolism 5, 346.
Alvi, M. Y. (1969). Brit. Med. J. 4, 237.
Ananth, J. V. (1971). Can. Med. Ass. J. 105, 1280.
Anderson, J. (1974). Practitioner 212, 536.
Asher, W. L., and Harper, H. W. (1973). Amer. J. Clin. Nutr. 26, 211.
Bagdade, J. D., Bierman, E. L., and Porte, D., Jr. (1967). J. Clin. Invest. 46, 1549.
Ball, M. F., Canary, J. J., and Kyle, L. H. (1970). Arch. Intern. Med. 125, 62.
Beckett, A. H., and Brookes, L. G. (1971). J. Pharm. Pharmacol. 23, 288.
Bell, J. D., Margen, S., and Calloway, D. H. (1969). Metabolism 18, 193.
Benoit, F. L., Martin, R. L., and Watten, R. H. (1965). Ann. Intern. Med. 63, 604.
Berger, H. J., Brown, C. C., and Krantz, J. C., Jr. (1973). J. Pharm. Sci. 62, 788.
Berry, M. J., Poyser, R. H., and Robertson, M. I. (1971). J. Pharm. Pharmacol. 23, 140.
Bernstein, R. A., Werner, L. H., and Rimm, A. A. (1973). Health Serv. Rep. 88, 925.
Blundell, J. E., and Leshem, M. B. (1973). Brit. J. Pharmacol. 47, 183.
Blundell, J. E., Latham, C. J., and Leshem, M. B. (1973). J. Pharm. Pharmacol. 25, 492.
Bortz, W. M., Wroldsen, A., Issekutz, B., and Rodahl, K. (1966). New Engl. J. Med. 274, 376.
Bradford, R. B., and Jourdan, M. H. (1973). Lancet ii, 640.
Brandon, S. (1969). Brit. Med. J. 4, 557.
Bray, G. A. (1970). Ann. Intern. Med. 73, 565.
Bray, G. A. (1972). J. Clin. Invest. 51, 537.
Bray, G. A. (1973). Bariatric Med. 2, 146.
Bray, G. A., and York, D. A. (1971). Physiol. Rev. 51, 598.
Bray, G. A., Melain, K. E., and Chopra, I. J. (1973). Amer. J. Clin. Nutr. 26, 715.
Carlson, L. A., and Bottiger, L. E. (1972). Lancet i, 865.
Cattabeni, F., Revuelta, A., and Costa, E. (1972). Neuropharmacology 11, 753.
Chhina, G. S., Kang, H. K., Singh, B., and Anand, B. K. (1971). Physiol. Behav. 7, 433.
Council on Foods and Nutrition (1973). J. Amer. Med. Ass. 224, 1415.
Court, J. M. (1972). S. Afr. Med. J. 46, 132.
Craig, L. S., Ray, R. E., Waxler, S. H., and Madigan, H. (1963). Amer. J. Clin. Nutr. 12, 230.
Crook, G. H., Bennett, C. A., Norwood, W. D., and Mahaffey, J. A. (1966). J. Amer. Med. Ass. 198, 39.
Danowski, T. S., Tsai, C. T., Morgan, C. R., Sieracki, J. C., Alley, R. A., Robbins, T. J., Sabeh, G., and Sunder, J. H. (1969), Metabolism 18, 811.
Datey, K. K., Kelkar, P. N., and Pandya, R. S. (1973). Brit. J. Clin. Pract. 27, 373.
DeFelice, E. A., Chaykin, L. B., and Cohen, A. (1973). Curr. Ther. Res. 15, 358.
Drenick, E. J., Simmons, F., and Murphy, J. F. (1970). New Engl. J. Med. 282, 829.
Duell, H. J., Jr., and Gulick, M. (1932). J. Biol. Chem. 96, 25.

Duncan, G. G., Jenson, W. K., Cristofori, E. C., and Shcless, G. L. (1963). *Amer. J. Med. Sci.* **245**, 515.

Dunn, J. P., Ipsen, J., Elsom, K. O., and Ahtani, M. (1970). *Amer. J. Med. Sci.* **259**, 309.

Durnin, J. U. G. A., and Womersley, J. (1973). *Brit. J. Pharmacol.* **49**, 115.

Dykes, J. R. W. (1973). *Postgrad. Med. J.* **49**, 314.

Editorial (1970). *Med. J. Aust.* **2**, 1052.

Editorial (1967). *J. Amer. Med. Ass.* **200**, 638.

Edison, G. R. (1971). *Ann. Intern. Med.* **74**, 605.

Elliott, B. W. (1970). *Curr. Ther. Res.* **12**, 502.

Ellis, G. (1960). *Brit. Med. J.* **4**, 558.

Elmaleh, M. K., and Miller, J. (1974). *Pa. Med.* **77**, 46.

Fabry, P., Fodor, J., Hejl, Z., Braun, T., and Zvolankova, K. (1964). *Lancet* **ii**, 614.

Fabry, P., Hejda, S., Cerny, K., Asancova, K., and Perkar, J. (1966). *Amer. J. Clin. Nutr.* **18**, 358.

Felig, P., Marliss, E., and Cahill, G. F., Jr. (1969). *New Engl. J. Med.* **281**, 811.

Fineberg, S. K. (1972). *J. Amer. Geriat. Soc.* **20**, 576.

Fink, M., and Shapiro, D. M. (1969). *Electroencephalogr. Clin. Neurophysiol.* **27**, 710.

Fink, M., Itil, T. M., and Shapiro, D. M. (1970). *Electroencephalogr. Clin. Neurophysiol.* **28**, 102.

Finkelstein, B., and Fryer, B. A. (1971). *Amer. J. Clin. Nutr.* **24**, 465.

Frank, B. W. (1964). *Amer. J. Clin. Nutr.* **14**, 133.

Funderburk, W. H., and Ward, J. W. (1970). *J. Pharm. Pharmacol.* **22**, 786.

Funderburk, W. H., Hazelwood, J. C., Ruckart, R. T., and Ward, J. W. (1971a). *J. Pharm. Pharmacol.* **23**, 468.

Funderburk, W. H., Hash, A. M., Jr., Hazelwood, J. C., and Ward, J. W. (1971b). *J. Pharm. Pharmacol.* **23**, 509.

Garrow, J. S., Belton, E. A., and Daniels, A. (1972). *Lancet* **ii**, 559.

Gastineau, C. F. (1972). *Med. Clin. N. Amer.* **56**, 1021.

Gershberg, H. (1972). *Postgrad. Med.* **51**, 135.

Glennon, J. A. (1966). *Arch. Intern. Med.* **118**, 1.

Golding, D. (1970). *Brit. Med. J.* **238**, 238.

Goldrick, R. B., Havenstein, N., and Whyte, H. M. (1973). *Aust. New Zealand J. Med.* **3**, 131.

Goldrick, R. B., Nestel, P. J., and Havenstein, N. (1974). *Med. J. Aust.* **1**, 882.

Grande, F. (1968). *Ann. Intern. Med.* **368**, 467.

Grey, N., and Kipnis, D. (1971). *New Engl. J. Med.* **285**, 827.

Gropetti, A., Misher, A., Naimzada, M., Reveulta, A., and Costa, E. (1972). *J. Pharmacol. Exp. Ther.* **182**, 464.

Gubner, R. S. (1957). *Nutr. Rev.* **15**, 353.

Gurtner, H. P., Gertsch, M., Selzmann, C., Scherrer, M., Stucki, P., and Wyss, F. (1968). *Schweiz. Med. Wochenschr.* **98**, 1579.

Gusman, H. A. (1969). *Amer. J. Clin. Nutr.* **22**, 686.

Hadler, A. J. (1969). *Curr. Ther. Res.* **11**, 750.

Hadler, A. J. (1970). *Curr. Ther. Res.* **12**, 639.

Hadler, A. J. (1971). *J. Clin. Pharmacol.* **11**, 52.

Hadler, A. J. (1972). *J. Clin. Pharmacol.* **12**, 453.

Harding, T. (1971). *Brit. Med. J.* **3**, 305.

Harding, T. (1972). *Brit. J. Psych.* **121**, 338.

Hart, A., and Cohen, H. (1970). *Brit. Med. J.* 1, 22.
Hejda, S., and Fabry, P. (1964). *Nutr. Dieta* 6, 216.
Hirsch, J. (1972). *Advan. Psychosomatic Med.* 7, 229.
Hirsch, J., and Knittle, J. L. (1970). *Fed. Proc. Fed. Amer. Soc. Exp. Biol.* 29, 1516.
Hirsch, J., and Van Italie, T. B. (1973). *Amer. J. Clin. Nutr.* 26, 1039.
Hoebel, B. G. (1971). *Annu. Rev. Physiol.* 33, 533.
Hollingsworth, D. R., Amatruda, T. T., Jr., and Scheig, R. (1970). *Metabolism* 19, 934.
Holtzman, S. G., and Jewett, R. E. (1971). *Psychopharmacology* 22, 151.
Hood, C. E. A., Goodhart, J. M., Fletcher, F., Gloster, J., Bertrand, P. V., and Crooke, A. C. (1970). *Brit. J. Nutr.* 24, 39.
Hooper, A. C. (1971). *Brit. Med. J.* 3, 305.
Hoyer, I., and Van Zwieten, P. A. (1972). *J. Pharm. Pharmacol.* 24, 452.
Imlah, N. W. (1970). *Brit. Med. J.* 2, 178.
Issekutz, B., Walter, J. N., Bortz, M., Miller, H. I., and Paul, P. (1967). *Metabolism* 16, 1001.
Jespersen, J., and Bonaccursi, A. (1969). *J. Pharm. Pharmacol.* 21, 776.
Jun, H. W., and Triggs, E. J. (1970). *J. Pharm. Sci.* 59, 306.
Kalkhoff, R. K., Kim, H. J., Cerletty, J., and Ferrou, C. A. (1971). *Diabetes* 20, 83.
Kannel, W. B., LeBauer, J. E., Dawbert, T. R., and McNamara, P. M. (1967). *Circulation* 35, 734.
Kaplan, N. M., and Jose, A. (1970). *Amer. J. Med. Sci.* 260, 105.
Kasper, H. (1970). *Digestion* 3, 321.
Kasper, H., Thiel, H., and Ehl, M. (1973). *Amer. J. Clin. Nutr.* 26, 197.
Kaufman, N. A., and Blondheim, S. H. (1973). *Lancet* i, 104.
Kekwick, A., and Pawan, G. L. S. (1956). *Lancet* ii, 155.
Kew, M. C. (1970). *S. Afr. Med. J.* 44, 421.
Keys, A., Aravanis, C., Blackburn, H., Van Buchem, F. S. P., Buzina, R., Bjordjevic, B. S., Fidanza, F. (1972). *Ann. Intern. Med.* 77, 15.
Kinsell, L. W., Gunning, B., Michaels, G. D., Richardson, J., Cox, S. E., and Lemon, C. (1964). *Metabolism* 13, 195.
Kraisberg, R. A., Boshell, B. R., Di Placido, J., and Roddam, R. F. (1967). *New Engl. J. Med.* 276, 314.
Lamki, K., Ezrin, C., Koven, I., and Steiner, G. (1973). *Metabolism* 22, 617.
Lawson, A. A., Strong, J. A., Peattie, P., Roscoe, P., and Gibson, A. (1970). *Lancet* ii, 437.
Leibowitz, S. F. (1970). *Proc. Nat. Acad. Sci. U.S.* 67, 1063.
Lepkovsky, S. (1973). *Amer. J. Clin. Nutr.* 26, 271.
Lewis, S. A., Oswald, I., and Dunleavy, D. L. (1971). *Brit. Med. J.* 3, 67.
Lullmann, H., Lullmann-Rauch, R., and Reil, G. H. (1973a). *Klin. Wochenschr.* 51, 284.
Lullmann, H., Rossen, E., and Seiler, K. U. (1973b). *J. Pharm. Pharmacol.* 25, 239.
Lullmann-Rauch, R. (1973). *Virchows Arch.* 12, 295.
McClelland, W. S., and DuBois, E. F. (1930). *J. Biol. Chem.* 87, 651.
Mace, P. M., Malcolm, A. D., Outar, K. P., and Pawan, G. L. (1972). *Proc. Nutr. Soc.* 21, 14A.
Malcolm, A. D., Mace, R. M., Outar, K. U., and Pawan, G. L. S. (1972). *Proc. Nutr. Soc.* 31, 12A.
Miettinen, T. A. (1971). *Circulation* 44, 842.
Montegriffo, V. M. E. (1968). *Ann. Hum. Genet.* 31, 389.

Mroczek, W. J., Lee, W. R., and Finnerty, F. A., Jr. (1974). Curr. Ther. Res. 16, 1197.

Nelson, R. A., Anderson, L. F., Gastineau, C. F., Hoyles, A. B., and Stannes, C. L. (1973). J. Amer. Med. Ass. 223, 627.

Noble, R. E. (1972). Curr. Ther. Res. 14, 162.

Nussey, A. M. (1973). Brit. Med. J. 1, 177.

Olefsky, J., Reaven, C. M., and Farquhar, J. W. (1974). J. Clin. Invest. 53, 1974.

Olesen, E. S., and Quaade, F. (1960). Lancet i, 1048.

Oswald, I. (1969). Exp. Med. Surg. 27, 68.

Oswald, I., Lewis, S. A., Dunleavy, D. L., Brezinova, V., and Briggs, M. (1971). Brit. Med. J. 3, 70.

Panksepp, K., and Booth, D. A. (1973). Pharmacologia 29, 45.

Pardridge, W. M., and Connor, J. D. (1973). Experientia 29, 302.

Pawan, G. L. S. (1969). Lancet i, 498.

Pawan, G. L. S., Payne, P. M., and Sheldrick, E. C. (1971a). Brit. J. Pharmacol. 41, 416P.

Pawan, G. L. S., Payne, P. M., and Sheldrick, E. C. (1971b). Proc. Nutr. Soc. 20, 8A.

Penick, S. G., and Stunkard, A. J. (1972). Advan. Psychosomatic Med. 7, 217.

Pennington, A. W. (1951). Del. State Med. J. 23, 79.

Pennington, A. W. (1955). Amer. J. Dig. Dis. 22, 33.

Pilkington, T. R. E., Gainsborough, H., Rosenoer, V. M., and Carey, M. (1960). Lancet i, 856.

Rabinowitz, D. (1970). Annu. Rev. Med. 21, 241.

Rabinowitz, D., and Zierler, K. (1962). J. Clin. Invest. 41, 2173.

Randle, P. J., Garland, P. B., Holes, C. N., and Newsholme, E. A. (1963). Lancet i, 785.

Rivlin, R. S. (1975). New Engl. J. Med. 292, 26.

Robertson, R. P., Gavareski, D. J., Henderson, J. D., Porte, D., Jr., and Bierman, E. L. (1973). J. Clin. Invest. 52, 1620.

Rogers, O. H. (1901). Proc. Ass. Life Insurance Med. Directors Amer. 11, 280.

Samanin, R., Ghezzi, D., Valzelli, L., and Garattini, S. (1972). Eur. J. Pharmacol. 19, 318.

Sandoval, R. G., Wang, R. I., and Rimm, A. A. (1971). J. Clin. Pharmacol. 11, 120.

Sapeika, N. (1972). Practitioner 208, 660.

Sapeika, N. (1973). Central Afr. J. Med. 19, 23.

Sapeika, N. (1974). S. Afr. Med. J. 48, 2027.

Sassoon, H. F. (1973). Amer. J. Clin. Nutr. 26, 776.

Schachter, S. (1968). Science 161, 751.

Schatz, H., Doci, S., and Hofer, R. (1972). Diabetologia 8, 1.

Seedat, Y. K., and Reddy, J. (1974). Curr. Ther. Res. 16, 398.

Sharma, R. K., Collipp, P. J., Rezvani, I., Strimas, J., Maddaiah, V. T., and Rezvani, E. (1973). Clin. Pediatr. 12, 145.

Silverman, M., and Okun, R. (1971). Curr. Ther. Res. 13, 648.

Silverstone, T. (1972). Psychopharmacologia 25, 315.

Simeons, A. T. W. (cited by Grusman, H. A.) (1969). Amer. J. Clin. Nutr. 22, 686.

Simon, E., and Kraicer, P. F. (1966). Isr. J. Med. Sci. 2, 785.

Sirtori, C., Hurwitz, A., and Azarnoff, D. L. (1971). Amer. J. Med. Sci. 261, 341.

Soc. Actuaries (1959). Build Blood Pressure Study, Chicago, The Soc. 1, 268.

Sohar, E., and Sneh, E. (1973). Amer. J. Clin. Nutr. 26, 845.

Southgate, P. J., Mayer, S. R., Boxall, E., and Wilson, A. B. (1971). *J. Pharm. Pharmacol.* **23**, 600.

Steel, J. M., and Briggs, M. (1972). *Brit. Med. J.* **3**, 26.

Stern, J. S., Batchelor, B. R., Hollander, N., Cohn, C. K., and Hirsch, J. (1972). *Lancet* **ii**, 948.

Stock, A. L., and Yudkin, J. (1970). *Amer. J. Clin. Nutr.* **23**, 948.

Stunkard, A., and McLaren-Hume, M. (1959). *Arch. Intern. Med.* **103**, 79.

Stunkard, A., Rickels, K., and Hesbacher, P. (1973). *Lancet* **i**, 503.

Tang, A. H., and Kirch, J. D. (1971). *Psychopharmacologia* **21**, 139.

Thorpe, P. C., Isaac, P. F., and Rodgers, J. (1975). *Curr. Ther. Res.* **17**, 149.

Tinklenberg, J. R. (1971). *Amer. Fam. Physician* **4**, 82.

Tognoni, G., Morselli, P. L., and Garattini, S. (1972). *Eur. J. Pharmacol.* **20**, 125.

Tolstoi, E. (1929). *J. Biol. Chem.* **83**, 753.

Tormey, J., and Lasagna, L. (1960). *J. Pharmacol. Exp. Ther.* **128**, 201.

Truant, A. P., Olon, L. P., and Cobb, S. (1972). *Curr. Ther. Res.* **14**, 726.

Turtle, J. R., and Burgess, J. A. (1973). *Diabetes* **22**, 858.

United States Public Health Service (1966). *Pub. Health Serv. Publ. No. 1485.* Washington, D.C.

Vallance-Owen, J. (1964). *Diabetes* **13**, 241.

Vermeulen, A., and Rottiers, R. (1972). *Diabetologia* **8**, 8.

Waal-Manning, H. J., and Simpson, F. D. (1969). *Lancet* **ii**, 1392.

Wall, J. R., Pyke, D. A., and Oakley, W. G. (1973). *Brit. Med. J.* **1**, 577.

Werner, S. C. (1955). *New Engl. J. Med.* **252**, 661.

Wirz, P., and Arbenz, U. (1970). *Schweiz. Med. Wochenschr.* **100**, 2147.

Worthington, B. S., and Taylor, L. E. (1974). *J. Amer. Diet. Ass.* **64**, 47.

Yelnosky, J., and Lawlor, R. B. (1970). *Arch. Int. Pharmacodyn. Ther.* **184**, 374.

Young, C. M., and Scanlan, S. S. (1971). *Amer. J. Clin. Nutr.* **24**, 290.

Young, C. M., Moore, N. S., Berresford, K., Einset, B. M., and Waldner, B. C. (1955). *J. Amer. Dietetic Ass.* **31**, 1111.

Yudkin, J., and Carey, M. (1960). *Lancet* **ii**, 939.

Chapter 6

ETHANOL AND LIPID METABOLISM

Laboratory of the Section of Liver Disease and Nutrition
Bronx Veterans Administration Hospital
Bronx, New York, and Department of Medicine, Mount Sinai
School of Medicine of the City University of New York,
New York, New York

I. Introduction

The most common disturbance in lipid metabolism produced by alcohol abuse is that of excessive accumulation of lipids in the liver resulting in a fatty liver, often associated with hyperlipemia. Attempts at elucidating the pathogenesis of the alcoholic fatty liver and hyperlipemia over the last decade have revealed a variety of effects of ethanol on lipid metabolism not only in the liver but in other organs as well, including the adipose tissue.

The immediate effects of the oxidation of ethanol on lipid and intermediary metabolism will be reviewed first, to be followed by an analysis of some persistent and, in part, adaptive changes following chronic ethanol consumption. Fatty liver and hyperlipemia associated with alcoholism will then be discussed.

II. Oxidation of Ethanol and Its Direct Effects on Hepatic and Extrahepatic Lipid and Intermediary Metabolism

A. Pathways of Ethanol Oxidation

Ethanol can be synthesized endogenously in trace amounts, but it is primarily an exogenous compound that is readily absorbed from the gastrointestinal tract. Only 2 to 10% of that absorbed is eliminated through the kidneys and lungs; the rest must be oxidized in the body, principally in the liver. This relative organ specificity probably explains why, despite the existence of intracellular mechanisms responsible for redox homeostasis, ethanol oxidation produces striking metabolic imbalances in the liver. These effects are aggravated by the lack of feedback mechanism to adjust the rate of ethanol oxidation to the metabolic state of the hepatocyte, and the inability of ethanol, unlike other major sources of calories, to be stored or metabolized to a significant degree in peripheral tissues. The main hepatic pathway for ethanol disposition involves alcohol dehydrogenase (ADH), an enzyme of the cell sap (cytosol) that catalyzes the conversion of ethanol to acetaldehyde. Hydrogen is transferred from ethanol to the cofactor nicotinamide adenine dinucleotide (NAD), which is converted to its reduced form (NADH) (Fig. 1A). The acetaldehyde produced again loses hydrogen and is converted to acetate, most of which is released into the bloodstream. As a net result, ethanol oxidation generates an excess of reducing equivalents in the liver, primarily as NADH. In addition, ethanol can also be metabolized by an accessory pathway which requires NADPH as a cofactor and is localized in the endoplasmic reticulum, which is gathered in the microsomal fraction upon subcellular fractionation and ultracentrifugation (Fig. 1B). There is also a debate on a possible role of catalase (Fig. 1C and D). The various metabolic effects of ethanol can be attributed either to the NADH generation by the ADH pathway or to the interaction with other microsomal functions in association with metabolism

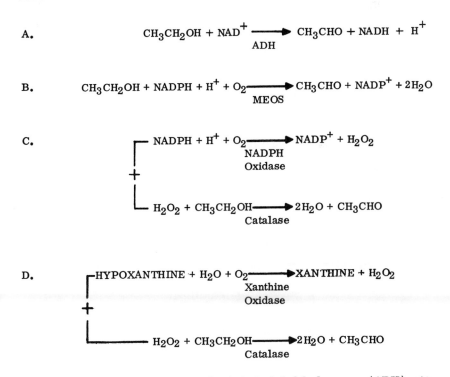

A. $$CH_3CH_2OH + NAD^+ \xrightarrow{\text{ADH}} CH_3CHO + NADH + H^+$$

B. $$CH_3CH_2OH + NADPH + H^+ + O_2 \xrightarrow{\text{MEOS}} CH_3CHO + NADP^+ + 2H_2O$$

C. $$NADPH + H^+ + O_2 \xrightarrow{\substack{\text{NADPH} \\ \text{Oxidase}}} NADP^+ + H_2O_2$$

$$+$$

$$H_2O_2 + CH_3CH_2OH \xrightarrow{\text{Catalase}} 2H_2O + CH_3CHO$$

D. $$HYPOXANTHINE + H_2O + O_2 \xrightarrow{\substack{\text{Xanthine} \\ \text{Oxidase}}} XANTHINE + H_2O_2$$

$$+$$

$$H_2O_2 + CH_3CH_2OH \xrightarrow{\text{Catalase}} 2H_2O + CH_3CHO$$

FIG. 1. Hepatic ethanol oxidation by (A) alcohol dehydrogenase (ADH); nicotinamide adenine dinucleotide (NAD); nicotinamide adenine dinucleotide, reduced form (NADH). (B) The microsomal ethanol-oxidizing system (MEOS); nicotinamide adenine dinucleotide phosphate, reduced from (NADPH); nicotinamide adenine dinucleotide phosphate (NADP). (C) A combination of NADPH oxidase and catalase; or (D) xanthine oxidase and catalase.

by the microsomal ethanol-oxidizing system (MEOS). Some of the effects are also due to the metabolites of ethanol, acetaldehyde, and acetate.

B. Effects of Excessive Hepatic NADH Generation

As shown in Fig. 2, the oxidation of ethanol results in the transfer of hydrogen to NAD. The resulting enhanced NADH/NAD ratio, in turn, produces a change in the ratio of those metabolites that are dependent for reduction on the NADH/NAD couple. It was therefore proposed that the altered NADH/NAD ratio is responsible for a number of metabolic abnormalities associated with alcohol abuse (Lieber and Davidson, 1962).

1. HYPERLACTACIDEMIA, HYPERURICEMIA, AND ACIDOSIS

The enhanced NADH/NAD ratio reflects itself in an increased lactate/pyruvate ratio that results in hyperlactacidemia (Lieber *et al.*, 1962a,b), because of both decreased utilization and enhanced production of lactate by the liver. The hyperlactacidemia contributes to acidosis and also re-

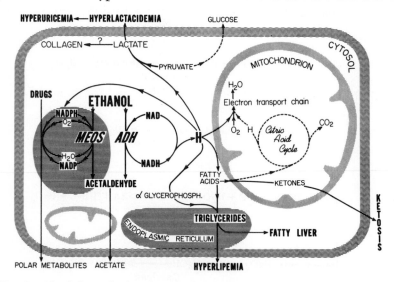

FIG. 2. Metabolism of ethanol in the hepatocyte and schematic representation of its link to fatty liver, hyperlipemia, hyperuricemia, hyperlactacidemia, ketosis, and hypoglycemia. Pathways which are decreased by ethanol are represented by dashed lines. ADH, alcohol dehydrogenase: MEOS, microsomal ethanol-oxidizing system; NAD, nicotinamide adenine dinucleotide; NADH, nicotinamide adenine dinucleotide, reduced form; NADP, nicotinamide adenine dinucleatide phosphate; NADPH, nicotinamide adenine dinucleotide phosphate, reduced form.

duces the capacity of the kidney to excrete uric acid, leading to secondary hyperuricemia (Lieber *et al.*, 1962a). Alcohol-induced ketosis (vide infra) may also promote the hyperuricemia. The latter may be related to the common clinical observation that excessive consumption of alcoholic beverages frequently aggravates or precipitates gouty attacks (Newcombe, 1972). Alcoholic hyperuricemia can be readily distinguished from the primary variety by its reversibility upon discontinuation of ethanol abuse (Lieber *et al.*, 1962a).

2. ENHANCED LIPOGENESIS AND DEPRESSED LIPID OXIDATION

The increased NADH/NAD ratio also raises the concentration of α-glycerophosphate (Nikkila and Ojala, 1963) that favors hepatic triglyceride accumulation by trapping fatty acids. In addition, excess NADH promotes lipogenesis (Lieber and Schmid, 1961; Gordon, 1972a) possibly by the elongation pathway or transhydrogenation to nicotinamide adenine dinucleotide phosphate (NADP) (Lieber, 1968), although *in vivo* acute administration of a high dose of ethanol did not enhance fatty acid synthesis (Guynn *et al.*, 1973). Theoretically, enhanced lipogenesis can be considered a means for disposing of the excess hydrogen. Some hydrogen equivalents can be transferred into the mitochondria by various "shuttle" mechanisms. Since the activity of the citric acid cycle is depressed (Forsander *et al.*, 1965; Lieber *et al.*, 1967), partly because of a slowing of the reactions of the cycle that require NAD, the mitochondria will use the hydrogen equivalents originating from ethanol, rather than from the oxidation through the citric acid cycle of two carbon fragments derived from fatty acids. Thus, fatty acids that normally serve as the main energy source of the liver are supplanted by ethanol. Decreased fatty acid oxidation by ethanol has been demonstrated in liver slices (Lieber and Schmid, 1961; Blomstrand *et al.*, 1973), perfused liver (Lieber *et al.*, 1967) (Fig. 3), isolated hepatocytes (Ontko, 1973), and *in vivo* (Blomstrand and Kager, 1973). This reduction results in the deposition in the liver of dietary fat, when available, or fatty acids derived from endogenous synthesis in the absence of dietary fat (Lieber *et al.*, 1966, 1969; Lieber and Spritz, 1966; Mendenhall, 1972).

As emphasized previously (Lieber and Davidson, 1962; Lieber, 1968, 1969), a number of metabolic effects of alcohol can be attributed to the generation of NADH; these include interference with glucose, galactose, serotonin, and norepinephrine metabolism, and the alteration of hepatic steroid metabolism in favor of the reduced compounds (Admirand *et al.*, 1970).

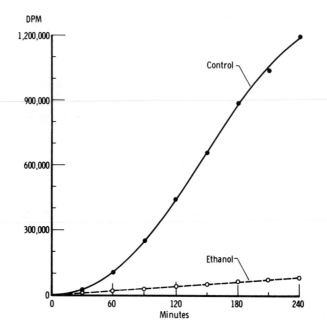

FIG. 3.   Effect of alcohol on total $^{14}CO_2$ production from ^{14}C-labeled chylomicrons in a pair of isolated perfused rat livers (Lieber *et al.*, 1967).

C. The Microsomal Ethanol-Oxidizing
System (MEOS) and Associated Interaction
of Ethanol with Drug Metabolism

1. NATURE OF THE NON-ADH PATHWAYS

The first indication of an interaction of ethanol with the microsomal fraction of the hepatocyte was provided by the morphological observation that in rats, ethanol feeding results in a proliferation of the smooth endoplasmic reticulum (SER), which will be discussed subsequently. This increase in SER resembles that seen after the administration of a wide variety of xenobiotic compounds including known hepatotoxins (Meldolesi, 1967), numerous therapeutic agents (Conney, 1967), and food additives (Lane and Lieber, 1967). Most of these substances which induce a proliferation of the SER are metabolized, at least in part, in the microsomal fraction of the hepatocyte which comprises the SER. The observation that ethanol also produces proliferation of the SER raised the possibility that, in addition to its oxidation by ADH in the cytosol, ethanol may be metabolized by the microsomes. A microsomal system capable of methanol oxidation was described (Orme-Johnson and Ziegler, 1965) but its capacity for ethanol oxidation was extremely low. Subsequently, a microsomal ethanol-oxidizing system with a rate of ethanol oxidation 10 times higher than reported by Orme-Johnson and Ziegler was described (Lieber and DeCarli, 1968, 1970a). Differentiation from alcohol dehydrogenase was achieved by subcellular localization, pH optimum *in vitro* cofactor requirements (Fig. 1A, B), and effects of inhibitors such as pyrazole (Lieber *et al.*, 1970; Lieber and DeCarli, 1973).

In 1945, Keilin and Hartree showed that in the presence of a H_2O_2-generating system, catalase is capable of oxidizing ethanol. Some H_2O_2-generating systems, such as hypoxanthine-xanthine oxidase (Fig. 1D), are present in the cytosol. In addition, hepatic microsomes also contain a H_2O_2-generating system, namely NADPH oxidase (Fig. 1C). It was, indeed, found that the addition of catalase in this system allows it to oxidize methanol (Gillette *et al.*, 1957). Catalase resides primarily in the microbodies which, upon ultracentrifugation, are separated with the mitochondrial fraction. Other organelles, however, including the microsomes, contain traces of catalase activity. Therefore, the question arose as to what extent a combination of NADPH oxidase–catalase (Fig. 1C) could account for MEOS activity (Fig. 1B). It has been reported that microsomes of acatalasemic mice fail to oxidize ethanol (Vatsis and Schulman, 1973), but this claim has now been retracted (Vatsis and Schulman, 1974); microsomes of these mice do indeed actively oxidize ethanol in the presence of NADPH (Lieber and DeCarli, 1974a; Teschke *et al.*, 1975). Differentiation of MEOS from catalase was achieved utilizing various inhibitors (Lieber and DeCarli, 1968, 1970a, 1973; Lieber *et al.*, 1970) but these studies became the subject of controversy (Thurman *et al.*, 1972). Actually, inhibitor studies are no longer relevant now that MEOS has been clearly differentiated and separated from ADH (Teschke *et al.*, 1972, 1974b; Mezey *et al.*, 1973) and from catalase (Teschke *et al.*, 1972, 1974b) by column chromatography and substrate specificity (Teschke *et al.*, 1974a).

The absence of catalase in the purified MEOS preparation was verified by the lack of catalatic activity (O_2 appearance) in the presence of a H_2O_2-generating system or the failure of H_2O_2 to disappear (Teschke *et al.*, 1972, 1974b). To rule out the possibility that the peroxidatic activity of catalase may still be present though catalatic activity may not be detectable, the effect of various H_2O_2-generating systems on the capacity of the purified fraction to oxidize ethanol was studied. The purified MEOS which actively oxidized ethanol in the presence of a NADPH-generating system had no activity with the H_2O_2-generating systems unless exogenous catalase was added (Teschke *et al.*, 1974b).

2. Respective Roles of the Various Ethanol
Oxidizing Systems

Even in the presence of pyrazole (a potent ADH inhibitor) ethanol metabolism was found to persist, both *in vivo* (Lieber and DeCarli, 1972) and *in vitro* in isolated perfused liver (Papenberg *et al.*, 1970), liver slices (Lieber and DeCarli, 1970a) and isolated liver cells (Thieden, 1971; Grunnet *et al.*, 1973). Furthermore, in the presence of pyrazole, glucose labeling from [1R-^{3}H]ethanol was nearly abolished, while ^{3}H$_2$O production was inhibted less than 50%. In view of the stereospecificity of ADH for [1R-^{3}H]ethanol, these findings again suggest "the presence of a significant pathway not mediated by cytosolic ADH" (Rognstad and Clark, 1974). The rate of this non-ADH medicated oxidation varied depending on the concentrations of ethanol used, from 20 to 25% (Lieber and DeCarli, 1970a, 1972; Papenberg *et al.*, 1970) to half or more (Thieden, 1971; Grunnet *et al.*, 1973) of the total ethanol metabolism. Additional evidence that this pyrazole-insensitive residual ethanol metabolism is not ADH mediated was derived from the fact that the cytosolic redox state was unaffected (Grunnet and Thieden, 1972). The striking increase in the non-ADH fraction of ethanol metabolism with increasing ethanol concentrations (Thieden, 1971; Grunnet *et al.*, 1973) is consistent with the known K_m for ADH and MEOS; whereas the former has a K_m varying from 0.5 to 2 mM (Reynier, 1969; Makar and Mannering, 1970), the latter has a value of 8 to 9 mM (Lieber and DeCarli, 1970a). It has been estimated in liver microsomes that the H$_2$O$_2$-dependent catalase-mediated ethanol oxidation has a K_m of the same order of magnitude as that of MEOS (Thurman *et al.*, 1972) whereas in perfused livers others found a low K_m of 0.5 to 0.6 mM for the dissociation of the catalase–H$_2$O$_2$ compound by ethanol (Theorell *et al.*, 1972). An indirect answer to the question of the respective roles of these two systems for the non-ADH-mediated ethanol oxidation can be derived from the generally accepted view that rates of H$_2$O$_2$-mediated peroxidation are limited by the amount of H$_2$O$_2$ generated rather than the amount of catalase itself. In perfused livers, the total capacity of the organ to generate H$_2$O$_2$ is very low and has been estimated at 40 to 70 nmoles/min/gm of liver (Oshino *et al.*, 1973) in the presence of physiological substrates. This is only 1/10 or even less than the estimated rates of non-ADH-mediated ethanol oxidation (Lieber and DeCarli, 1972). Moreover, acute ethanol administration was found to inhibit the activity of NADPH oxidase (Hasumura *et al.*, 1975a). Even in the presence of unusual substrates for H$_2$O$_2$ generation, this pathway still would only account for a minor fraction of the non-ADH-mediated ethanol oxidation in control rats. This interpretation is consistent with the results of Thurman and Scholz (1973) who found that menadione, although it strikingly increased H_2O_2 generation and ethanol oxidation by microsomes *in vitro*, nevertheless failed to affect the alcohol dehydrogenase independent rates of ethanol utilization by perfused livers. By contrast, the MEOS could account for the bulk of the non-ADH-mediated ethanol oxidation (Lieber and DeCarli, 1972).

3. INTERACTION OF ETHANOL WITH DRUG METABOLISM

Interactions of the effects of ethanol and various drugs have been widely recognized (Forney and Hughes, 1968). Intoxicated individuals are more susceptible to several medications (Soehring and Schüppel, 1966). These various effects are usually attributed to additive or synergistic effects of alcohol and various drugs on the central nervous system. However, the existence of an at least partially common microsomal system for ethanol and drug metabolism sheds new light on their interaction. The increased susceptibility of the inebriated individual could be explained, at least in part, by the effect of ethanol or microsomal drug-detoxifying enzymes. It has, indeed, been found that ethanol inhibits the metabolism of a variety of drugs *in vitro* (Rubin and Lieber, 1968a; Rubin *et al.*, 1970a,b; Ariyoshi *et al.*, 1970; Schüppel, 1971). With some systems, such as aniline hydroxylase, this inhibition is of a competitive nature (Rubin *et al.*, 1970b; Cohen and Mannering, 1973). These *in vitro* effects may explain the observation that *in vivo*, simultaneous administration of ethanol and drugs slows the rate of drug metabolism (Rubin *et al.*, 1970b; Schüppel, 1971). Conversely, in the presence of drugs, there is inhibition of ethanol metabolism (Lieber and DeCarli, 1972). Drugs inhibit ethanol oxidation by microsomes *in vitro* in a way which has been considered as strong evidence for the catalase-independent nature of MEOS (Hildebrandt *et al.*, 1974). In addition, some drugs inhibit alcohol dehydrogenase (Sutherland *et al.*, 1960). The interaction of ethanol with drug metabolism may have some important practical consequences. Indeed, in the United States, more than 50% of all lethal road accidents are associated with an elevated blood alcohol level (Voas, 1973). One may wonder to what extent the loss of control on the road may be due not only to ethanol itself, but to an ethanol–drug interaction considering that a large segment of the population is given sedatives and tranquilizers.

D. Effects of the Metabolites of Ethanol:
   ## Acetaldehyde and Acetate

Acetaldehyde is the first major "specific" oxidation product of ethanol, whether the latter is oxidized by the classic alcohol dehydrogenase of the cytosol or by the more recently described microsomal system. Except after Antabuse administration, acetaldehyde concentrations after alcohol ingestion are low, but it has long been speculated that they may contribute to the complications of alcoholism (Truitt and Duritz, 1966). Moreover, it has been found recently that at high ethanol blood levels, blood acetaldehyde is higher in alcoholics than in nonalcoholics. This possibly reflects enhanced MEOS activity, since the blood acetaldehyde concentration drops precipitously at ethanol levels which correspond to MEOS desaturation (Korsten et al., 1975). Although the exact pathway of its metabolism is still unknown, it is generally accepted that aldehyde oxidation proceeds via aldehyde dehydrogenase, 80% of the activity of which is located in the mitochondria (Marjanen, 1972; Grunnet, 1973). Whether ethanol consumption alters the activity of acetaldehyde dehydrogenase is the subject of debate, with both increases (Dajani et al., 1963; Horton, 1971) and no changes (Raskin and Sokoloff, 1972; Redmond and Cohen, 1971) reported. Recently, it was found that the increase is limited to the activity of the unphysiologically high K_m enzyme; total mitochondrial acetaldehyde dehydrogenation was decreased after ethanol consumption (Hasumura et al., 1975b). Since metabolism of acetaldehyde via aldehyde dehydrogenase results in the generation of NADH, some of the acetaldehyde effects could be attributed to the NADH generation, as discussed before in the case of ethanol. Acetaldehyde, however, is a very reactive compound which may exert some toxic effects of its own, such as the reduction of the activity of the shuttles for the transport of reducing equivalents into the mitochondria (Cederbaum et al., 1973) and inhibition of site I of oxidative phosphorylation (Cederbaum et al., 1974). In addition, acetaldehyde could produce its effects through catecholamine release (Eade, 1959) whose excretion is reported to be increased after ethanol (Perman, 1958). Several ethanol effects upon the brain have been attributed to acetaldehyde (Davis and Walsh, 1970; Truitt and Walsh, 1973). As reviewed before (Lieber, 1967), ethanol exerts various cardiovascular effects, including increased splanchnic blood flow and cardiac output (Stein et al., 1963); a number of these could be secondary to the action of acetaldehyde. In addition, acetaldehyde was shown to interfere with myocardial protein synthesis (Schreiber et al., 1972), which may relate to the cardiotoxicity of ethanol.

The exact fate of acetaldehyde is still the subject of debate. That acetyl-CoA is formed from ethanol is indicated by the observation that [^{14}C]ethanol can be traced to a variety of metabolites of which acetyl-CoA is a precursor, such as fatty acids and cholesterol, as reviewed elsewhere (Lieber, 1967). It is noteworthy that a large fraction of the carbon skeleton of ethanol is incorporated in hepatic lipids after ethanol administration (Scheig, 1971; Brunengraber et al., 1973). The acetaldehyde which results from the oxidation of ethanol could be converted to acetyl-CoA via acetate. The reverse possibility, namely that ethanol is converted directly to acetyl-CoA which in turn could be either incorporated into various metabolites or yield acetate, has not been ruled out. In any event, acetate has been found to markedly increase in the blood after ethanol administration (Lundquist et al., 1962; Crouse et al., 1968).

Although in vitro, the liver can readily utilize acetate, in vivo most of the acetate is metabolized in peripheral tissues (Katz and Chaikoff, 1955). The effects of a rise of circulating acetate on intermediary metabolism in various tissues have not been defined, except for adipose tissue where it was found to be responsible, at least in part, for the decreased release of free fatty acids (FFA) and the fall of circulating FFA (Crouse et al., 1968), which will be discussed subsequently. A fall in FFA, a major fuel for peripheral tissues, may have significant metabolic consequences.

III. Adaptive Metabolic Changes following Prolonged Ethanol Intake

It is common knowledge that chronic alcohol consumption produces increased tolerance to ethanol. This is generally attributed to central nervous system adaptation. In addition, recent studies have shown the development of metabolic adaptation, that is an accelerated clearance of alcohol from the blood. Furthermore, there is an associated increased capacity to metabolize other drugs as well. Moreover, the liver acquires an enhanced capacity to rid itself of lipids through lipoprotein secretion into the blood stream. It is noteworthy that these functions which adaptively increase after chronic ethanol feeding involve to a large extent the activity of the hepatic smooth endoplasmic reticulum, which undergoes significant change after chronic alcohol consumption. It was indeed observed 10 years ago that ethanol feeding results in a proliferation of the smooth membranes of the hepatic endoplasmic reticulum (Iseri et al., 1964, 1966). This ultramicroscopic finding was subsequently confirmed (Lane and Lieber, 1966; Rubin et al., 1968; Lieber and Rubin, 1968; Carulli et al., 1971) and was established on a biochemical basis by the demonstration of an increase in both phospholipids and total protein content of the smooth membranes (Ishii et al., 1973). Its functional counterparts include accelerated metabolism of drugs (including ethanol) and lipoprotein production.

A. Accelerated Ethanol Metabolism after Chronic Ethanol Consumption

Regular drinkers tolerate large amounts of alcoholic beverages, mainly because of central nervous system adaptation. In addition, alcoholics develop increased rates of blood ethanol clearance, so-called metabolic tolerance (Kater *et al.*, 1969a; Ugarte *et al.*, 1972). Experimental ethanol administration also results in an increased rate of ethanol metabolism (Lieber and DeCarli, 1970a; Tobon and Mezey, 1971; Misra *et al.*, 1971). The mechanism of this acceleration is the subject of debate.

1. INCREASE OF ETHANOL METABOLISM RELATED TO THE ADH PATHWAY

There is a controversy over whether ethanol consumption affects activities of hepatic ADH, with most investigators reporting no change or even decreases (Lieber, 1973). In alcoholics, liver ADH was found to be lowered even in the absence of liver damage (Ugarte *et al.*, 1967). Extra-hepatic ADH, particularly the gastric one, has been reported to increase after alcohol feeding (Mistilis and Garske, 1969) but this has not been confirmed either after acute or chronic ethanol administration (de Saint-Blanquat *et al.*, 1972).

Actually, the question of whether there is a moderate change in hepatic ADH activity may not have direct bearing on the problem of rates of alcohol metabolism, since it is generally recognized that ADH activity is usually not the rate-limiting factor in that pathway. There are numerous examples of the lack of correlation between rates of ethanol oxidation and hepatic ADH activity which supports the view that in the process of ADH-mediated ethanol oxidation, ADH itself is not rate-limiting but that velocities may depend on availability of the co-factor NAD, especially the speed of the reoxidation of the ADH–NADH complex.

The mechanism which could contribute to the acceleration of ADH dependent-ethanol metabolism after ethanol consumption is based on increased NADH reoxidation, for instance, because of enhanced ATPase activity (Bernstein *et al.*, 1973). Mitochondrial mechanisms which have been postulated include enhanced shuttling of the H equivalent from the cytosol to the mitochondria after chronic ethanol feeding. We failed to find evidence in favor of this possibility (Cederbaum *et al.*, 1973). In general, it must be pointed out that if, following chronic ethanol consumption, changes affecting the ADH pathway (such as ATPase activity) were responsible exclusively for the acceleration of ethanol metabolism, the latter should be fully abolished by pyrazole treatment, but this was not the case (Lieber and DeCarli, 1970a, 1972). This raises the possibility of the involvement of non-ADH pathways.

2. NON-ADH RELATED ACCELERATION OF ETHANOL METABOLISM

Calculations show that when corrected for microsomal losses during the preparative procedure, the rise in MEOS activity can account for one-half to two-thirds of the increase in blood ethanol clearance (Lieber and DeCarli, 1972). The unaccounted for difference may actually result from a secondary increase in oxidation via ADH, a pathway limited by the rate of NADH reoxidation. This indeed could be accelerated by an increase in MEOS activity, since the latter is associated with enhanced NADPH utilization and the NADPH/NADP and NADH/NAD systems are linked (Veech et al., 1969). Moreover, evidence is accumulating that NADH may serve as partial electron donor for microsomal drug-detoxifying systems (Cohen and Estabrook, 1971). Interestingly, upon addition of ethanol to microsomes, a modified type II binding spectrum appears, the magnitude of which is tripled by ethanol treatment (Rubin et al., 1971).

Indirect evidence that MEOS activity may play a role in vivo can be derived from the fact that other drugs (such as barbiturates) which increase total hepatic MEOS activity (Lieber and DeCarli, 1970b) were also found to enhance rates of blood ethanol clearance (Lieber and DeCarli, 1972; Fischer, 1962; Mezey and Robles, 1974). Some other studies failed to verify this effect (Tephly et al., 1969; Klaassen, 1969). In the latter investigations, however, long-acting barbiturates were used and ethanol clearance was tested in close association with barbiturate administration, at a time when blood barbiturate levels were probably elevated. Under these conditions, it was found that barbiturates interfere with blood ethanol clearance (Lieber and DeCarli, 1972).

There is also some debate over whether, in rats, ethanol feeding enhances catalase activity. Both an increase (Carter and Isselbacher, 1971) and no change (Lieber and DeCarli, 1970a; Hawkins et al., 1966; von Wartburg and Rothlisberger, 1961) have been reported. In man there was no increase (Ugarte et al., 1972). This question, however, may not be fully relevant to the rate of ethanol metabolism since peroxidative metabolism of ethanol in the liver is probably limited by the rate of hydrogen peroxide formation rather than by the amount of available catalase (Boveris et al., 1972). Ethanol consumption does, however, enhance the activity of hepatic NADPH oxidase (Carter and Isselbacher, 1971; Lieber and DeCarli, 1970c; Thurman, 1973) which, as illustrated in Fig. 1C, can participate in H_2O_2 generation. It is conceivable that this mechanism contributes to ethanol metabolism *in vivo* (and to its increase after chronic ethanol consumption) by furnishing the H_2O_2 needed for peroxidative oxidation of ethanol. As discussed before, however, the amount of H_2O_2 generated by the liver is small (Oshino *et al.*, 1973; Boveris *et al.*, 1972) and even when increased by ethanol consumption could not account for the rate of ethanol clearance observed (Lieber, 1973; Israel *et al.*, 1973). Catalase appears to participate primarily in the oxidation of methanol, at least in the rat, whereas in the monkey alcohol dehydrogenase may play a greater role in that respect (Makar *et al.*, 1968).

B. Stimulation of the Microsomal Drug Metabolizing Enzymes

1. ENHANCED DRUG METABOLISM (DRUG TOLERANCE)

The proliferation of hepatic SER induced by ethanol has a functional counterpart: an increased activity of a variety of microsomal drug-detoxifying enzymes (Rubin and Lieber, 1968a; Ariyoshi *et al.*, 1970; Carulli *et al.*, 1971; Misra *et al.*, 1971; Joly *et al.*, 1973b). Ethanol also increases the content of microsomal cytochrome P450 and of NADPH-cytochrome P450 reductase (Rubin *et al.*, 1968; Joly *et al.*, 1973b). These increases occur in the smooth membranes (Ishii *et al.*, 1973; Joly *et al.*, 1973b). Furthermore, ethanol feeding raises the hepatic phospholipid content (Lieber *et al.*, 1965) including that of the smooth microsomal membranes (Ishii *et al.*, 1973). Moreover, it has been shown that microsomal cytochrome P450, a reductase, and phospholipids play a key role in the microsomal hydroxylation of various drugs (Lu *et al.*, 1969). Therefore, the increase in the activity of hepatic microsomal drug-detoxifying enzymes and in the content of cytochrome P450 induced by ethanol ingestion offers a likely explanation for the recent observation that ethanol consumption enhances the rate of drug clearance *in vivo*. The tolerance of the alcoholic to various drugs has been generally attributed to central nervous system adaptation (Kalant *et al.*, 1970). In addition to central nervous system adaptation, metabolic adaptation must now be considered. Indeed, it has been shown recently that the rate of drug clearance from the blood is enhanced in alcoholics (Kater *et al.*, 1969b). Of course, this could be due to a variety of factors other than ethanol, such as the congeners and the use of other drugs so commonly associated with alcoholism. Our studies showed, however, that administration of pure ethanol with nondeficient diets under metabolic ward conditions resulted in a striking increase in the rate of blood clearance of meprobamate and pentobarbital (Misra et al., 1971). Similarly, an increase in the metabolism of aminopyrine (Vesell et al., 1971) and tolbutamide (Carulli et al., 1971) was found. Furthermore, the capacity of liver slices from animals fed ethanol to metabolize meprobamate was also increased (Misra et al., 1971), which clearly shows that ethanol consumption affects drug metabolism in the liver itself, independent of drug excretion or distribution.

2. INCREASED CCl_4 TOXICITY IN ALCOHOLICS

The stimulation of microsomal enzyme activities also applies to those which convert exogenous substrates to toxic compounds. For instance, CCl_4 exerts its toxicity only after conversion in the microsomes. Alcohol pretreatment remarkably stimulates the toxicity of CCl_4 (Hasumura et al., 1974). These experiments were carried out at a time when the ethanol had disappeared from the blood to rule out the increase of the toxicity of CCl_4 due to the presence of ethanol (Traiger and Plaa, 1972). The potentiation of the CCl_4 toxicity by ethanol pretreatment may account for the clinical observation of the enhanced susceptibility of alcoholics to the hepatotoxic effects of CCl_4 (Moon, 1950). Thus, these studies demonstrate that chronic alcohol consumption is associated with enhanced toxicity of some noxious compounds and it is likely that a larger number of toxic agents will be found to display a selective injurious action in the alcoholic. This side effect is possibly an undesirable consequence of the "adaptive" response to chronic ethanol consumption.

Microsomal functions related to lipid metabolism are also increased and these will be alluded to in connection with the discussion of the pathogenesis of the alcoholic fatty liver and hyperlipemia.

IV. Pathogenesis of the Alcoholic Fatty Liver

A. Etiologic Role of Ethanol in the Pathogenesis of the Alcoholic Fatty Liver

Until a decade ago, the concept prevailed that malnutrition is primarily responsible for the development of the alcoholic fatty liver. This notion was based largely on experimental work in rats given ethanol in drinking water (Best et al., 1949). With this new technique, ethanol consumption usually does not exceed 10–25% of the total caloric intake of the animal. A comparable amount of alcohol, when given with an adequate diet, resulted in negligible ethanol levels in the blood (Lieber et al., 1965). By incorporating ethanol in a totally liquid diet, the amount of ethanol consumed was increased to 36% of total calories, a proportion comparable to moderate alcohol intake in man. With these nutritionally adequate diets, isocaloric replacement of sucrose or other carbohydrate by ethanol consistently produces a 5- to 10-fold increase in hepatic tri-glycerides (Lieber *et al.*, 1963, 1965; DeCarli and Lieber, 1967; Porta *et al.*, 1965). As shown in Fig. 4, isocaloric replacement of fat by ethanol also produced steatosis whereas isocaloric replacement of carbohydrate by fat did not. Hepatic lipid accumulation developed progressively over the first month of alcohol administration and persisted thereafter for at least 1 year in the rat (Lieber and DeCarli, 1970d) and 3 years in the baboon (Lieber *et al.*, 1972). This contrasts with results of Porta *et al.* (1965) who found hepatic fat accumulation to be only transient during the first 4–8 weeks of ethanol consumption; this may be due to the fact that the amount of ethanol consumed by animals studied for 4 months (which did not develop steatosis) was only one-half of the short-term group. Similarly, when in our studies in the rat the ethanol intake was

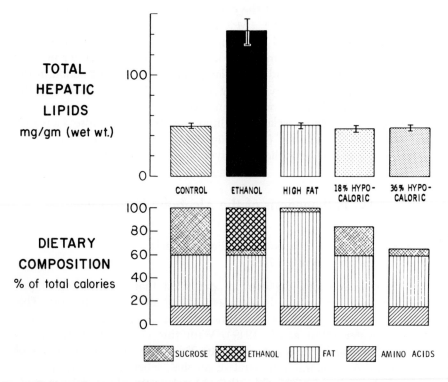

Fɪɢ. 4.　Effect on total hepatic lipids of five types of liquid diets fed to rats for 24 days (Lieber, 1967).

decreased from 36 to 20% of the total calories, no fatty liver was observed (Lieber *et al.*, 1965).

An etiological role for ethanol in the pathogenesis of human liver disease was suggested by the parallel changes in alcohol consumption and death rate from cirrhosis as discussed elsewhere (Lieber *et al.*, 1975a). Epidemiological studies also indicated that alcohol, rather than malnutrition, is the determining factor (Lelbach, 1966; Pequignot, 1962). The importance of the degree of alcohol abuse is illustrated by the fact that of those with a daily consumption in excess of 160 gm of ethanol per day, 75% displayed severe liver damage, whereas only 17% of those who consumed a lesser amount were so affected (Lelbach, 1966). The duration of alcohol abuse is also important: after 15 years of excessive alcohol consumption the incidence of severe liver damage was eight times greater than after 5 years. Furthermore, ethanol itself rather than the congeners of the alcoholic beverages was implicated (Lelbach, 1967).

Undernutrition during World War II (Sherlock and Walshe, 1948) and starvation as a treatment for obesity (Rozental *et al.*, 1967) or as a consequence of anorexia (Solbach and Franken, 1968) were not associated with hepatic steatosis.

Volwiler *et al.* (1948) and Summerskill *et al.* (1957) failed to detect any deleterious effects from alcohol administration in patients recovering from alcoholic fatty liver. In these studies, however, the amounts of alcohol given were less than the usual intake of alcoholics. With larger amounts of alcohol, Menghini (1960) found that the clearance of fat from the alcoholic fatty liver was prevented. Moreover, individuals with a morphologically normal liver (with or without a history of alcoholism) developed a fatty liver when given a variety of nondeficient diets under metabolic ward conditions, with ethanol either as a supplement to the diet or as an isocaloric substitution for carbohydrates (Lieber *et al.*, 1963, 1965; Lieber and Rubin, 1968; Rubin and Lieber, 1968b). This was evident both by morphological examination and by direct measurement of the lipid content of the liver biopsies which revealed up to a 25-fold rise in triglyceride concentration. Even with a high protein, vitamin-supplemented diet, there was a significant increase in hepatic triglycerides, as measured in percutaneous biopsies (Fig. 5). This increase was apparent already after a few (Rubin and Lieber, 1968b) or even 1 (Wiebe *et al.*, 1971) day. This steatosis, though reversible, was accompanied by striking ultrastructural changes, as discussed subsequently. Alcohol interferes with intestinal absorptive functions (Lindenbaum and Lieber, 1969, 1971; Barboriak and Meade, 1969), and its integrity (Rubin *et al.*, 1972b; Baraona and Lieber, 1975; Baraona *et al.*, 1975a). Impairment of lipid absorption, pancreatic function, and intraluminal fat digestion have been

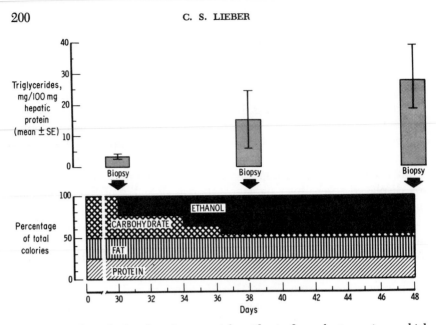

Fig. 5. Effect of ethanol on hepatic triglycerides in five volunteers given a high protein, low fat diet (Lieber, 1967).

reported in alcoholics (Roggin *et al.*, 1969, 1972; Mezey *et al.*, 1970) or after ethanol administration (Bayer *et al.*, 1972; Mott *et al.*, 1972). Furthermore, in rats, deficiency states can aggravate the effect of alcohol upon the liver (Klatskin *et al.*, 1954; Lieber *et al.*, 1969). However, it is unlikely that a deficiency state developed over the short period of some of the studies conducted in volunteers (Lieber *et al.*, 1963; Lieber and Rubin, 1968; Rubin and Lieber, 1968b; Wiebe *et al.*, 1971) especially in view of the liberal enrichment of the diets (Lieber and Rubin, 1968; Rubin and Lieber, 1968b).

B. The Influence of Dietary Factors

1. ROLE OF DIETARY FAT

As discussed subsequently, alcohol ingestion leads to the deposition in the liver of dietary fat. This observation prompted an investigation into the role of the amount and kind of dietary fat in the pathogenesis of alcohol-induced liver injury. Rats were given liquid diets containing an adequate amount of protein for rodents (18% of total calories), with varying amounts of fat (Fig. 6). In the 2% fat diet, the only lipid given was linoleate, to avoid essential fatty acid deficiency. Reduction in dietary fat

to a level of 25% (or less) of total calories was accompanied by a significant decrease in the steatosis induced by ethanol (Lieber and DeCarli,
1970d). These results obtained with alcohol differ from the fatty liver
resulting from choline deficiency, the degree of which was found to be
independent of the amount of dietary fat (Iwamoto et al., 1963). The
importance of dietary fat was confirmed in volunteers: for a given alcohol
intake, much more steatosis developed with diets of normal fat content
than with a low fat diet (Lieber and Spritz, 1966). In addition to the
amount, the chain length of the dietary fatty acid is also important for
the degree of fat deposition in the liver. Replacement of dietary triglycerides containing long-chain fatty acids (LCT) by fat containing
medium-chain fatty acids (MCT) reduces the capacity of alcohol to
produce a fatty liver in rats (Lieber and DeCarli, 1966). The propensity
of medium-chain fatty acids to undergo oxidation rather than esterification probably explains this phenomenon (Lieber et al., 1967).

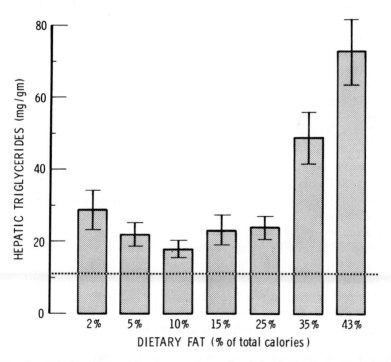

Fig. 6. Hepatic triglycerides in seven groups of rats given ethanol (36% of
calories) with a diet normal in protein (18% of calories) but varying fat content.
Average hepatic triglyceride concentration in the control animals is indicated by a
dotted line (Lieber and DeCarli, 1970d).

2. ROLE OF PROTEIN AND LIPOTROPIC FACTORS (CHOLINE AND METHIONINE)

In growing rats, deficiencies in dietary protein and lipotropic factors (choline and methionine) can produce fatty liver (Best et al., 1949), but primates are far less susceptible to protein and lipotropic deficiency than rodents (Hoffbauer and Zaki, 1965). Clinically, treatment with choline of patients suffering from alcoholic liver injury has been found to be ineffective in the face of continued alcohol abuse and, experimentally, massive supplementation with choline failed to prevent the fatty liver produced by alcohol in volunteer subjects (Rubin and Lieber, 1968b). This is not surprising, since there is no evidence that a diet which is deficient in choline is deleterious to adult man. Unlike rat liver, human liver contains very little choline oxidase activity which may explain the species difference with regard to choline deficiency. The phospholipid content of the liver represents another key difference between the fatty liver produced by ethanol and that caused by choline deficiency. After the administration of ethanol, hepatic phospholipids increase (Lieber et al., 1965) whereas in the fatty liver produced by choline deficiency, they decrease (Ashworth et al., 1961). Similarly, hepatic carnitine is decreased by choline deficiency (Corredor et al., 1967) but increased after ethanol feeding (Kondrup and Grunnet, 1973). Hepatic injury induced by choline deficiency appears to be primarily an experimental disease of rats with little, if any, relevance to human alcoholic liver injury. Even in rats, massive choline supplementation failed to prevent fully the ethanol-induced lesion, whether alcohol was administered acutely (DiLuzio, 1958) or chronically (Lieber and DeCarli, 1966). Ultrastructurally, the two types of fatty liver also differ (Iseri et al., 1966). Furthermore, orotic acid, which reduces the choline deficiency fatty liver, has no such effect on the ethanol variety (Edreira et al., 1974).

The effect of protein deficiency has not yet been clearly delineated in human adults. In children, protein deficiency leads to hepatic steatosis, one of the manifestations of kwashiorkor. In adolescent baboons, however, protein restriction to 7% of total calories did not result in conspicuous liver injury either by biochemical analysis or by light and electron microscopic examination even after 19 months.

Significant steatosis was observed only when the protein intake was reduced to 4% of total calories (Lieber et al., 1972). Furthermore, an excess of protein (25% of total calories, or twice the recommended amount) did not prevent ethanol from producing fat accumulation in human volunteers (Fig. 5). Thus, in man, ethanol is capable of producing striking changes in the liver even in the absence of protein deficiency.

When protein deficiency is present, it may potentiate the effect of ethanol. In the rat, a combination of ethanol and a diet deficient in both protein and lipotropic factors leads to more pronounced hepatic steatosis than either deficiency alone (Klatskin *et al.*, 1954; Lieber *et al.*, 1969).

C. Origin of the Fatty Acids Which Are Deposited in the Alcoholic Fatty Liver and Mechanism for Their Accumulation

Lipids which accumulate in the liver can originate from three main sources: dietary lipids, which reach the bloodstream as chylomicrons, adipose tissue lipids, which are transported to the liver as free fatty acids (FFA), and lipids synthesized in the liver itself (Fig. 7). These fatty acids can accumulate in the liver because of a variety of metabolic disturbances (Lieber, 1969). The four major mechanisms which have been proposed are: (a) decreased lipid oxidation in the liver, (b) enhanced hepatic lipogenesis, (c) decreased hepatic release of lipoproteins, and (d) increased mobilization of peripheral fat. Depending on the experimental conditions, any of the three sources and the four mechanisms can be implicated.

After consumption of ethanol with lipid-containing diets, the fatty acids which accumulate in the liver are derived primarily from dietary fatty acids, whereas when ethanol is given with a low fat diet, en-

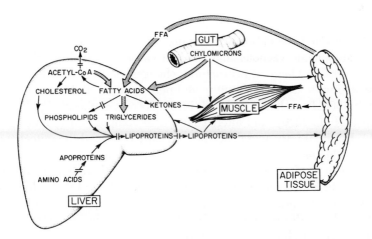

FIG. 7. Possible mechanisms of fatty liver production through either increase (———→) or decrease (———| |———→) of lipid transport and metabolism (Lieber, 1971).

dogenously synthesized fatty acids are deposited in the liver (Lieber *et al.*, 1966, 1969; Lieber and Spritz, 1966·). Some of these effects can be considered as consequences of the metabolism of ethanol in the liver. Indeed, as discussed before both decreased lipid oxidation and enhanced lipogenesis can be linked to ethanol oxidation and the associated increased generation of NADH.

In addition to the changes in mitochondrial functions, which are a direct consequence of the metabolism of ethanol, chronic ethanol abuse results in more persistent changes in the mitochondria. Indeed, alcoholics are known to have profound mitochondrial changes in their liver (Svoboda and Manning, 1964; Kiessling and Pilström, 1971), including the swelling and disfiguration of mitochondria, disorientation of the cristae, and intramitochondrial crystalline inclusions. Increased serum activity of the intramitochondrial enzyme glutamate dehydrogenase was also reported in alcoholics (Konttinen *et al.*, 1970). From these clinical observations, however, it was impossible to assess whether the mitochondrial changes were a direct result of chronic ethanol intake or were secondary to other factors such as dietary deficiencies. This question was resolved by the observation of Iseri *et al.* (1966) who showed that in the rat, isocaloric substitution of ethanol for carbohydrates in otherwise adequate diets leads to enlargement and alterations of the configuration of the mitochondria. Mitochondrial changes similar to those seen in chronic alcoholics were also produced by isocaloric substitution of ethanol for carbohydrate in baboons (Lieber *et al.*, 1972) and in man, both in alcoholics (Lane and Lieber, 1966, Lieber and Rubin, 1968) and in nonalcoholics (Rubin and Lieber, 1968b). Degenerated mitochondria were conspicuous and the debris of these degraded organelles was also found within autophagic vacuoles and residual vacuolated bodies (Rubin and Lieber, 1967). These ultrastructural changes in the mitochondria are associated with increased fragility and permeability (Rubin *et al.*, 1970a; French, 1968; French and Todoroff, 1970), decreased phospholipid content (French *et al.*, 1970), and altered fatty acid composition (French *et al.*, 1971). Essential fatty acid deficiency partially protected the mitochondria from developing the increased fragility induced by chronic ethanol feeding (French *et al.*, 1969). The mechanism of the alteration of mitochondrial membranes is unknown but could possibly be linked to depression of mitochondrial protein synthesis by ethanol consumption (Rubin *et al.*, 1970a). These altered mitochondria have a reduction in cytochrome *a* and *b* content (Rubin *et al.*, 1970a) and in succinic dehydrogenase activity (Rubin *et al.*, 1970a; Oudea *et al.*, 1970). In one study (Videla and Israel, 1970) succinic dehydrogenase activity measured in total liver homogenates was reported to be increased in ethanol-fed rats. The striking structural changes of the mitochondria are associated with corresponding functional abnormalities. Indeed, the respiratory capacity of the mitochondria was found to be depressed (Pilström and Kiessling, 1972; Kiessling and Pilström, 1968; Rubin et al., 1972a; Gordon, 1973), using pyruvate and succinate as substrates. Other substrates were also found to have reduced oxidation by the mitochondria of ethanol-fed rats, except for α-glycerophosphate, the oxidation of which was reported by some to be increased (Kiessling, 1968) or unchanged (Pilström and Kiessling, 1972) whereas others found it to be decreased (Banks et al., 1970), or variable depending on the substrate (Pilström and Kiessling, 1972) or perhaps on the duration of treatment. The major function of the mitochondria is fatty acid oxidation. As discussed before, ethanol depresses hepatic fatty acid oxidation and this inhibition can be attributed to the associated redox change and the inhibition of the activity of the citric acid cycle. This redox change has now been demonstrated in mitochondria of rats fed ethanol chronically (Gordon, 1973), but this alteration returns rapidly to normal after cessation of alcohol intake (Gordon, 1972b, 1973). Since the structural changes of the mitochondria persist, however, the question arose as to whether these in turn could be responsible for some alterations in lipid metabolism beyond those produced by the altered redox change. The first indication that ethanol consumption may result in more persistent metabolic changes arose from the observation that alcohol ingestion is associated with a progressive increase in ketonemia and ketonuria, which was most pronounced in the fasting state, in the absence of ethanol (Lefevre et al., 1970). Moreover, in experimental animals this is associated with enhanced ketogenesis in liver slices, in the absence of ethanol (Lefevre et al., 1970), which contrasts with the inhibition of ketogenesis reported in the presence of ethanol (Ontko, 1973). The ketonemia may aggravate the acidosis and hyperuricemia resulting from hyperlactacidemia (as discussed previously) and on occasion may lead to severe alcoholic ketoacidosis (Jenkins et al., 1971). The capacity for ethanol to produce ketones was found to be greater than that of fat itself, provided, however, that fat was present in the diet. Thus fat seems to play a permissive role (Lefevre et al., 1970). Recent observations indicate that mitochondria obtained from ethanol-fed rats, when incubated in vitro, even in the absence of ethanol, display decreased capacity to oxidize fatty acid but enhanced β-oxidation, possibly responsible for the increased ketogenesis (Toth et al., 1973; Cederbaum et al., 1975). In any event, decreased fatty acid oxidation, whether as a function of the reduced citric acid cycle activity (secondary to the altered redox potential) or whether as a consequence of permanent changes in mitochondrial structure, offers the most likely explanation for the deposition of fat in the liver after chronic alcohol ingestion, especially fat derived from the diet.

V. Pathogenesis of Alcoholic Hyperlipemia

A. Hepatic Lipoprotein Production and Interaction of Ethanol with Hepatic Microsomal Functions

The initial phase of hepatic lipid deposition after ethanol is accompanied by an increased release of lipoproteins into the blood; this tends to counteract lipid accumulation in the liver. However, this "adaptive" mechanism is generally insufficient to prevent fully the development of hepatic steatosis. In any event, decreased hepatic lipoprotein secretion and/or release, which has been proposed as an explanation for the fatty liver produced by a variety of toxic agents, does not apparently play a role in initiating the development of the fatty liver produced by the usual ethanol abuse. The effect is dose dependent. *In vitro*, high ethanol concentration may decrease hepatic lipoprotein release (Schapiro *et al.*, 1964). More recently, when livers were perfused with ethanol in concentrations more in keeping with *in vivo* conditions, no inhibition of lipoprotein secretion was found (Gordon, 1972a). Similarly, contrasting with the hyperlipemia which is commonly associated with the administration of moderate to large amounts of ethanol (Lieber *et al.*, 1963; Jones *et al.*, 1963; Baraona and Lieber, 1970), an extremely high dose has been reported to decrease serum triglyceride (Dajani and Kouyoumjian, 1967), very low-density lipoproteins (Madsen, 1969), high-density lipoproteins (Koga and Hirayama, 1968), and the incorporation of glucosamine into the carbohydrate moiety of serum lipoproteins (Mookerjea and Chow, 1969) in the rat. However, both in man (Lieber *et al.*, 1963; Jones *et al.*, 1963) and in rats (Baraona and Lieber, 1970) ethanol administration usually produces hyper- rather than hypolipemia, and the most striking changes take place in the very low-density lipoprotein fraction. The alcohol-induced hyperlipemia can occur in the fasting state (Jones *et al.*, 1963); it is markedly exaggerated, however, when alcohol is given with a fat-containing diet (Brewster *et al.*, 1966; Barboriak and Meade, 1968; Wilson *et al.*, 1970). This alcohol effect does not result solely from caloric overload, since no comparable hyperlipemia was produced by isocaloric amounts of either carbohydrate or lipids (Losowsky *et al.*, 1963). Incorporation into lipoprotein of intragastrically administered [³H]palmitate and intravenously injected [¹⁴C]lysine was significantly increased by alcohol administration (Baraona and Lieber, 1970). These as well as other studies suggested that the ethanol-induced hyperlipemia results from enhanced lipoprotein production. This contrasts with choline (Oler and Lombardi, 1970) and protein (Flores et al., 1970; Seakins and Waterlow, 1972) deficiencies which produce the opposite effect. Ethanol could act by enhancing the availability of fatty acids, which, in turn, can induce hepatic synthesis of lipoproteins (Alcindor et al., 1970). Furthermore, fatty acids are esterified (Stein and Shapiro, 1958) and lipoproteins are formed (Jones et al., 1967) in the endoplasmic reticulum. Moreover, ethanol consumption enhances the activity of hepatic microsomal L-α-glycerophosphate acyltransferase (Joly et al., 1973a), as well as that of other acyltransferases, depending upon the dietary conditions (Mendenhall et al., 1969). The mechanism of the alterations of these microsomal functions produced by ethanol has not been clarified. It could be linked directly to the fact that ethanol can be oxidized at this key metabolic site and results in proliferation of the membranes of the smooth endoplasmic reticulum, the morphological counterpart of the microsomal fraction obtained by ultracentrifugation. Ethanol feeding was also found to enhance the activity of glycosyltransferase in the Golgi apparatus (Gang et al., 1973).

B. Potentiating Factors

Ethanol-induced hyperlipemia is usually moderate. However, some alcoholic patients develop marked hyperlipemia, which suggests that other factors in addition to ethanol itself contribute to this alteration. A possible role of postheparin lipoprotein lipase activity (PHLA) in alcoholic hyperlipemia has been suggested on the basis of the finding that 6 of the 8 patients with marked hyperlipemia reported by Losowsky et al. (1963) has a decreased PHLA. Furthermore, a mild decrease in the fractional turnover rate of intravenously injected exogenous triglycerides has been reported in alcoholics who develop marked hyperlipemia (Chait et al., 1972). However, in most of the subjects reported by Losowsky et al. (1963), the PHLA remained low after the hyperlipemia had subsided and after alcohol had been withdrawn for weeks or months. This alteration could not be reproduced either by ethanol in vitro (Losowsky et al., 1963) or by administration of ethanol in vivo (Wilson et al., 1970; Verdy and Gattereau, 1967; Kudzma and Schonfeld, 1971). Thus one factor in the development of hyperlipemia could be the existence of a defective removal of serum lipids in some patients. Furthermore, some of the reported alcoholic patients with marked hyperlipemia had other conditions which can contribute to the hyperlipemia, such as diabetes (Losowsky et al., 1963; Kudzma and Schonfeld, 1971) or pancreatitis (Albrink and

Klatskin, 1957). The latter condition has been reported to be associated with the production of an inhibitor of PHLA (Kessler et al., 1963). Type IV hyperlipidemia may also predispose to alcoholic hyperlipemia (Mendelson and Mello, 1973; Ginsberg et al., 1974).

Another possible mechanism could be an increased capacity to secrete serum lipoproteins when challenged with alcohol. This may account for the observation that some alcoholic patients appear to have an unusual sensitivity to the hyperlipemic effect of ethanol (Kudzma and Schonfeld, 1971). Thus, patients, with normal PHLA have been shown to develop hyperlipemia with doses of ethanol (120–160 gm/day) that do not produce hyperlipemia in normal subjects or individuals with endogenous hypertriglyceridemia (type IV). The mechanism for the increased capacity of these patients to develop alcoholic hyperlipemia remains unknown. Since ethanol consumption results in an increased capacity to secrete lipoproteins in response to a lipid load (Baraona et al., 1973), one may wonder whether the difference in response to ethanol between some alcoholics and some individuals with type IV hyperlipemia may be secondary, at least in part, to a difference in prior alcohol consumption.

C. Site of the Ethanol Effects

It is noteworthy that both in men and in rats, ethanol-induced hyperlipemia results in increased concentrations of the various serum lipoprotein fractions, but the main change occurs in the lipoproteins of low density. In the postprandial state, this fraction includes very low-density lipoproteins and chylomicrons. In patients with alcoholic hyperlipemia, chylomicronlike particles have been observed in the fasting state (Chait et al., 1972). In the rat rendered hyperlipemic by ethanol feeding, the lipid/protein ratio of the $d < 1.006$ lipoproteins approaches that of chylomicrons (Baraona et al., 1973). However, the site of origin of these particles cannot be deduced with certainty from physical or chemical characteristics. Indeed, in other states of accelerated lipoprotein production, such as carbohydrate-induced hyperlipemia, the lipid/protein ratio and particle size of the $d < 1.006$ lipoproteins increases even in the absence of dietary fat. The increase in serum lipoproteins of higher density both in man (Wilson et al., 1970) and in rats (Baraona and Lieber, 1970) indicates that the hyperlipemia is not merely of intestinal origin and that the liver participates in this process.

The possibility still remains that after alcohol feeding the intestine releases more lipid into the lymph, either by decreasing oxidation of fatty acid or by increasing the synthesis of lipids from sources other than dietary fat (Windmueller and Levy, 1968; Ockner et al., 1969). A decreased production of $^{14}CO_2$ from labeled fatty acids by intestinal slices after an acute load of alcohol (Baraona et al., 1973) and an increased incorporation of these fatty acids into intestinal triglycerides by slices obtained from rats fed ethanol (Carter et al., 1971) have been reported. To what extent these alterations contribute to alcoholic hyperlipemia is unknown. Recently, Mistilis and Ockner (1972) have shown that intraduodenal infusion of 10% ethanol to the fasted rat, in a dose of 5 gm/kg, produces a mild increase in the very low-density lipoprotein output in the lymph. They postulated that this increase in nondietary lymph lipid could contribute to the hyperlipemia although the peak serum rise actually preceded the maximum increase in intestinal lymph lipids. Furthermore, lymph lipoproteins can derive in part from plasma lipoproteins (Windmueller et al., 1973). Moreover, although a single intragastric administration of a diet containing ethanol (3 gm/kg) increased both intestinal lymph flow and lipid output in rats not previously fed alcohol, postprandial hyperlipemia was not produced under these conditions (Baraona et al., 1973). Actually, the acute load of an ethanol-containing diet did not increase lymph lipid output in rats fed alcohol for several weeks compared to their pair-fed controls; however, marked hyperlipemia developed in these alcohol-fed rats. In addition, when a similar lymph lipid load was infused intravenously to alcohol pretreated and control rats with diversion of intestinal lymph, the alcohol-fed rats developed hyperlipemia. If lymph depletion was not prevented by intravenous replacement, hepatic and plasma lipids decreased and alcoholic hyperlipemia did not occur. This indicates that, although an adequate supply of dietary lipids represents a permissive factor needed to induce alcoholic hyperlipemia in the rat, changes in lymph lipid output do not seem to play a major role in the lipemic effect of ethanol, and that the site of origin of the increased production of serum lipoprotein is a nonintestinal one, most likely hepatic. Similarly, the contribution of lymph lipids to the steatosis appears to be a minor one (Ockner et al., 1973). Consistent with this interpretation is the observation that the blood lipid increase produced by Triton in rats treated with ethanol was prevented by the administration of orotic acid (Hernell and Johnson, 1973).

D. Lipids Other than Triglycerides

The mechanism for the increase in lipids other than triglycerides in the course of alcoholic hyperlipemia remains unknown. This is partly due to the fact that the role of cholesterol and phospholipids in serum lipoproteins has not been clarified. The changes in the plasma concentration of these lipids could be a reflection of variations in the mass of serum lipoprotein secondary to changes in triglyceride transport. Ethanol also increases cholesterogenesis in the liver (Lefevre et al., 1972) and in the small intestine (Middleton et al., 1971). Furthermore, ethanol feeding decreases bile acid excretion (Lefevre et al., 1972) though after cessation of ethanol administration, the opposite effect was observed (Boyer, 1972).

After the initial development of fatty liver associated with hyperlipemia, the blood lipids return toward normal (Lieber et al., 1963). Progressive deterioration of liver function, including lipoprotein production and secretion, could be responsible, and may secondarily aggravate fat accumulation in the liver. In any event, whenever there is fat deposition in the liver, removal is obviously inadequate to offset the increased availability of lipids in the liver.

E. Fatty Acid Mobilization from Adipose Tissue; Effect of Ethanol and Acetate on FFA Metabolism

In rats given one large, sublethal dose of ethanol, it was observed that fatty acids resembling those of adipose tissue accumulate in the liver (Brodie et al., 1961; Lieber et al., 1966). Experimental procedures or agents which reduce the normal rate of peripheral fat mobilization (i.e. adrenalectomy, spinal cord transection, or ganglioplegic drugs) prevent or decrease this type of hepatic fat accumulation (Brodie et al., 1961; Mallov, 1957; Rebouças and Isselbacher, 1961). More direct approaches, however, such as studies in rats with prelabeled epididymal fat pads yielded conflicting information, with evidence for increased (Kessler and Yalovsky-Mishkin, 1966) or unchanged (Poggi and DiLuzio, 1964) fatty acid mobilization. Similarly, in rats one large dose of ethanol has been reported to result in increased (Brodie et al., 1961) or unchanged (Elko et al., 1961) circulating levels of FFA. In man, even with amounts of ethanol as large as 300 gm/day, the concentration of circulating FFA did not increase; it rose only after ingestion of very large doses of ethanol (400 gm/day) (Lieber et al., 1963). In short-term studies, ethanol administration produced a fall in the level of circulating FFA in man (Lieber et al., 1962b; Jones et al., 1963) with reduced peripheral venous-arterial differences in FFA (Lieber et al., 1962b), decreased FFA turnover (Jones et al., 1965), and concomitant reduction in circulating glycerol (Feinman and Lieber, 1967). This effect of ethanol on FFA mobilization from adipose tissue was found to be mediated by acetate (Crouse et al., 1968). Acetate is the end product of ethanol metabolism in the liver (Fig. 2) and is released into the bloodstream. Since stressful doses of ethanol probably both stimulate fatty acid mobilization (via catecholamine release) and depress it (via the acetate produced), the net effect may depend on the particular experimental conditions. This may account for some of the apparent contradictions of the literature.

Actually, whether enhanced peripheral fat mobilization is responsible for hepatic fat accumulation after one large sublethal dose of ethanol in the rat is of little clinical relevance after chronic ethanol consumption. Under the latter conditions, the fatty acids deposited in the liver do not derive primarily from adipose tissue (Lieber and Spritz, 1966; Lieber *et al.*, 1966).

VI. Progression of Alcoholic Fatty Liver to Hepatitis and Cirrhosis

Rats fed alcohol, although they get a fatty liver, do not develop the more severe forms of liver injury seen in alcoholics, namely, hepatitis and cirrhosis. We wondered whether this failure might be due to the fact that in the rat, even when alcohol is given as part of a liquid diet, its intake does not exceed 36% of total calories, which corresponds to moderate consumption in man. Also of potential importance is the fact that whereas development of cirrhosis in man requires 5 to 20 years of steady drinking, the rat lives only about 2 years. To overcome this difficulty, we turned to the baboon, a species that is long-lived and phylogenetically closer to man than the rat. Baboons were given a dose of ethanol that is 50% of the total caloric intake, taking advantage of the liquid diet technique first developed in the rat and now applied to the baboon (Lieber and DeCarli, 1974b). With this diet, alcohol intake was sufficient to result in periods of obvious inebriation. Upon interruption of alcohol administration, some withdrawal symptoms (such as seizures) were observed. Fifteen baboons fed the isocaloric control diet maintained normal livers, and all the animals given ethanol developed excessive fat accumulation. In addition, 5 showed typical alcoholic hepatitis; cirrhosis evolved in 6 baboons studied for 2 to 4 years. The features of cirrhosis and alcoholic hepatitis were comparable to the human variety (Lieber *et al.*, 1975b). These included hyaline sclerosis in the central zones of the lobules, inflammation, necrosis, and occasional hyaline bodies. Thus, for the first time an experimental model has now been developed that reproduces all the liver lesions observed in the alcoholic, namely, fatty liver, hepatitis, and cirrhosis, and shows their sequential development. It is usually accepted that cirrhosis (characterized by extensive scarring or fibrosis) may be, at least in part, a consequence of the necrosis and inflammation associated with the alcoholic hepatitis. The idea that the fatty liver is a precursor of the hepatitis has been well accepted. It must be pointed out, however, that although hepatic fat accumulation by itself may be harmless, it reflects a severe metabolic imbalance in the liver. It is possible that this disturbance, when exaggerated, may eventually engender irreversible damage of the hepatocyte. Necrosis in turn could lead to inflammation, resulting in "alcoholic hepatitis." Indeed, comparable electron microscopic changes of the mitochondria accompany alcoholic hepatitis (Svoboda and Manning, 1964) and the fatty liver produced experimentally by the administration of alcohol (Lane and Lieber, 1966; Lieber and Rubin, 1968) were seen. Alteration of the rough endoplasmic reticulum was also found in patients with alcoholic hepatitis (Svoboda and Manning, 1964). This corresponds to the reduction in rough endoplasmic reticulum seen by electron microscopy after ethanol consumption (Lane and Lieber, 1966; Iseri et al., 1966). Although the alcoholic fatty liver is not an inflammatory condition and is distinguishable from alcoholic hepatitis by light microscopy, the remarkable similarity of the ultrastructural features in the hepatocytes suggests that the former may be a precursor of the latter. This concept has been substantiated more recently by the demonstration of additional signs of hepatoxicity already present at the fatty liver stage. The fatty liver is commonly associated with hepatomegaly, which is generally attributed to fat accumulation. However, there is no evidence that fat by itself can account for the total increase in liver mass. Therefore, a study was recently undertaken in our group (Baraona et al., 1975b) to determine whether there are liver constituents other than fat which contribute to ethanol hepatomegaly. Livers from ethanol-fed rats increased 30% both in volume and in wet weight. The specific gravity of the liver was unchanged despite doubling of hepatic lipids in ethanol-fed rats. The increase in liver fat accounted for only half of the increase in liver dry weight. In addition to the increase in liver lipid, there was a concomitant increase in liver protein. The increase in protein accounted for almost all of the other half of the increased hepatic dry weight. Hepatic protein concentration did not change. This similarity in protein concentration indicates that water also increased in proportion to the increase in protein. Liver wet weight to dry weight ratios were similar in both groups of animals.

In contrast to the increase in protein, hepatic DNA content remained unchanged. This dissociation suggested that hepatomegaly was due to an increase in cell size rather than to an increase in cell number. This was verified in histologic sections of the liver. Hepatocytes occupied a significantly larger area in livers from ethanol-fed rats than in those from pair-fed controls, even in zones where there was little visible fat. The 36% increase in size of the hepatocytes of ethanol-fed rats can account for the hepatomegaly, since these cells contribute to the bulk of the liver volume. The increase in hepatocyte size associated with increased intracellular protein and water may be the basis for the "ballooning" of the hepatocytes commonly observed in alcoholic liver disease.

To determine the subcellular distribution of the increase in liver protein mitochondria, microsomes, and cytosol were prepared. Protein increased in all three fractions, but the total increase in mitochondrial and microsomal proteins accounted for less than half of the total increase in liver protein. The major protein increase occurred in the 100,000 g supernatant, or cytosolic fraction. Thus, in addition to the known increase in organelle protein (Ishii et al., 1973), there is an even greater increase in soluble protein.

To determine whether the increase in soluble protein could be due, at least in part, to hepatic accumulation of proteins that are primarily secreted into the plasma, albumin and transferrin were chosen as typical examples of export proteins. The concentration of intrahepatic albumin increased in the ethanol-fed rats compared to controls. Hepatic transferrin concentration also increased in the ethanol-fed rats. Plasma concentrations of these proteins were similar in both groups of animals. The increases in hepatic albumin and transferrin concentrations indicate that intrahepatic deposition of export proteins could contribute to the increase in soluble protein. By contrast, the concentration of nonexport proteins of the cytosol, namely, ferritin and tubulin, showed a significant decrease in the ethanol-treated animals. Tubulin is the constituent protein of microtubules, which play a key role in protein export.

Accumulation of export proteins in the liver could be due either to increased production or to decreased transport into the blood. Albumin and transferrin production were found to decrease after acute administration of ethanol, whereas albumin production was unaffected by chronic ethanol feeding (Rothschild et al., 1971). By contrast, our findings are consistent with the hypothesis that ethanol feeding may decrease the ability of the liver to export proteins. This defect in the export of proteins has been substantiated recently by the observation of delayed appearance of labeled amino acids in serum protein of alcohol-fed rats, whereas intrahepatic labeling was not retarded (Baraona et al., 1975c). The alteration of this fundamental hepatic function and the hitherto unrecognized accumulation of proteins in the liver reveal a potentially important new site of the hepatotoxic action of ethanol.

ACKNOWLEDGMENT

Original studies reported in this paper were supported, in part, by the Veterans Administration and by United States Public Health Service Grants AM12511 and AA00224.

References

Admirand, W. H., Cronholm, T., and Sjovall, J. (1970). *Biochim. Biophys. Acta* **202**, 343.

Albrink, M. J., and Klatskin, G. (1957). *Amer. J. Med.* **23**, 26.

Alcindor, L. G., Infante, R., Soler-Argilaga, C., Raisonnier, A., Polonovski, J., and Caroli, J. (1970). *Biochim. Biophys. Acta* **210**, 483.

Ariyoshi, T., Takabatake, E., and Remmer, H. (1970). *Life Sci.* **9**, 361.

Ashworth, C. T., Wrightsman, F., and Buttram, V. (1961). *Arch. Pathol.* **72**, 620.

Banks, W. L., Kline, E. S., and Higgins, E. S. (1970). *J. Nutr.* **100**, 581.

Baraona, E., and Lieber, C. S. (1970). *J. Clin. Invest.* **49**, 769.

Baraona, E., and Lieber, C. S. (1975). *Gastroenterol.* **68**, 495.

Baraona, E., Pirola, R. C., and Lieber, C. S. (1973). *J. Clin. Invest.* **52**, 296.

Baraona, E., Pirola, R. C., and Lieber, C. S. (1975a). *Biochim. Biophys. Acta* **388**, 19.

Baraona, E., Leo, M., Borowsky, S. A., and Lieber, C. S. (1975b). *Science* **190**, 794.

Baraona, E., Leo, M., Borowsky, S. A., and Lieber, C. S. (1975c). *Gastroenterol.* **69**, 806.

Barboriak, J. J., and Meade, R. C. (1968). *Amer. J. Med. Sci.* **255**, 245.

Barboriak, J. J., and Meade, R. C. (1969). *J. Nutr.* **98**, 373.

Bayer, M., Rudick, J., Lieber, C. S., and Janowitz, H. D. (1972). *Gastroenterology* **63**, 619.

Bernstein, J., Videla, L., and Israel, Y. (1973). *Biochem. J.* **134**, 515.

Best, C. H., Hartroft, W. S., Lucas, C. C., and Ridout, J. H. (1949). *Brit. Med. J.* **2**, 1001.

Blomstrand, R., and Kager, L. (1973). *Life Sci.* **13**, 113.

Blomstrand, R., Kager, L., and Lantto, O. (1973). *Life Sci.* **13**, 1131.

Boveris, A., Oshino, N., and Chance, B. (1972). *Biochem. J.* **128**, 617.

Boyer, J. L. (1972). *Gastroenterology* **62**, 294.

Brewster, A. C., Lackford, H. G., Schwartz, M. G., and Sullivan, J. F. (1966). *Amer. J. Clin. Nutr.* **19**, 255.

Brodie, B. B., Butler, W. M., Horning, M. G., Maickel, R. P., and Maling, H. M. (1961). *Amer. J. Clin. Nutr.* **9**, 432.

Brunengraber, H., Boutry, M., Lowenstein, L., and Lowenstein, J. M. (1973). *In* "Alcohol and Aldehyde Metabolizing Systems" (R. G. Thurman, T. Yonetani, J. R. Williamson, and B. Chance, eds.), pp. 329. Academic Press, New York.

Carter, E. A., Drummey, G. D., and Isselbacher, K. J. (1971). *Science* **174**, 1245.

Carter, E. A., and Isselbacher, K. J. (1971). *Ann. N.Y. Acad. Sci.* **179**, 282.

Carulli, N., Manenti, F., Gallo, M., and Salvioli, G. F. (1971). *Eur. J. Clin. Invest.* **1**, 421.

Cederbaum, A. I., Lieber, C. S., Toth, A., Beattie, D. S., and Rubin, E. (1973). *J. Biol. Chem.* **248**, 4977.

Cederbaum, A. I., Lieber, C. S., and Rubin, E. (1974). *Arch. Biochem. Biophys.* **161**, 26.

Cederbaum, A. I., Lieber, C. S., Beattie, D. S., and Rubin, E. (1975). *J. Biol. Chem.* **250**, 5122.

Chait, A., Mancini, M., February, A. W., and Lewis, B. (1972). *Lancet* ii, 62.

Cohen, B. S., and Estabrook, R. W. (1971). *Arch. Biochem. Biophys.* **143**, 54.

Cohen, G. M., and Mannering, G. J. (1973). *Mol. Pharmacol.* **9**, 383.

Conney, A. H. (1967). *Pharmacol. Rev.* **19**, 317.

Corredor, C., Mansbach, C., and Bressler, R. (1967). *Biochim. Biophys. Acta* **125**, 244.
Crouse, J. R., Gerson, C. D., DeCarli, L. M., and Lieber, C. S. (1968). *J. Lipid Res.* **9**, 509.
Dajani, R. M., and Kouyoumjian, C. (1967). *J. Nutr.* **91**, 535.
Dajani, R. M., Danielski, J., and Orten, J. M. (1963). *J. Nutr.* **80**, 196.
Davis, V. E., and Walsh, M. J. (1970). *Science* **167**, 1005.
DeCarli, L. M., and Lieber, C. S. (1967). *J. Nutr.* **91**, 331.
de Saint-Blanquat, G., Fritsch, P., and Derache, R. (1972). *Pathol. Biol.* **20**, 249.
DiLuzio, N. R. (1958). *Amer. J. Physiol.* **194**, 453.
Eade, N. R. (1959). *J. Pharmacol. Exp. Ther.* **127**, 29.
Edreira, J. G., Hirsch, R. L., and Kennedy, J. A. (1974). *Quart. J. Stud. Alc.* **35**, 20.
Elko, E. E., Wooles, W. R., and DiLuzio, N. R. (1961). *Amer. J. Physiol.* **201**, 923.
Feinman, L., and Lieber, C. S. (1967). *Amer. J. Clin. Nutr.* **20**, 400.
Fischer, H. -D. (1962). *Biochem. Pharmacol.* **11**, 307.
Flores, H., Pak, N., Maccioni, A., and Monckeberg, F. (1970). *Brit. J. Nutr.* **24**, 1005.
Forney, R. B., and Hughes, F. W. (1968). "Combined Effects of Alcohol and Other Drugs." Thomas, Springfield, Illinois.
Forsander, O. A., Maenpaa, P. H., and Salaspuro, M. P. (1965). *Acta Chem. Scand.* **19**, 1770.
French, S. W. (1968). *Gastroenterology* **54**, 1106.
French, S. W., and Todoroff, T. (1970). *Arch. Pathol.* **89**, 329.
French, S. W., Sheinbaum, A., and Morin, R. J. (1969). *Proc. Soc. Exp. Biol. Med.* **130**, 781.
French, S. W., Ihrig, T. J., and Morin, R. J. (1970). *Quart. J. Stud. Alcohol* **31**, 801.
French, S. W., Ihrig, T. J., Shaw, G. P., Tanaka, T. T., and Norum, M. L. (1971). *Res. Commun. Chem. Pathol. Pharmacol.* **2**, 567.
Gang, H., Lieber, C. S., and Rubin, E. (1973). *Nature (London)* **243**, 123.
Gillette, J. R., Brodie, B. B., and La Du, B. N. (1957). *J. Pharmacol. Exp. Ther.* **119**, 532.
Ginsberg, H., Olefsky, J., Farquhar, J. W., and Reaven, G. M. (1974). *Ann. Intern. Med.* **80**, 143.
Gordon, E. R. (1972a). *Biochem. Pharmacol.* **21**, 2291.
Gordon, E. R. (1972b). *Can. J. Biochem.* **50**, 949.
Gordon, E. R. (1973). *J. Biol. Chem.* **248**, 8271.
Grunnet, N. (1973). *Eur. J. Biochem.* **35**, 236.
Grunnet, N., and Thieden, H. I. D. (1972). *Life Sci.* (Part II) **11**, 983.
Grunnet, N., Quistorf, B., and Thieden, H. I. D. (1973). *Eur. J. Biochem.* **40**, 275.
Guynn, R. W., Velsoso, D., Harris, R. L., Lawson, J. W. R., and Veech, R. L. (1973). *Biochem. J.* **136**, 639.
Hasumura, Y., Teschke, R., and Lieber, C. S. (1974). *Gastroenterology* **66**, 415.
Hasumura, Y., Teschke, R., and Lieber, C. S. (1975a). *Science* **184**, 727.
Hasumura, Y., Teschke, R., and Lieber, C. S. (1975b). *J. Pharmacol. Exp. Ther.* **194**, 469.
Hawkins, R. D., Kalant, H., and Khanna, J. M. (1966). *Can. J. Physiol. Pharmacol.* **44**, 241.
Hernell, O., and Johnson, O. (1973). *Lipids* **8**, 503.
Hildebrandt, A. G., Speck, M., and Roots, I. (1974). *Naunyn-Schmiedebergs Arch. Pharmakol. Exp. Pathol.* **281**, 371.
Hoffbauer, F. W., and Zaki, F. G. (1965). *Arch. Pathol.* **79**, 364.

Horton, A. A. (1971). *Biochim. Biophys. Acta* **253**, 514.
Iseri, O. A., Gottlieb, L. S., and Lieber, C. S. (1964). *Fed. Proc. Fed. Amer. Soc. Exp. Biol.* **23**, 579.
Iseri, O. A., Lieber, C. S., and Gottlieb, L. S. (1966). *Amer. J. Pathol.* **48**, 535.
Ishii, H., Joly, J. -G., and Lieber, C. S. (1973). *Biochim. Biophys. Acta* **291**, 411.
Israel, Y., Videla, L., MacDonald, A., and Bernstein, J. (1973). *Biochem. J.* **134**, 523.
Iwamoto, A., Hellerstein, E. E., and Hegsted, D. M. (1963). *J. Nutr.* **79**, 488.
Jenkins, D. W., Eckel, R. W., and Craig, J. W. (1971). *J. Amer. Med. Ass.* **217**, 177.
Joly, J. -G., Feinman, L., Ishii, H., and Lieber, C. S. (1973a). *J. Lipid Res.* **14**, 337.
Joly, J. -G., Ishii, H., Teschke, R., Hasumura, Y., and Lieber, C. S. (1973b). *Biochem. Pharmacol.* **22**, 1532.
Jones, A. L., Ruderman, N. B., and Herrera, M. G. (1967). *J. Lipid Res.* **8**, 429.
Jones, D. P., Losowsky, M. S., Davidson, C. S., and Lieber, C. S. (1963). *J. Lab. Clin. Med.* **62**, 675.
Jones, D. P., Perman, E. S., and Lieber, C. S. (1965). *J. Lab. Clin. Med.* **66**, 804.
Kalant, H., Khanna, J. M., and Marshman, J. (1970). *J. Pharmacol. Exp. Ther.* **175**, 318.
Kater, R. M. H., Carulli, N., and Iber, F. L. (1969a). *Amer. J. Clin. Nutr.* **22**, 1608.
Kater, R. M. H., Roggin, G., Tobon, F., Zieve, P., and Iber, F. L. (1969b). *Amer. J. Med. Sci.* **258**, 35.
Katz, J., and Chaikoff, I. L. (1955). *Biochim. Biophys. Acta* **18**, 87.
Keilin, D., and Hartree, E. F. (1945). *Biochem. J.* **39**, 293.
Kessler, J. I., and Yalovsky-Mishkin, S. (1966). *J. Lipid Res.* **7**, 772.
Kessler, J. I., Kniffen, J. C., and Janowitz, H. D. (1963). *New Eng. J. Med.* **269**, 943.
Kiessling, K.-H. (1968). *Acta Pharmacol,* **26**, 245.
Kiessling, K.-H., and Pilström, L. (1968). *Quart. J. Stud. Alc.* **29**, 819.
Kiessling, K.-H., and Pilström, L. (1971). *Cytobiologie* **4**, 339.
Klaassen, C. D. (1969). *Proc. Soc. Exp. Biol. Med.* **132**, 1099.
Klatskin, G., Krehl, W. A., and Conn, H. O. (1954). *J. Exp. Med.* **100**, 605.
Koga, A., and Hirayama, C. (1968). *Experientia* **24**, 438.
Kondrup, J., and Grunnet, N. (1973). *Biochem. J.* **132**, 373.
Konttinen, A., Härtel, G., and Louhija, A. (1970). *Acta Med. Scand.* **188**, 257.
Korsten, M., Matsuzaki, S., Feinman, L., and Lieber, C. S. (1975). *New Engl. J. Med.* **292**, 386.
Kudzma, D. J., and Schonfeld, G. (1971). *J. Lab. Clin. Med.* **77**, 384.
Lane, B. P., and Lieber, C. S. (1966). *Amer. J. Pathol.* **49**, 593.
Lane, B. P., and Lieber, C. S. (1967). *Lab. Invest.* **16**, 342.
Lefevre, A., Adler, H., and Lieber, C. S. (1970). *J. Clin. Invest.* **49**, 1775.
Lefevre, A., DeCarli, L. M., and Lieber, C. S. (1972). *J. Lipid Res.* **13**, 48.
Lelbach, W. K. (1966). *Acta Hepatosplenol.* **13**, 321.
Lelbach, W. K. (1967). *Deut. Med. Wochenschr.* **92**, 233.
Lieber, C. S. (1967). *Fed. Proc. Fed. Amer. Soc. Exp. Med.* **26**, 1443.
Lieber, C. S. (1968). *Advan. Intern. Med.* **14**, 151.
Lieber, C. S. (1969). *In* "The Biological Basis of Medicine" (E. E. Bittar, ed.), Vol. 5, Chapter 9, p. 317. Academic Press, New York.
Lieber, C. S. (1971). *Quad. Sclavo Diagn. Clin. Lab.* **7**, 861.
Lieber, C. S., and Davidson, C. S. (1962). *Amer. J. Med.* **33**, 319.
Lieber, C. S. (1973). *Gastroenterology* **65**, 821.
Lieber, C. S., and DeCarli, L. M. (1966). *Gastroenterology* **50**, 316.

Lieber, C. S., and Decarli, L. M. (1968). *Science* **162**, 917.

Lieber, C. S., and DeCarli, L. M. (1970a). *J. Biol. Chem.* **245**, 2505.

Lieber, C. S., and DeCarli, L. M. (1970b). *Life Sci.* **9**, 267.

Lieber, C. S., and DeCarli, L. M. (1970c). *Science* **170**, 78.

Lieber, C. S., and DeCarli, L. M. (1970d). *Amer. J. Clin. Nutr.* **23**, 474.

Lieber, C. S., and DeCarli, L. M. (1972). *J. Pharmacol. Exp. Ther.* **181**, 279.

Lieber, C. S., and DeCarli, L. M. (1973). *Drug Metabol. Disposition* **1**, 428.

Lieber, C. S., and DeCarli, L. M. (1974a). *Biochem. Biophys. Phys. Commun.* **60**, 1187.

Lieber, C. S., and DeCarli, L. M. (1974b). *J. Med. Primatology* **3**, 153.

Lieber, C. S., and Rubin, E. (1968). *Amer. J. Med.* **44**, 200.

Lieber, C. S., and Schmid, R. (1961). *J. Clin. Invest.* **40**, 394.

Lieber, C. S., and Spritz, N. (1966). *J. Clin. Invest.* **45**, 1400.

Lieber, C. S., Jones, D. P., Losowsky, M. S., and Davidson, C. S. (1962a). *J. Clin. Invest.* **41**, 1863.

Lieber, C. S., Leevy, C. M., Stein, S. W., George, W. S., Cherrick, G. R., Abelmann, W. H., and Davidson, C. S. (1962b). *J. Lab. Clin. Med.* **59**, 826.

Lieber, C. S., Jones, D. P., Mendelson, J., and DeCarli, L. M. (1963). *Trans. Ass. Amer. Physicians* **76**, 289.

Lieber, C. S., Lefevre, A., Spritz, N., Feinman, L., and DeCarli, L. M. (1967). *J. Clin. Invest.* **46**, 1451.

Lieber, C. S., Jones, D. P., and DeCarli, L. M. (1965). *J. Clin. Invest.* **44**, 1009.

Lieber, C. S., Spritz, N., and DeCarli, L. M. (1966). *J. Clin. Invest.* **45**, 51.

Lieber, C. S., Spritz, N., and DeCarli, L. M. (1969). *J. Lipid Res.* **10**, 283.

Lieber, C. S., Rubin, E., and DeCarli, L. M. (1970). *Biochem. Biophys. Res. Commun.* **40**, 858.

Lieber, C. S., Rubin, E., DeCarli, L. M., Gang, H., and Walker, G. (1972). *In* "Medical Primatology" (E. I. Goldsmith and J. Moor-Jankowski, eds.), Part 3, p. 270. S. Karger, Basel.

Lieber, C. S., Feinman, L., and Rubin, E. (1975a). *In* "Gastroenterology" (H. L. Bockus, ed.), Vol. III, 3rd ed., Chapter 106, in press. Saunders, Philadelphia, Pennsylvania.

Lieber, C. S., DeCarli, L. M., Rubin, E. (1975b). *Proc. Nat. Acad. Sci. U.S.* **72**, 437.

Lindenbaum, J., and Lieber, C. S. (1969). *Nature (London)* **224**, 806.

Lindenbaum, J., and Lieber, C. S. (1971). *In* "Biological Aspects of Alcohol" (M. K. Roach, W. M. McIsaac, and P. J. Craven, eds.), Vol. 3, p. 27. Univ. of Texas Press, Austin, Texas.

Losowsky, M. S., Jones, D. P., Davidson, C. S., and Lieber, C. S. (1963). *Amer. J. Med.* **35**, 794.

Lu, A. Y. H., Strobel, H. W., and Coon, M. J. (1969). *Biochem. Biophys. Res. Commun.* **36**, 545.

Lundquist, F., Tygstrup, N., Winkler, K., Mellemgaard, K., and Munck-Petersen, S. (1962). *J. Clin. Invest.* **41**, 955.

Madsen, N. P. (1969). *Biochem. Pharmacol.* **18**, 261.

Makar, A. B., and Mannering, G. J. (1970). *Biochem. Pharmacol.* **19**, 2017.

Makar, A. B., Tephly, T. R., and Mannering, G. J. (1968). *Mol. Pharmacol.* **4**:471.

Mallov, S. (1957). *Amer. J. Physiol.* **189**, 428.

Marjanen, L. (1972). *Biochem. J.* **127**, 633.

Meldolesi, J. (1967). *Biochem. Pharmacol.* **16**, 125.

Mendelson, J. H., and Mello, N. K. (1973). *Science* **180**, 1372.

Mendenhall, C. L. (1972). *J. Lipid Res.* 13, 177.

Mendenhall, C. L., Bradford, R. H., and Furman, R. H. (1969). *Biochim. Biophys. Acta* 187, 510.

Menghini, G. (1960). *Bull. Acad. Suisse Sci. Med.* 16, 36.

Mezey, E., and Robles, E. A. (1974). *Gastroenterology* 66, 248.

Mezey, E., Jow, E., Slavin, R. E., and Tobon, F. (1970). *Gastroenterology* 59, 657.

Mezey, E., Potter, J. J., and Reed, W. D. (1973). *J. Biol. Chem.* 248, 1183.

Middleton, W. R. J., Carter, E. A., Drummey, G. D., and Isselbacher, K. J. (1971). *Gastroenterology* 60, 880.

Misra, P. S., Lefevre, A., Ishii, H., Rubin, E., and Lieber, C. S. (1971). *Amer. J. Med.* 51, 346.

Mistilis, S. P., and Garske, A. (1969). *Aust. Ann. Med.* 18, 227.

Mistilis, S. P., and Ockner, R. K. (1972). *J. Lab. Clin. Med.* 80, 34.

Mookerjea, S., and Chow, A. (1969). *Biochim. Biophys. Acta* 184, 83.

Moon, H. D. (1950). *Amer. J. Pathol.* 26, 1041.

Mott, C., Sarles, H., Tiscornia, O., and Gullo, L. (1972). *Amer. J. Dig. Dis.* 17, 902.

Newcombe, D. S. (1972). *Metabolism* 21, 1193.

Nikkila, E. A., and Ojala, K. (1963). *Proc. Soc. Exp. Biol. Med.* 113, 814.

Ockner, R., Hughes, F. B., and Isselbacher, K. J. (1969). *J. Clin. Invest.* 48, 2079.

Ockner, R. K., Mistilis, S. P., Poppenhausen, R. B., and Stiehl, A. F. (1973). *Gastroenterology* 64, 603.

Oler, A., and Lombardi, B. (1970). *J. Biol. Chem.* 245, 1282.

Ontko, J. A. (1973). *J. Lipid Res.* 14, 78.

Orme-Johnson, W. H., and Ziegler, D. M. (1965). *Biochem. Biophys. Res. Commun.* 21, 78.

Oshino, N., Chance, B., Sies, H., and Bucher, T. (1973). *Arch. Biochem. Biophys.* 154, 117.

Oudea, M. C., Launay, A. N., Queneherve, S., and Oudea, P. (1970). *Rev. Eur. Etudes Clin. Biol.* 15, 748.

Papenberg, J., von Wartburg, J. P., and Aebi, H. (1970). *Enzymol. Biol. Clin.* 11, 237.

Pequignot, G. (1962). *Muench. Med. Wochenschr.* 103, 1464.

Perman, E. S. (1958). *Acta Physiol. Scand.* 44, 241.

Pilström, L., and Kiessling, K. -H. (1972). *Histochemie* 32, 329.

Poggi, M., and DiLuzio, N. R. (1964). *J. Lipid Res.* 5, 437.

Porta, E. A., Hartroft, W. S., and de la Iglesia, F. D. (1965). *Lab. Invest.* 14, 1437.

Raskin, N. H., and Sokoloff, L. (1972). *Nature (London)* 236, 138.

Rebouças, G., and Isselbacher, K. J. (1961). *J. Clin. Invest.* 40, 1355.

Redmond, G., and Cohen, G. (1971). *Science* 171, 387.

Reynier, M. (1969). *Acta Chem. Scand.* 23, 1119.

Roggin, G. M., Iber, F. L., Kater, R. M. H., and Tobon, F. (1969). *Johns Hopkins Med. J.* 125, 321.

Roggin, G. M., Iber, F. L., and Linscheer, W. G. (1972). *Gut* 13, 107.

Rognstad, R., and Clark, D. G. (1974). *Eur. J. Biochem.* 42, 51.

Rothschild, R., Oratz, M., Mongelli, J., Schreiber, S. S. (1971). *J. Clin. Invest.* 50, 1812.

Rozental, P., Biava, C., Spencer, H., and Zimmerman, H. J. (1967). *Amer. J. Dig. Dis.* 12, 198.

Rubin, E., and Lieber, C. S. (1967). *Gastroenterology* 52, 1.

Rubin, E., and Lieber, C. S. (1968a). *Science* 162, 690.

Rubin, E., and Lieber, C. S. (1968b). *New Eng. J. Med.* **278,** 869.
Rubin, E., Hutterer, F., and Lieber, C. S. (1968). *Science* **159,** 1469.
Rubin, E., Beattie, D. S., and Lieber, C. S. (1970a). *Lab. Invest.* **23,** 620.
Rubin, E., Gang, H., Misra, P. S., and Lieber, C. S. (1970b). *Amer. J. Med.* **49,** 801.
Rubin, E., Lieber, C. S., Alvares, A. P., Levin, W., Kuntzman, R. (1971). *Biochem. Pharmacol.* **20,** 229.
Rubin, E., Beattie, D. S., Toth, A., and Lieber, C. S. (1972a). *Fed. Proc. Fed. Amer. Soc. Exp. Biol.* **31,** 131.
Rubin, E., Rybak, B., Lindenbaum, J., Gerson, C. D., Walker, G., and Lieber, C. S. (1972b). *Gastroenterology* **63,** 801.
Schapiro, R. H., Drummey, G. D., Shimizu, Y., and Isselbacher, K. J. (1964). *J. Clin. Invest.* **43,** 1338.
Scheig, R. (1971). *Gastroenterology* **60,** 751.
Schreiber, S. S., Briden, K., Oratz, M., and Rothschild, M. A. (1972). *J. Clin. Invest.* **51,** 2820.
Schüppel, R. (1971). *In* "Alcohol and the Liver" (W. Gerok, K. Sickinger, and H. H. Hennekeuser, eds.) p. 227. Schattauer Verlag, New York.
Seakins, A., and Waterlow, J. C. (1972). *Biochem. J.* **129,** 793.
Sherlock, S., and Walshe, V. (1948). *Nature (London)* **161,** 604.
Soehring, K., and Schüppel, R. (1966). *Deut. Med. Wochenschr.* **91,** 1892.
Solbach, H. G., and Franken, F. H. (1968). *Deut. Med. Wochenschr.* **93,** 1990.
Stein, S., Lieber, C. S., Cherrick, G. R., Leevy, C. M., and Abelmann, W. H. (1963). *Amer. J. Clin. Nutr.* **13,** 68.
Stein, Y., and Shapiro, B. (1958). *Biochim. Biophys. Acta* **30,** 271.
Summerskill, W. H. J., Wolfe, S. J., and Davidson, C. S. (1957). *Lancet* i, 335.
Sutherland, V. C., Burbridge, T. N., Adams, J. E., and Simon, A. (1960). *J. Appl. Physiol.* **15,** 189.
Svoboda, D. J., and Manning, R. T. (1964). *Amer. J. Pathol.* **44,** 645.
Tephly, T. R., Tinelli, F., and Watkins, W. D. (1969). *Science* **166,** 627.
Teschke, R., Hasumura, Y., Joly, J. -G., Ishii, H., and Lieber, C. S. (1972). *Biochem. Biophys. Res. Commun.* **49,** 1187.
Teschke, R., Hasumura, Y., and Lieber, C. S. (1974a). *Biochem. Biophys. Res. Commun.* **60,** 851.
Teschke, R., Hasumura, Y., and Lieber, C. S. (1974b). *Arch. Biochem. Biophys.* **163,** 404.
Teschke, R., Hasumura, Y., and Lieber, C. S. (1975). *Molec. Phar.* **11,** 841.
Theorell, H., Chance, C., Yonetani, T., and Oshino, N. (1972). *Arch. Biochem. Biophys.* **151,** 434.
Thieden, H. I. D. (1971). *Acta Chem. Scand.* **25,** 3421.
Thurman, R. G. (1973). *Mol. Pharmacol.* **9,** 670.
Thurman, R. G., and Scholz, R. (1973). *Drug Metabol. Disposition* **1,** 441.
Thurman, R. G., Ley, H. G., and Scholz, R. (1972). *Eur. J. Biochem.* **25,** 420.
Tobon, F., and Mezey, E. (1971). *J. Lab. Clin. Med.* **77,** 110.
Toth, A., Lieber, C. S., Cederbaum, A. I., Beattie, D. S., and Rubin, E. (1973). *Gastroenterology* **64,** 198.
Traiger, G. J., and Plaa, G. L. (1972). *J. Pharmacol. Exp. Ther.* **183,** 481.
Truitt, E. B., and Duritz, G. (1966). *In* "Biochemical Factors in Alcoholism" (P. P. Maickel, ed.), p. 61. Pergamon Press, New York.
Truitt, E. B., and Walsh, M. J. (1973). *In* "Proceedings of the First Annual Alcoholism Conference of the National Institute on Alcohol Abuse and Alcoholism" (M. E.

Chafetz, ed.), p. 100. DHEW (NIH) 74-675, U.S. Government Printing Office, Washington, D.C.

Ugarte, G., Pino, M. E., and Insunza, I. (1967). *Amer. J. Dig. Dis.* **12**, 589.

Ugarte, G., Pereda, T., Pino, M. E., and Iturriaga, H. (1972). *Quart. J. Stud. Alc.* **33**, 698.

Vatsis, K. P., and Schulman, M. P. (1973). *Biochem. Biophys. Res. Commun.* **52**, 588.

Vatsis, K. P., and Schulman, M. P. (1974). *Fed. Proc. Fed. Amer. Soc. Exp. Biol.* **33**, 554.

Veech, R. L., Eggleston, L. V., and Krebs, H. A. (1969). *Biochem. J.* **115**, 609.

Verdy, M., and Gattereau, A. (1967). *Amer. J. Clin. Nutr.* **20**, 997.

Vesell, E. S., Page, J. G., and Passananti, G. T. (1971). *Clin. Pharmacol. Ther.* **12**, 192.

Videla, L., and Israel, Y. (1970). *Biochem. J.* **118**, 275.

Voas, R. B. (1973). *In* "Proceedings of the First Annual Alcoholism Conference of the National Institute on Alcohol Abuse and Alcoholism" (M. E. Chafetz, ed.), p. 324. DHEW (NIH) 74-675. U. S. Government Printing Office, Washington, D.C.

Volwiler, W., Jones, C. M., and Mallory, T. B. (1948). *Gastroenterology* **11**, 164.

von Wartburg, J. P., and Rothlisberger, M. (1961). *Helv. Physiol. Acta* **19**, 30.

Wiebe, T., Lundquist, A., and Belfrage, P. (1971). *Scand. J. Clin. Lab. Invest.* **27**, 33.

Wilson, D. E., Schreibman, P. H., Brewster, A. C., and Arky, R. A. (1970). *J. Lab. Clin. Med.* **75**, 264.

Windmueller, H. G., and Levy, R. I. (1968). *J. Biol. Chem.* **243**, 4878.

Windmueller, H. G., Herbert, P. N., and Levy, R. I. (1973). *J. Lipid Res.* **14**, 215.

Chapter 7

CHOLESTYRAMINE AND ION-EXCHANGE RESINS

H. RICHARD CASDORPH

Lipid Research Foundation,
Long Beach, California

I. Cholestyramine and Ion-Exchange Resins

During the past decade cholestyramine, a bile acid sequestering resin, has proved to be a very effective and useful agent in the treatment of patients with elevated serum cholesterol. In 1973, it was released by the Food and Drug Administration in the United States for the treatment of type II hyperlipoproteinemias, although it was used even prior to this throughout the western world.

It has quite an interesting history inasmuch as the original uses of the resin were unrelated to its effect on serum cholesterol. The initial reports (1–5) indicated that it was useful in the treatment of pruritus due to partial biliary obstruction and the initial studies with this resin indicated that patients with partial biliary obstruction, who frequently had elevated serum cholesterols, would have a decline in their levels of serum cholesterol shortly after being placed on cholestyramine therapy.

The initial product had the trade name Cuemid (Merck, Sharp & Dohme Co.). It was subsequently developed and made more palatable in the form of Questran (Mead Johnson). In its present form it is readily acceptable to a majority of patients who are motivated to be treated for their hyperlipidemias. It is available in an orange-flavored powdered form, and is quite effective and consistent in its cholesterol-reducing properties. The efficacy of cholestyramine in reducing the serum cholesterol (and β-lipoprotein levels) has been confirmed by many investigators (6–23, 24a).

II. Other Ion Exchangers

Although cholestyramine is the only resin that has been released by the FDA in the United States, there are two other agents in the investigative phase of development. The Upjohn Company has developed a resin, Colestid formerly called Colestipol (*U-26, 597A*). The hypolipidemic effect of this resin has been documented by numerous investigators (24–51, 79). This agent acts as a bile acid sequestrant. It is also assumed to act by increasing low-density lipoprotein (LDL) catabolism. Indications, side effects, and drug interactions may be similar to those noted for cholestyramine (133).

A third resin is in the process of development and evaluation by Pharmacia Laboratories Inc. This resin is called Secholex (PDX-Chloride), formerly called DEAE-Sephadex. In order to avoid confusion and the possibility that physicians might use the biochemical grade material, Pharmacia has asked all authors to drop the old name of DEAE Sephadex and use the name Secholex or PDX-chloride in referring to this resin. As of this writing there are no clinical trials being conducted in the United States with this agent, however, some animal studies are being conducted (52–58), and other reports indicate the effectiveness of PDX-chloride (DEAE Sephadex) in its cholesterol-reducing properties (59–62).

Both (PDX-Chloride) DEAE Sephadex and Colestid are tertiary amines. Cholestyramine (Questran) is a quaternary amine. The tertiary amines are reported to bind bile acids less strongly than quaternary amines (63). Whether this will have an advantage or disadvantage so far as the use of the resins in the treatment of the hyperlipidemias remains to be seen.

Since most of our experience during the past 10 years has been with cholestyramine, it will of necessity serve as the basic reference point for discussion in this chapter. When experimental evidence is sufficient for an adequate comparison, the advantages or disadvantages of the other resins will be compared with cholestyramine.

Dr. A. N. Howard has reported that he finds no difference between the efficacy of Secholex and cholestyramine (Questran) when given alone. He has also indicated that he has seen the unpublished data of Lars Carlson and Esmo Nikkila in which a comparison of the two resins was made. Pooling their data with those of Howard (at least 30 patients), there was no statistical difference (64) in the effect of the two resins. Dr. Howard also has indicated that in his hands when combined with clofibrate, Secholex was superior to Questran (59).

Dr. Scott M. Grundy has indicated that he has initiated studies to evaluate the combination of Colestipol and clofibrate. As of this writing, Dr. Grundy indicates that he has given the combination to only a few patients and the results are insufficient to draw any definite conclusions (65). Dr. Grundy has been particularly interested in the possibility that the combined use of these two drugs may give rise to lithogenic bile, i.e., that these patients may develop cholesterol gallstones.

III. Other Agents Acting by the Ion-Exchange Mechanism

Several agents appear to have a slight degree of cholesterol-reducing properties and act in a manner possibly similar to the ion-exchange mechanism. These will be briefly mentioned with appropriate bibliographic references when available.

Pectin, ingested orally, is reported to lower serum cholesterol in man (66, 67) as do other mucilaginous polysaccharides in chickens (68) and rats (69–72). The exact mechanism of action of these compounds is not known, however, it appears to be related to a reduction or interference with cholesterol absorption.

Parkinson has written a comprehensive article describing the various cellulose and dextran anion exchangers which bind bile salts *in vitro* under conditions of pH and ion exchange resembling those in the lumen of the small intestine (52). These substances, diethylaminoethyl DEAE cellulose, guanidoethyl cellulose, and DEAE Sephadex reduce hypercholesteremia when added to the diet of cholesterol-fed cockerels.

Cellulose ion exchangers, cross-linked by hydrogen bonds, and DEAE-Sephadex, a chemically modified dextran cross-linked by ether bonds, possess a loose hydrophilic molecular structure that permits free diffusion of even large molecules, such as proteins and nucleic acids, through a three-dimensional network of polysaccharide chains (52). Accessibility of the binding sites of these anion exchangers to large ionic species may be important in determining their affinity for bile acids.

Hypocholesterolemic activity of DEAE cellulose, GE cellulose, and DEAE Sephadex, parallels bile acid binding activity *in vitro*. Lack of activity of the cholate salt of DEAE Sephadex and increased fecal excretion of bile acids in cockerels treated with DEAE Sephadex suggest that this compound and the cellulose exchangers enhance the excretion of bile acids as insoluble salts of polymeric amines, with concomitant increase in the oxidation of cholesterol to cholic acid in the liver.

Lignin is an amorphous insoluble phenyl propane polymer present in vegetable fiber. Although it is reported to be a bile salt binding agent effective *in vitro*, Heaton et al. (73) found no significant effect on sequestration of bile salts. Lignin has also been evaluated by others (74–76).

Cayen has studied the effect of neomycin and cholesterol metabolism and concluded that the hypocholesteremic action of neomycin in man or chick may be mediated by precipitation of dihydroxy bile acid conjugates (77).

Metamucil and cellulose increase bile salt excretion significantly but the effect of cholestyramine is several times as great according to Stanley et al. (78). In this article it was pointed out that Metamucil and cellulose resemble frequently ingested food components which may be important determinants of variations in bile salt turnover under normal circumstances.

IV. Ion-Exchange Mechanism

In 1850, J. Thomas Way performed experiments on the exchange of cations that occur when soils and other silicates are treated with various electrolytes. Since the time of his studies, other investigators have shown a similar phenomenon in many other systems (81). Phosphates, humus, cellulose, wood, protein, carbon, AL_2O_3, resins, lignin, living cells, $BaSO_4$, AgCl, and other inorganic precipitates have been shown to exhibit the property of ion exchange.

In ion-exchange resins the exchange of ions occur throughout the entire ion-exchange particle. As indicated above, in addition to the polystyrene resins of the cholestyramine type, various cellulose and dextran anion exchangers have been shown to bind bile salts and lower serum cholesterol in experimental animals (52).

V. Mode of Action of Ion Exchangers

In the discussion that follows, cholestyramine will be used as the prototype for the class of resins acting by the ion-exchange mechanism. In the discussion of mode of action, unless otherwise stated, the information presented will pertain to cholestyramine. The large majority of the research data that is available pertains to this resin.

Cholestyramine is a strongly basic ion-exchange resin composed of quaternary ammonium exchange groups attached to a styrene-divinyl benzene polymer lattice. The resin is insoluble in water and is not absorbed from the intestinal tract. It is used as its chloride salt and exchanges chloride in the intestine for bile acids (80). Thus chlorides are released in the exchange and absorbed from the intestine. In adult patients there has been no problem with acidosis in our experience. Even after long-term administration, serum chlorides remain within normal limits or at the upper limits of normal. Experience in children is limited. Extremely large doses of resin in experimental animals has produced a metabolic acidosis.

The equivalent weight of this resin is about 230. It has a high molecular weight, over 1 million (82). As the chloride salt of an anhydrous basic exchange resin, it is slightly acid in reaction (pH 5.6) (83).

VI. Influence on Absorption from the Gastrointestinal Tract

A. Bile Acids

The normal enterohepatic recirculation of bile acid is shown in Fig. 1. This illustration also shows the action of cholestyramine in binding

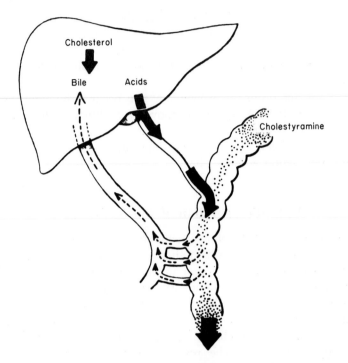

Fig. 1. This figure shows the normal enterohepatic circulation of bile acids with the presence of cholestyramine. In the intestinal tract, bile acids are bound to the resin and thus carried out into the stool, decreasing the amount reabsorbed and returned to the liver. As a consequence of this, more cholesterol is oxidized to bile acids as shown.

bile acids in the intestinal lumen resulting in less reabsorption of bile acids from the intestine. The bile acids are bound to the resin and carried out in the feces thus resulting in a net loss of bile acids from the body.

The liver recycles daily about 25 gm of bile acids, most of which are reabsorbed by the intestine into the portal circulation and thus carried to the liver. They then are reexcreted through the bile into the intestine. Reabsorption of bile acids is from the ileum and is normally 90–95% efficient (84).

Bergstrom has estimated that about 0.7 gm cholesterol is oxidized to bile acids daily (84). This amount enters a pool of 3 to 5 gm which circulates and recirculates through the enterohepatic circulation (85). A corresponding amount of bile acids is lost daily in the stools. At the same time, about 0.4 to 0.5 gm/day of neutral sterols is excreted in the stools. Combined the fecal sterols and bile acids account for nearly all the cholesterol loss from the body each day.

Bile acids are formed in the liver by the oxidation of cholesterol; as bile acids are lost from the body, more cholesterol is oxidized to bile acids in the liver. This results in a loss of body cholesterol and a lowering of serum cholesterol in man.

Studies with radioiodinated low-density lipoproteins (132) indicate that cholestyramine reduces plasma cholesterol levels by increasing the fractional catabolic rate of these lipoproteins, whereas synthesis is unchanged. How increased bile acid excretion brings about this increased catabolism remains unexplained.

B. Cholesterol

With the presence of cholestyramine in the intestinal tract and the consequent binding of bile acids, fewer bile acids are available for cholesterol absorption and there is presumably a diminished intestinal absorption of cholesterol.

It has been shown that cholesterol absorption stops completely in the total absence of bile in the experimental animal (86). It should be noted, however, that the primary mode of action of cholestyramine lowering plasma cholesterol is in the intestinal binding of bile acids and enhanced fecal elimination of these bile acids.

Cholestyramine administration increased the fecal loss of steroidal substances in all species studied. This loss is reflected in a decrease in plasma cholesterol in some species such as the chicken, dog, and man. However, in other species, such as the rat, mouse, and pig the feeding of cholestyramine does not lower the plasma cholesterol (7). The apparent explanation is that the rat, mouse, and pig can increase sterol synthesis sufficiently to compensate for increased fecal loss of bile acids and other steroidal substances.

Studies in lymph-fistula rats (8) have demonstrated reduced intestinal absorption of [4-^{14}C]cholesterol with cholestyramine administration. Other studies in rats suggest cholestyramine may also lower serum cholesterol by preventing cholesterol synthesized by the intestine from reaching circulation (87).

VII. [4-^{14}C]Cholesterol Turnover

Goodman and Noble (88) have performed studies in which [4-^{14}C]-cholesterol was injected intravenously into a series of normal men, untreated hyperlipidemic patients, and hyperlipidemic patients being treated with cholestyramine. The specific activity of plasma total cholesterol was measured during the ensuing 10 weeks. Analysis of the turnover curves of plasma cholesterol revealed that the turnover of plasma cholesterol conformed to a two-pool model. In this study blood samples were collected from 8 hours to 10 weeks after isotope injection. By contrast, Casdorph et al. (89) reported that the turnover of plasma cholesterol in dogs was best fitted to a four-compartment system. In the study by Casdorph, blood samples were obtained within minutes after injection of $[4-^{14}C]$cholesterol, most likely indicating evidence of at least two additional pools having rapid exchange with the plasma pool.

Goodman and Noble calculated several parameters including the constants C_A, C_B, a, and β, the size of the first pool (M_A), the rate constants for the total rate of removal of cholesterol from each pool $(K_{AA}$ and $K_{BB})$, and the production rate in A (PR_A). In two normal men and five untreated patients the average size of pool A was 25 gm.

The effect of cholestyramine was assessed by comparing the results obtained with and without therapy in five subjects studied under both conditions. Cholestyramine therapy produced a large increase in PR_A, from 0.98 to 1.98 gm/day, and in the rate of removal of cholesterol from pool A. The resin did not significantly alter the size of pool A under the conditions of this study. It seems likely that the plasma, red cell, and liver cholesterol together comprise about 15 gm of cholesterol in the average adult, and that they are all a part of the pool with rapid turnover, i.e., pool A. Assuming that the production rate provides a valid estimate of the metabolic turnover rate then these results indicate that cholestyramine produces a 100% increase in the rate of cholesterol degradation and excretion from the body.

In this study cholestyramine at a dose level of 12 gm/day, produced an average lowering of the plasma cholesterol of 14% (range 2–28%). In another larger series, a dose of 24 gm/day produced an average decrease of 20%.

In another study, Goodman and Noble (90) reported that cholestyramine administration resulted in an accelerated turnover of plasma cholesterol. Turnover was measured by $[^{14}C]$cholesterol. The metabolic turnover of cholesterol averaged 1.17 gm/day in the untreated and 1.95 gm/day in the treated subjects.

VIII. Fat-Soluble Vitamins

Because cholestyramine sequesters bile acids there has been some concern that it might interfere with normal fat absorption and thus prevent or impede the absorption of fat-soluble, vitamins such as A, D, E, and K. The general recommendation has been that if cholestyramine is to be given for long periods of time, supplemental fat-soluble vitamins should be given daily in a water-miscible form. It has been stated that chronic use of cholestyramine may be associated with increased bleeding tendencies due to hypoprothrombinemia associated with vitamin K deficiency and that this will usually respond promptly to parenteral vitamin K_1. Recurrences can be prevented by oral administration of vitamin K_1. Gross and Brotman (138) reported the first case of cholestyramine induced hemorrhagic complications secondary to malabsorption of vitamin K. This occurred in a 57-year-old woman who was being treated with cholestyramine for cholerheic enteropathy. See below for further discussion of possible vitamin K deficiency while taking cholestyramine.

The real risk of developing a deficiency of the fat-soluble vitamins is as yet undetermined. In order to assess this potential danger we have drawn, at approximately 12-month intervals, blood levels for vitamins A, E, and carotene. Prothrombin times have been obtained as an indirect measure of vitamin K absorption. Serum calcium levels likewise determined with other parameters at 6-month intervals have been used as a possible indirect measure of calcium absorption. Similarly, folate blood levels as discussed below have been determined at approximately 1-year intervals. Our findings are preliminary and will be briefly summarized.

We have performed 67 determinations of blood levels of vitamin E in 45 hyperlipidemic patients while taking cholestyramine. Sixty-six of these blood levels were within established normal limits, and one was borderline low.

One hundred seventy-two measurements of vitamin A blood level have been obtained in 72 hyperlipidemic patients. Of these determinations 162 were within normal limits, and ten determinations performed in nine patients were below the lower limits of normal. In several of these patients after the low level was obtained, subsequent determinations of vitamin A blood level were normal, without the administration of supplemental vitamins.

One hundred and six determinations of serum carotene in 63 hyperlipidemic patients have been obtained. Of these determinations 99 were within normal limits, and seven values obtained in three different patients were low, and in all three of these patients subsequent to the low values for serum carotene, normal values were obtained on recheck of the serum without the administration of vitamins.

One hundred seventy prothrombin times have been obtained on 71 patients during the course of cholestyramine therapy. Twenty-nine of

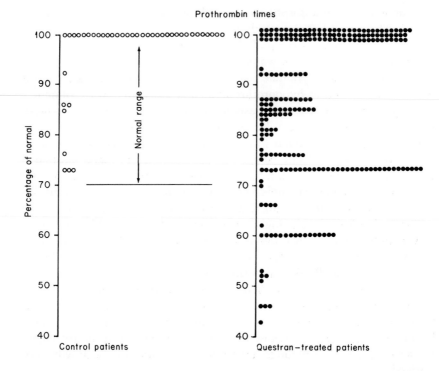

FIG. 2.   170 prothrombin times have been obtained on 71 patients during the course of cholestyramine therapy. 29 of these were below 70% (range 43–66%). These values below 70% suggest some interference with vitamin K absorption. Prothrombin times obtained on 41 control patients all fell within the "normal range," and the large majority of these prothrombin times were 100% of normal. Not shown in the graph is a recent finding of a prothrombin time of 17% in a Questran-treated patient also taking four aspirin tablets daily.

these were below 70% (range 43–66%). These values below 70% may represent some interference with vitamin K absorption. Figure 2 illustrates that prothrombin times obtained on Questran-treated patients fell below the so-called normal range, whereas prothrombin times obtained on 41 control patients all fell within the normal range and the large majority of these prothrombin times were 100% of normal.

As stated above, no bleeding phenomena have been observed in our large group of cholestyramine-treated patients. However, recently a routine recheck of a prothrombin time on a 52-year-old male who had taken the cholestyramine for 5 years revealed a prothrombin time of 17% (control, 11.5 seconds; patient, 32 seconds). Further questioning revealed that he was taking approximately four aspirin tablets daily for mild headaches. Recheck prothrombin time was 78% after cessation of aspirin intake.

Robinson and associates (91) performed studies to determine whether cholestyramine would interfere with the intestinal absorption of fat-soluble vitamins in dogs. These studies were designed to give the resin an optimal opportunity to interfere with vitamin K absorption. Administration of the resin in the dose range recommended for humans had no effect on vitamin K_1 absorption. Large doses of resin administered before and shortly after feeding the vitamin did delay and decrease the absorption of the vitamin.

Kelley et al. (92) have obtained similar results in studies of vitamin K_1 absorption in dogs. They concluded that cholestyramine showed no inhibitory effect on the absorption of vitamin K_1 in dogs when administered at a dosage comparable to that used clinically in human subjects. However, when the resin was given at a dose of 10 gm/day or greater, interference with the absorption of vitamin K_1 occurred. Similar results have been reported by others (91, 93).

Basu has studied in detail the effect of cholestyramine on absorption of vitamin A (94). In this study the addition of 4 gm of cholestyramine to the test meal had no significant effect, while twice this dose or 8 gm, reduced vitamin A absorption significantly. The recommendation of the study was that for prolonged use, or particularly in case of high dose treatment, the patient should safeguard his vitamin A status by the addition of vitamin A supplement to the diet.

Bressler et al. (95) measured vitamin A and carotene serum levels in a group of patients before administration of cholestyramine and again at intervals of 5 to 9 months of therapy. There were no significant depressions of either serum vitamin A or carotene levels.

Longenecker and Basu (96) studied the absorption of vitamin A in four young men. These subjects were fed a test meal (control) or with the addition of 4 to 8 gm of cholestyramine. Vitamin A (250,000 U) was added to the test meal. During a 9-hour postprandial period the high level, 8 gm of cholestyramine, significantly reduced plasma vitamin A levels below values obtained with the control meals. The 4 gm addition of the resin had no significant effect.

Whiteside and associates (97) studied the effects of cholestyramine on the utilization of vitamins A and K in animals. They used 2% cholestyramine in the diet which is equivalent (as a percentage of food intake) to a daily intake in man of about 12 to 15 gm of the resin. In this study cholestyramine was found to have a moderate effect on the absorption of vitamin A and a slight effect on vitamin K.

In adults subsisting on a diet adequate in fat-soluble vitamins, it seems unlikely that the addition of cholestyramine will have a marked effect on vitamin nutrition. When given over a long period especially at a large dosage, it is probably wise to supplement the diet with fat-soluble vitamins (16, 98).

IX. Folic Acid

West and Lloyd (108) have reported decreased serum folate levels in nineteen children treated with cholestyramine for periods up to 20 months. All of these subjects had decreased serum folate levels and six out of twelve tested had subnormal red blood cell folate. The reason that some malabsorption of folic acid might be expected with cholestyramine administration is because the folate ion "will be bound to the resin." They also reported that this abnormality can be corrected by the addition of oral folic acid "which should probably be given to all patients receiving long-term cholestyramine therapy." They found no evidence of malabsorption of other vitamins or minerals. Their study consisted of nineteen children from eleven families, 1–14 years of age, with heterozygous familial hypercholesteremia, treated daily in a dose of 8 to 24 gm daily. Their serum cholesterols were reduced by a mean of 36% (range 27–47%). Following their report we have obtained serum folate levels on 15 control patients, and on 34 patients being treated with long-term cholestyramine therapy. Statistical analysis[1] indicates no statistically significant difference in the folate blood levels obtained on control as well as the Questran-treated patients.

The normal range for serum folate is 5–21 mg/ml. Values between 3–5 mg/ml are considered borderline or are suggestive of deficiency, but an occasional normal will fall into this category. Values below 3 mg/ml are reported to be indicative of folate deficiency (109).

Of our control folate blood values, five out of fifteen were below the "normal range," i.e., below 5 mg/ml. Of the Questran-treated patients, 16 out of 34 had values below the supposed lower limit of normal of 5 mg/ml. In addition the lowest values of blood folates were found in the cholestyramine-treated patients with several having values below 3 mg/ml, i.e., 2.4, 2.8, and 2.4. Although the difference may not be statistically significant, the findings here warrant further investigation and observation of the folic acid levels, especially in patients on long-term cholestyramine therapy. Should patients show signs suggestive of folic acid deficiency, a folic acid level should be obtained at that time especially if anemia or evidence of peripheral neuropathy should develop. It is part of our study

---

[1] Analysis performed by Mead-Johnson Research Center.

protocol to obtain periodic determinations of the serum folic acid level as well as other vitamin blood levels as indicated above.

Table I contains the average and the standard deviation for the control and treatment groups in addition to raw data. It also contains the

TABLE I

VALUES ARE SERUM FOLATE LEVELS (NG/ML) FOR
CONTROL PATIENTS OR QUESTRAN PATIENTS[a, b]

Control		Questran			
Initials	Value	Initials	Value	Initials	Value
D.A.	6.2	R.A.	12.0	E.L.	5.2
M.C.	9.0	G.B.	5.4	C.L.	4.4*
R.C.	8.0	O.B.	2.4*	S.M.	6.0
L.D.	9.0	M.C.	6.0	J.M.	3.0*
C.H.	3.0*	J.C.	13.0	M.N.	3.7*
P.H.	12.0	A.C.	4.0*	A.N.	2.4*
F.J.	15.0	C.D.	7.8	V.P.	6.8
A.L.	4.0*	D.D.	6.0	F.R.	3.4*
L.N.	3.0*	M.E.	5.0	V.R.	4.6*
C.O.	3.0*	G.F.	12.6	E.S.	3.0*
R.R.	4.0*	R.G.	2.8*	R.S.	10.2
D.S.	12.0	W.G.	3.0*	D.S.	8.6
M.S.	8.0	E.H.	8.2	E.S.	11.0
H.T.	9.0	N.H.	2.4*	N.T.	4.2*
W.W.	10.0	B.H.	18.0	M.T.	3.0*
		R.H.	5.0	L.T.	3.0*
		F.J.	5.0	A.W.	4.0*

	Mean	S.D.	"t"		Abnormal	Normal	Total
Control	7.68	3.75		Control	5	10	15
Questran	6.03	3.72		Questran	16	18	34
			1.426		21	28	49
$df = 47$				$\chi^2 = 0.80$			
$P = 0.16$				$P = 0.63$			

[a] In addition to raw data Table I contains the average and the standard deviation for the control and Questran-treated groups. It also contains the results of statistical analysis. The first consideration was to determine whether the average serum folate levels were different in the two groups. A two-sample t test was used to determine that the means did not differ significantly, ($P = 0.16$). Another consideration was to determine whether there was a difference between the proportions of patients with abnormal serum folate levels in the two groups. By counting the number of abnormal values recorded for each group and subjecting this data to a χ^2, it was determined that the proportion did not differ significantly ($P = 0.63$) from one group to the other. The limited analysis to which these data were subjected did not determine that serum folate levels were different in a group of Questran patients than in a group of control patients. Asterisk indicates that the value is outside of the laboratory normal range of 5 to 21 mg/ml.

[b] Statistical analysis, courtesy of Mead Johnson Research Center.

results of the analysis. The first consideration was to determine whether the average serum folate levels were different in the two groups. A two-sample t test was used to determine that the mean did not differ significantly, $P = 0.16$. Another consideration was to determine whether there was a difference between the proportions of patients with abnormal serum folate levels in the two groups. By counting the number of abnormal values recorded for each group and subjecting this data to a χ^2 it was determined that the proportion did not differ significantly, $P = 0.63$, from one group to the other.

The limited analysis to which these data were subjected did not determine that serum folate levels were different in a group of Questran patients than in a group of control patients.[2]

X. Amino Acids

Longenecker and Basu (96) studied the possible effect of cholestyramine on the absorption of amino acids in two young men. Their conclusion was that the relative postprandial concentration changes of plasma amino acid showed that cholestyramine had little effect on the absorption of amino acids.

XI. Fats

Bile acids are necessary for the intestinal digestion and absorption of dietary fat. One would naturally wonder about fat absorption when administering a resin capable of binding bile acids. Studies have shown normal fat absorption, i.e., no increase in fecal fat with the usual dose of cholestyramine. It is possible to induce steatorrhea with extremely large doses of the resin, and in fact some patients taking very large doses note an increase in the frequency of stools.

It has also been observed that in patients who already have steatorrhea, the administration of 4 to 6 gm of cholestyramine daily has been reported to increase the fecal fat excretion (99).

Similar results have been reported by Danhof (100). At a dosage level of 12 gm cholestyramine per day, no significant impairment in the absorption of neutral fat or fatty acid occurred. At a level of 24 gm daily, some impairment in neutral fat was observed, but oleic acid absorption

---

[2] Statistical analysis was kindly performed by Mead Johnson Research Center, Evansville, Indiana.

was considered essentially normal. At 36 gm/day significant impairment in both neutral fat and fatty acid absorption occurred.

Hashim *et al.* (101, 102) have studied fecal fat excretion in patients receiving this resin. In two subjects maintained on constant dietary intake and receiving 15 gm daily of resin, there was no significant difference in daily fecal excretion of fat when experimental periods were compared with control periods. Steatorrhea, however, did occur in two subjects receiving 30 gm daily of resin. All subjects noted borborygmi, increased frequency, and greater bulk of stools during the experimental period.

In five subjects given ^{131}I-labeled triolein before and during 30 gm/day resin treatment, there was depression of level of blood radioactivity over the 8-hour sampling period and significant increase in fecal radioactivity during the period of resin administration. As expected there was no interference in the absorption of ^{131}I-labeled oleic acid.

Steatorrhea induced by cholestyramine administration to normal subjects maintained on formula diets containing long-chain triglyceride (LCT) as the fat source was abolished during isocaloric substitution of medium-chain triglycerides (MCT) (103). These observations suggest that MCT is absorbed efficiently despite diminution in the quantity of available bile acids.

These findings have generally been reproduced in experimental animals (104–106).

XII. Calcium Absorption

In our own series, 275 serum calcium determinations were obtained on 84 hyperlipidemic patients during the course of long-term cholestyramine administration. Out of this group, 270 were within normal limits, and 5 were below the lower limits of normal. None of these 5 patients experienced any symptoms of hypocalcemia and in most of these the serum calcium was only slightly below the limits of normal.

The influence of cholestyramine on calcium has been studied by Briscoe and Ragan (139) in three human subjects under metabolism ward conditions. In seven studies on the three subjects, the amount of calcium retained in the body increased during the period of cholestyramine administration in spite of increased urinary output of calcium because of greater decrease in fecal excretion. This effect of the drug was reproducible at two different levels of calcium intake and two levels of physical activity.

Datta and Sherlock (107) found no change in pre- and posttreatment serum calcium levels in fifteen patients with intrahepatic cholestasis treated with cholestyramine.

XIII. Drug Absorption and Drug Interaction with Cholestyramine

In clinical practice over the past decade we have not encountered problems arising from the administration of cholestyramine to medical patients taking a variety of other drugs. However, because of the nature of the resin and its binding tendency the potential is there for interference with the absorption of some drugs. The resin may have a strong affinity for acidic materials and may also absorb neutral or less likely basic materials to some extent. Cholestyramine may delay or reduce the absorption of concomitant oral medication such as phenylbutazone, warfarin, chlorothiazide, as well as tetracycline, phenobarbital, thyroid and thyroxine preparations, and digitalis. The discontinuance of cholestyramine could pose a hazard to health if a potentially toxic drug such as digitalis had been titrated to a maintenance level while the patient was taking cholestyramine (98).

Cholestyramine will apparently bind some digitalis preparations and it has even been suggested that this resin might be used in the treatment of digitalis intoxication, in particular that due to the administration of digitoxin. Although void of carboxyl groups, cardiac glycosides have a sterol structure similar to bile acid. Saral and Spratt (134) showed that the oral administration of cholestyramine to mice, significantly lowered the toxicity of orally administered digitoxin but did not alter the toxicity of intraperitoneally administered digitoxin.

Caldwell et al. (110, 135) found that the administration of cholestyramine to digitalized human subjects accelerated the metabolic disposition of digitoxin and abbreviated the physiological response to the glycoside. The effect is presumably mediated by interruption of the enterohepatic circulation of digitoxin by cholestyramine. Cholestyramine treatment resulted in reduction in half-life to total serum radioactivity from 11.5 to 6.6 days and in chloroform-extractable radioactivity from 6.0 to 4.5 days as compared to controls.

Bazzano and Bazzano (34) suggest that Colestipol binds more digitalis glycosides in vivo than cholestyramine and more binding of digitoxin than digoxin. They administered Colestipol to five patients with digoxin or digitoxin intoxication and found that Colestipol shortened the plasma half-life of digitoxin from 9.3 days to an average of 2.75 days and that some signs and symptoms of digitalis intoxication and plasma glycoside levels subsided in a much shorter time in patients treated with Colestipol. "In vitro experiments demonstrated that the presence of duodenal juice may decrease the binding capacity of cholestyramine although it does not seem to have a significant effect on the binding effect of Colestipol."

The potential problems posed by the interaction of digitoxin and other drugs including cholestyramine have been summarized by Solomon and Abrams (136).

Gallo and associates (111, 112) studied the possible interaction between the resin and several therapeutic agents including examples of anionic, cationic, and neutral drugs of those most commonly used. Basic and neutral drugs were bound only weakly if at all *in vitro*. Many anionic drugs were not strongly bound and the gastrointestinal absorption of these drugs which were strongly bound was only moderately delayed by concurrent administration of a single dose of resin.

The binding of anionic drugs was compared to the binding of sodium cholate, a representative bile acid to cholestyramine. Warfarin and phenylbutazone were bound more strongly to the resin than was cholate. Chlorothiazide and hydrochlorothiazide were bound about the same extent as cholate, while other anionic drugs such as phenobarbital, aspirin, and tetracycline were bound to a lesser extent. The results of these studies seemed to indicate that the gastrointestinal absorption of some agents was retarded by the concomitant adminsitration of the resin, but the total amount absorbed was not significantly reduced.

Bergman *et al.* (82) have indicated that thyroxine or its iodine-containing derivatives are excreted faster in hamsters fed cholestyramine than in controls and that it enhances the excretion of these substances in the feces. Thyroxine is known to be excreted in the bile and reabsorbed from the intestine in the enterohepatic circulation.

Northcutt *et al.* (113) have reported the malabsorption of administered thyroxine in human patients induced by cholestyramine. According to their studies, this interference with thyroid hormone absorption can be minimized by allowing a time interval of at least 4 hours between the ingestion of the two agents.

The following summarizes the drug effect of cholestyramine: It has a strong affinity for acidic compounds and may decrease the absorption of concomitant oral administration of phenylbutazone, thiazide, tetracycline, phenobarbital, thyroid preparations, digitalis alkaloids, and warfarin. Their administration should be separated from that of cholestyramine by 1 or more hours and, where possible, their blood concentrations or clinical effects should be monitored and their dosage adjusted accordingly (133).

It does seem prudent to advise patients not to take other drugs at the same time that cholestyramine is ingested. Note that the single dose or b.i.d. dose regimen discussed below should help minimize the potential for drug interference. The usual recommendations are that chlorothiazide, phenylbutazone or warfarin should be ingested 1 hour before the anionic-exchange resin·(83). Other drugs should be taken at least 30 minutes to 1 hour before cholestyramine. In patients taking thyroid hormone, an interval between the two drugs of 4 or more hours may be necessary. In the single-dose method of administration discussed below, there would be less of a problem separating the time of administration of the resin from other drugs taken concomitantly by the patient.

Levy reported difficulty regulating anticoagulation with warfarin during cholestyramine therapy (118). He stated, "despite trials in several patients, anticoagulation with warfarin is nearly impossible to regulate during cholestyramine therapy, and if a hypercholesteremic patient has a clear-cut indication for anticoagulation, he should probably not receive cholestyramine." We have not had similar difficulties. As a matter of fact we have encountered no difficulty in regulating prothrombin times in patients anticoagulated with warfarin while taking cholestyramine concomitantly. We have treated 21 patients with warfarin anticoagulation while taking cholestyramine and have not encountered any unusual difficulty in maintaining an adequate degree of anticoagulation. Six of these patients have continued on long-term anticoagulant therapy over a period of several years without difficulty.

XIV. Effects on Serum and Fecal Bile Acid Concentrations

Carey (4, 114) measured the serum and fecal bile acid concentrations demonstrating that the administration of cholestyramine is associated with a fall in the level of serum bile acids and a rise in fecal bile acids.

The bile acid pool is confined almost entirely to the enterohepatic circulation. Only small amounts escape into the feces and even smaller amounts occur normally in systemic blood. To investigate the altered distribution and excretion of bile acids which occur in pruritic jaundice, and the effect of removing them from the enterohepatic circulation, total serum bile acids were measured spectrophotometrically and fecal deoxycholic acid was quantitated after isolation by column chromatography (2). After the administration of cholestyramine at 10 gm daily, the fecal excretion of deoxycholic acid was increased from 54 to 500 mg/day in a healthy adult.

The resin administered to six jaundiced patients with severe pruritus, hyperbileacidemia, lowered serum bile acid and relieved pruritus in each patient. A 5-year-old boy with congenital biliary atresia and unrelenting pruritus had total serum bile acids of 115 μg/ml and excreted less than 1 mg deoxycholic acid per day. Resin feeding lowered serum bile acids to 20 μg/ml and increased fecal deoxycholic acid to 15 mg/day. Pruritus ceased for the first time in his life.

Harkins et al. (105) have performed similar studies in rats. Two percent cholestyramine in the diet increased the fecal excretion of bile acids thirtyfold over control levels, although 5% cholestyramine did not further increase bile acid excretion. Huff et al. (115) also reported a rise in fecal bile acid excretion in rats administered cholestyramine.

Beher et al. (116) reported that bile acid turnover rate in cholestyramine-treated mice was four times as fast as normal control mice. Bile acid pool size was reduced in the treated animals. Similar results have been confirmed by others (117).

XV. Dosage and Administration

The usual adult dose of cholestyramine is 12 to 24 gm daily in three or four divided doses. Levy has reported using doses as high as 32 gm daily (range 16–32 gm daily) given four times daily (118, 133).

Traditionally the resin has been given in three divided doses as a suspension in water or fruit juice before meals. The last 2 years we have also given the drug in a simplified mode of administration either once a day as a single dose or as a "twice a day" program discussed below. In the usual therapeutic range the cholesterol-lowering effect is dose-dependent, i.e., the larger degree of cholesterol lowering is obtained by larger doses of the resin. For many patients with milder forms of hypercholesteremia, doses of 12 gm daily are adequate to normalize the serum cholesterol level and the larger doses are frequently reserved for the more severe forms of hypercholesteremia. A dosage larger than 24 gm daily has been used, however, dosage of resin approaching 30 gm, has been shown to increase fecal fat (101, 102). Up to 36 gm have been used for limited periods of time (9). This dose was associated with marked reduction in serum cholesterol with increase in the amount of feces, fecal fat, and fecal nitrogen. We also have treated a small selected group of patients with doses in this range. Caution is indicated here. These patients must be observed carefully with full knowledge of the possible side effects which might occur, including malabsorption of fat-soluble vitamins, etc.

Cholestyramine is in powder form and is taken orally only after mixing with a liquid. It is prepared in 9-gm packets containing 4 gm of active drug under the trade name of Questran.

XVI. Single-Dose Method and Twice a Day Program of Administering Cholestyramine

During the past 2 years we have conducted studies designed to simplify the method of administration of cholestyramine. Doses of 12 and 16 gm have been administered as a single dose once a day without reference to meals. Larger doses, i.e., 20 or 24 gm daily, are more difficult to take at one time because of the bulk that is produced and for this reason we generally administer these doses divided in half, morning and evening, without reference to food intake (137). We have had one patient take 20 gm daily and another patient take 24 gm as a single dose once a day. Our findings have been that our patients have obtained the same degree of cholesterol lowering (or greater) when the resin was taken as a single daily dose, as when the same dose was taken in the traditional manner of three times a day before meals. Likewise, the patients taking an equivalent dose morning and evening without reference to meals have obtained the same or greater cholesterol lowering than the same dose taken in the traditional manner before meals.

XVII. Results of Single-Dose Method

We have studied twenty-one patients by administering Questran in varying doses of 12 to 16 gm (one patient took 20 gm) administered as a single daily dose in a glass of water or orange juice. These results are shown in Table II. The twenty-one patients in Table II who received the single daily dose of cholestyramine had an average reduction in cholesterol values of 21.58% over baseline values. These results compare favorably with the experience of administering Questran three times a day to a larger group of hyperlipidemic patients.

Twenty patients were treated by administering Questran twice daily without reference to meals. These results are shown in Table III. Most of these patients received 24 gm daily (12 gm b.i.d.). The b.i.d. dose patients obtained a 27.57% reduction of the serum cholesterol.

There were twenty-two patients in Tables II and III, who at different times had received Questran on a t.i.d. regimen, and these patients had a 25.61% reduction of the serum cholesterol. All three percentage reductions compared to 0 are highly significant (P is less than 0.0001). (Statistical analysis performed by Mead-Johnson Research Center.)

Comparison of the q.d. versus t.i.d. regimen for six patients in Table II, who received both indicates that difference in percentage reduction is -4.5%, $P = 0.82$. Similarly, a comparison of the sixteen patients in Table

TABLE II

SINGLE-DOSE METHOD[a]

Patient	Daily dose (gm/day)	Control cholesterol[b]	Milligram reduction	Treatment cholesterol[b]	Percentage reduction	Average t.i.d.[b]	t.i.d. % reduction
1. P.B.	16	250(4)	30	220(3)	12.00	–	–
2. H.B.	16	261(4)	45	216(15)	17.24	227(5)	13
3. O.B.	12	302(5)	67	235(4)	22.19	219(9)	27
4. J.C.	16	256(24)	24	232(3)	9.38	–	–
5. E.E.	16	228(2)	23	205(8)	10.09	–	–
6. M.G.	12	288(4)	43	245(4)	14.93	–	–
7. R.H.	16	290(4)	63	227(3)	21.72	–	–
8. E.H.	12	257(5)	65	192(6)	25.29	195(7)	24
9. N.H.	20	267(10)	60	207(4)	22.47	215(7)	19
10. J.H.	12	290(3)	55	235(1)	18.97	–	–
11. F.J.	16	276(5)	76	200(2)	27.54	–	–
12. H.K.	16	251(14)	40	211(2)	15.94	–	–
13. I.K.	16	278(2)	58	220(5)	20.86	–	–
14. A.L.	16	271(15)	61	210(24)	22.51	–	–
15. J. McN.	16	254(5)	57	197(2)	22.44	–	–
16. E.M.	16	309(5)	46	263(6)	14.89	241(2)	22
17. M.N.	16	257(4)	77	180(3)	29.96	–	–
18. D.S.	12	297(6)	106	191(4)	35.69	195(12)	30
19. E.S.	16	275(2)	127	148(3)	46.18	–	–
20. M.S.	12	313(5)	59	254(1)	18.85	–	–
21. L.T.	16	286(3)	69	217(5)	24.13	–	–

[a] The results of the single-dose method of administering cholestyramine are shown in twenty-one patients. Six of these patients had previously received cholestyramine three times daily, before meals, and their results are shown in the right-hand column of the table.

[b] Numbers in parentheses indicate the number of different cholesterol values used to calculate average cholesterol.

TABLE III
b.i.d. Dose[a]

	Patient	Daily dose (gm/day)	Control cholesterol[b]	Milligram reduction	Treatment cholesterol[b]	Percentage reduction	Average t.i.d.[b]	t.i.d. % reduction
1.	N.B.	16	248(7)	33	215(1)	13.31	—	—
2.	B.C.	24	399(6)	105	294(3)	26.32	309(3)	23
3.	M.C.	24	293(7)	96	197(8)	32.76	181(5)	38
4.	A.C.	24	316(4)	95	221(2)	30.06	223(5)	36
5.	C.D.	16	318(5)	70	248(7)	22.01	229(7)	28
6.	G.F.	24	420(6)	160	260(3)	38.10	—	—
7.	W.G.	24	324(5)	56	268(3)	17.28	233(10)	28
8.	E.H.	24	389(3)	165	224(8)	42.42	253(8)	35
9.	E.L.	24	328(12)	46	282(5)	14.02	230(6)	18
10.	M. McK.	24	484(2)	218	266(5)	45.04	—	—
11.	J.M.	24	312(8)	88	224(3)	28.21	—	—
12.	A.N.	16	300(5)	69	231(4)	23.00	274(15)	9
13.	V.P.	24	255(5)	50	205(1)	19.61	212(5)	17
14.	A.R.	24	373(5)	42	331(4)	11.26	304(5)	18
15.	F.R.	24	269(4)	69	200(5)	25.65	188(4)	30
16.	V.R.	16	282(7)	75	207(2)	26.60	222(12)	21
17.	N.T.	24	284(6)	61	223(10)	21.48	240(10)	15
18.	M.T.	24	374(6)	150	224(4)	40.11	258(7)	31
19.	D.T.	24	349(9)	107	242(2)	30.66	269(1)	23
20.	D.S.	24	413(4)	180	233(5)	43.58	219(6)	47

[a] Twenty patients were treated by administering cholestyramine twice daily without reference to meals. Most of these patients received 24 gm daily (12 gm b.i.d.). The majority of these patients had previously been treated by the administration of cholestyramine three times daily, before meals. For comparison the percentage of cholesterol reduction by the t.i.d. program is shown in the right-hand column.

[b] Numbers in parentheses indicate the number of different cholesterol values used to calculate average cholesterol.

III who had both a b.i.d. and t.i.d. dose administration shows a difference in percentage reduction of 0.24%, $P = 9.1$. Therefore there is no significant difference between the effectiveness of the three dose regimens. It appears from these results that cholestyramine whether administered q.d., b.i.d., or t.i.d. is equally effective in reducing cholesterol levels. Although there was no statistically significant difference in the results obtained by the three different modes of administration, it is to be noted that patients receiving the largest dose, i.e., 24 gm a day, were on the b.i.d. regimen, and the maximal degree of reduction was obtained (27.57%) using this method.

Figure 3 shows daily cholesterol curves on hospitalized patients to whom Questran was administered as a single daily dose. These figures are essentially similar to the curves obtained on other patients receiving Questran by the traditional t.i.d. regimen, also obtained on hospitalized patients, shown on Fig. 4.

XVIII. Daily Cholesterol Curves

Figures 3 and 4 show daily cholesterol curves obtained on inpatients before and after the onset of cholestyramine therapy. By obtaining daily cholesterols we have determined, as shown in these illustrations, that the serum cholesterol falls immediately after the onset of cholestyramine therapy and generally levels off after 7 to 10 days of therapy at the new lower level. Hence the advantage of resin therapy is that one gets an immediate effect, within 7 to 10 days. In the majority of our patients this has represented more or less the maximum effect to be obtained. However, in the article by Levy et al. (23), they noted some additional further decline in the cholesterol during the 4 weeks of their study.

XIX. Results of Cholestyramine Therapy
Immediate and Long-Term

In most studies cholestyramine administration to hyperlipidemic patients and specifically those with type II hyperlipoproteinemia, has resulted in an average reduction of the serum cholesterol of 20 to 25% in various studies. This amount of reduction can be expected in addition to that already obtained by dietary manipulation. (See below for further discussion.) The effects on serum triglycerides have been inconsistent.

During the past 10 years we have conducted long-term studies on the safety and efficacy of cholestyramine. Our total experience includes 246

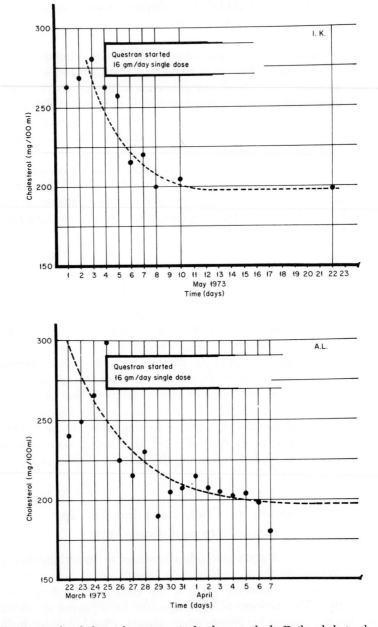

Fig. 3. Daily cholesterol curves—single dose method. Daily cholesterol curves were obtained on hospitalized patients to whom cholestyramine was administered as a single daily dose. These curves are essentially similar to those obtained on patients receiving cholestyramine by the traditional t.i.d. regimen.

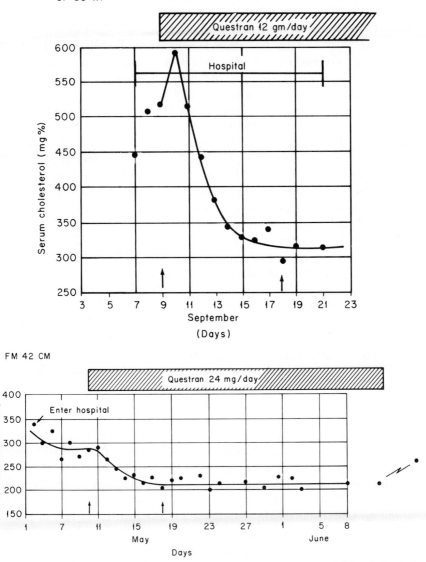

FIG. 4. Daily cholesterol curves—t.i.d. administration program. Daily cholesterol curves were obtained on hospitalized patients to whom cholestyramine was administered three times daily before meals in the usual manner. Generally from the first or second day of administration of cholestyramine the serum cholesterols drop and level off at a new low level approximately 7–10 days after administration. 80 WF, 80-year-old, Caucasian female; FM, patient's initials; 42 CM, 42-year-old Negro male. These curves are essentially similar to those obtained by the single-dose method of administration of cholestyramine.

hyperlipidemic subjects representing approximately 3564 patient-months of therapy. Serum cholesterols were reduced an average of 22% (statistical significance at the P 0.01 level). Serum triglycerides have not been significantly altered by this drug in our series. Results on 69 patients showed an average rise of 5.9 mg/100 ml in the serum triglycerides.[3] One hundred type II patients obtained an average reduction of the serum cholesterol of −20% (range 7–57%).

Cholestyramine has been shown since 1959 to be effective in lowering plasma cholesterol levels. Many additional studies since then have confirmed its effectiveness in patients with essential or familial hypercholesteremia or type II hyperlipoproteinemia (16). Its effect is principally the reduction of β-cholesterol levels secondary to the oxidation of cholesterol to bile acids in the liver, as a consequence of the intestinal sequestering and increased elimination of bile acids. This in turn results in an increased rate of catabolism of β-lipoproteins (LDL).

Levy and associates (23) have summarized the results of the first double-blind comparison of cholestyramine and placebo therapies in an outpatient population of subjects with primary familial type II hyperlipoproteinemia. In this study the efficacy of cholestyramine (16 gm/day) was compared with that of placebo in 47 outpatients with primary type II hyperlipoproteinemia (39 with type IIa), on isocaloric low cholesterol diets, over 14 weeks by a random cross-over, double-blind design. Cholestyramine significantly (P less than 0.005) lowered the plasma levels of cholesterol and low-density lipoprotein cholesterol from means of 333 ± 54 and 265 ± 49 mg/100 ml, respectively, to 264 ± 48 and 193 ± 45 mg/100 ml; these were reductions of 20.6 and 27.3%. Cholestyramine had its greatest lipid-lowing effect during the first week of therapy, although cholesterol levels continued to fall slightly to week four. When cholestyramine was withdrawn the cholesterols returned to within 10% of baseline levels by week one, but some effect persisted until week four. There was a direct relation (P less than 0.05) between the initial cholesterol and low-density lipoprotein levels and the absolute cholesterol fall. From this study an initial statistically significant (P less than 0.05) rise in plasma triglyceride values was observed at the end of the first week of cholestyramine therapy when placebo preceded cholestyramine, but this did not persist while the patient continued to take the drug. (See Fig. 5 and Table IV.)

Fallon and Woods (119) reported that fourteen patients with type II hyperlipoproteinemia responded with a 24% decline in serum cholesterol level. However, seven patients with other types of hyperlipoproteinemia

---

[3] Calculations of our data performed by the Mead-Johnson Research Center.

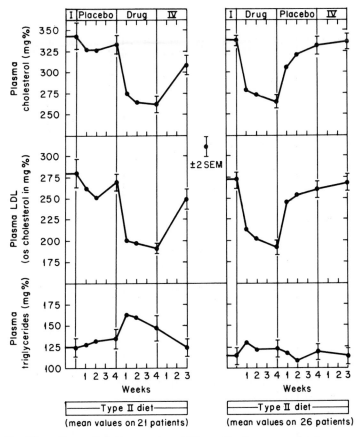

FIG. 5. The effect of cholestyramine (16 gm/day) versus placebo on plasma lipids and lipoproteins. Levels are in milligrams per 100 ml. LDL, low-density lipoproteins. Period I, no drug; period II, placebo; period III, 16 gm/day cholestyramine; and period IV, no drug. With permission of R. I. Levy and "Annals of Internal Medicine" from: Cholestyramine in type II hyperlipoproteinemia, Ann. Int. Med. 79, 51–58 (1973).

had no significant decrease in cholesterol. Serum triglycerides remained unchanged in all patients at the dose used (average 12 gm daily).

Connor has reported results (10) of treatment in fifty hypercholesterolemic adults and seven children with inherited hypercholesteremia type II. Thirty of the hypercholesteremic adults received an average dosage of 22 gm/day of cholestyramine for 12 to 24 months. The average reduction in the serum cholesterol was 23%. Twenty hypercholesteremic adults with tendon xanthomas had a 26% average decrease in the serum cholesterol, taking cholestyramine in a dose of 24 gm/day for 12 months. In

TABLE IV

Effect of Cholestyramine Versus Placebo on Blood Lipid and Lipoprotein Fractions[a, b]

Period[c] (47 patients)	Plasma cholesterol	Plasma triglyceride	(HDL) (cholesterol)	(LDL) (cholesterol)	(VLDL) (cholesterol)
I	341 ± 65.6	120 ± 55.1	44.3 ± 11.4	275 ± 63.2	21.7 ± 12.3
II On placebo[d]					
Week 1	317 ± 50.1	123 ± 49.6	44.7 ± 11.8	253 ± 48.3	20.8 ± 12.3
Week 2	325 ± 49.6	120 ± 51.4	45.2 ± 11.5	258 ± 48.2	21.2 ± 12.9
Week 4	333 ± 54.2	126 ± 50.6	46.9 ± 11.6	265 ± 49.4	21.8 ± 12.8
III On cholestyramine					
Week 1	277 ± 55.0	144 ± 75.1	45.1 ± 11.7	207 ± 55.3	23.9 ± 17.5
Week 2	270 ± 50.8	139 ± 67.0	45.3 ± 12.9	201 ± 51.5	24.1 ± 13.1
Week 4	264 ± 47.5	133 ± 64.5	47.1 ± 11.9	193 ± 44.5	24.2 ± 15.1
IV	325 ± 53.8	120 ± 49.1	45.1 ± 11.1	261 ± 52.2	19.3 ± 12.9

[a] With permission of R. I. Levy and "Annals of Internal Medicine" from: Cholestyramine in type II hyperlipoproteinemia. *Ann. Int. Med.* **79**, 51–58 (1973).

[b] Mean ± S.D.

[c] Patients on type II diet throughout. Period I, no drug; period II, placebo; period III, 16 gm/day cholestyramine; and period IV, no drug.

[d] In each double-blind study, a placebo period of 4 weeks randomly preceded or followed a 4-week period on drug. Placebo and drug values represent the mean values ± S.D. regardless of sequence.

seven children, 4–17 years of age, cholestyramine at an average dose of 14 gm/day reduced the serum cholesterols from 319 to 271 mg/100 ml (−15% reduction).

Fuson and associates (120) treated 65 patients, many with atherosclerotic sequelae using an average of 24 gm of cholestyramine daily in divided doses. Significant mean serum lipid depression occurred in all of these during therapy: cholesterol decreased 43%, 308–173 mg/100 ml, triglycerides decreased 55%, 324–150 mg/100 ml, and phospholipids decreased 25%, 293–220 mg/100 ml.

The cholesterol-lowering effect of cholestyramine has been confirmed by many other studies (9, 11, 121–125).

Effect on Triglycerides

The apparent effect of cholestyramine on triglycerides has been variable, different studies revealing no effect, elevation, or decline in triglycerides. From our own study, triglyceride results were available on 69 hyperlipidemic patients. In these patients an average rise of 5.9 mg/100 ml in the serum triglycerides was recorded, and this was not considered to be statistically significant.[4]

As stated above, Fuson and associates (120) reported a decline in triglycerides of 55% in 65 patients with atherosclerotic sequelae.

Other investigators have demonstrated no significant plasma triglyceride increase with cholestyramine (9, 95). In addition, cholestyramine treatment has apparently produced triglyceride elevation in subjects with normal lipids (126) and in patients with hypertriglyceridemia (127) or hypercholesteremia (123). So it is difficult, if not impossible, to draw valid conclusions regarding triglyceride effect of cholestyramine at this time. It does seem indicated to periodically monitor the serum triglycerides in patients on cholestyramine therapy and particularly in patients with pretreatment elevation of the triglyceride components.

XX. Side Effects

Systemic side effects have been minimized by the fact that the resin is water insoluble and is not absorbed from the gastrointestinal tract. Studies with [^{14}C]cholestyramine have confirmed that it is not absorbed from the intestine (128, 129).

---

[4] Analysis and calculations of triglyceride data performed by Mead-Johnson Research Center.

By far the most common side effect is that of constipation which has been encountered in about one-third to one-half of patients taking cholestyramine. Every patient to whom this resin is administered must be forewarned of the possibility of constipation and must be encouraged to maintain his normal regularity so far as bowel movements are concerned. In many of these patients it is necessary to administer stool softeners or laxatives and occasionally enemas. Even rectal impactions have been encountered in an occasional individual. In some of these patients the constipation seems to lessen with the continued administration of the resin. Levy (118) has reported a similar experience regarding the frequency with which constipation has been encountered in his patients. He stated that the side effect did not prevent the effective use of the drug as long as the usual precautions regarding constipation are taken. They have routinely prescribed hydrophilic stool softeners in order to minimize this problem.

Side effects encountered less frequently have included nausea and epigastric burning which have usually been relieved by antacids. In our experience these have usually occurred in the first few weeks of cholestyramine administration.

Rarely, patients have complained of abdominal fullness or distention, crampy abdominal discomfort, or even diarrhea. Hyperchloremic acidosis in children and gastrointestinal tract obstruction is reported as a rare side effect (133). Bleeding tendency due to hypoprothrombinemia (vitamin K deficiency), vitamin A (one case of night blindness reported) and vitamin D deficiency, rash, and irritation of the skin, tongue, and perianal area have been reported (98). One 10-month-old baby with biliary atresia had an impaction presumably due to cholestyramine resin after 3 days administration of 9 gm daily. She developed acute intestinal sepsis and died (98).

Rectal irritation and aggravation of rectal hemorrhoids has occurred as might be expected in an individual developing constipation. Proctitis and rectal bleeding have also occurred.

Several out of our series have developed acute pancreatitis while taking cholestyramine. This apparently has not been observed by others. Although no direct cause and effect relationship has been established, it is possible that in a patient predisposed to the development of pancreatitis, the resin may serve as a nonspecific irritant of the gastrointestinal tract and may have served as the triggering factor in initiating the episodes of pancreatitis. There were no deaths among this group of patients, however, one patient was seriously ill with necrotizing pancreatitis.

Several patients on long-term cholestyramine therapy for relief of pruritus resulting from primary biliary cirrhosis have developed calcification of intraabdominal organs primarily in the right upper quadrant. The relationship to the cholestyramine therapy has not been established (16).

Schaffner (93, 130) was the first to report the appearance of calcified material in the biliary tree. In his original report he stated that calcified material had been observed in the biliary tree in three cases for unknown reasons. The gallbladder and sometimes the extrahepatic ducts became totally opacified.

In a subsequent report, Schaffner (130) reported that one of the four patients with primary biliary cirrhosis treated with cholestyramine from 18 to 42 months developed peculiar calcifications in the biliary tree. The relationship to cholestyramine in these patients is most difficult to ascertain inasmuch as multiple medications are usually administered, including vitamin D and, at times, calcium tablets in an effort to suppress the osteoporosis, which is frequently severe and symptomatic.

Wells et al. (131) reported the first such case to include a postmortem study. This was the case of a 56-year-old Caucasian woman with primary biliary cirrhosis who developed a right upper quadrant calcification during the course of long-term cholestyramine therapy. At autopsy a single calcium-containing stone was found in the gallbladder. In addition, calcification was found in the pancreas within the ducts of the head of the pancreas, and this was associated with a smoldering acute pancreatitis. Disseminated coccidioidomycosis manifested by multiple pulmonary abscesses and suppuration of hilar and tracheal lymph nodes was also present in this patient at autopsy. Similar lesions were found in the spleen and bone marrow.

The patient described by Wells et al. also received vitamin D and calcium supplement as treatment for osteoporosis. This illustrates the complexity of these cases and the difficulty in relating these calcifications to cholestyramine therapy. Their conclusion was that there was insufficient evidence to implicate cholestyramine as the cause of the phenomenon and that it remained a valuable therapeutic measure.

XXI. Summary

During the past decade, cholestyramine, a bile acid sequestering resin, has found a useful place in the treatment of patients with elevated serum cholesterol. Hence, it is used in the treatment of type II hyperlipoproteinemia. The resin binds bile acids and decreases their reabsorption into the enterohepatic circulation. As a consequence of this loss of bile acids from the body, more cholesterol is oxidized to bile acids in the liver, and there is a net loss of cholesterol from the body and a decreased serum concentration. Other resins apparently acting in a similar manner are being tested and have been discussed.

Because bile acids are necessary for fat absorption there is some concern that prolonged use might interfere with the absorption of nutrients such as fat-soluble vitamins, and some cases of reduced concentration of these fat-soluble vitamins have been found. Reduced prothrombin concentrations have been encountered as well as reduced concentrations of serum folate. The general recommendation has been that if cholestyramine is to be given for long periods of time, supplemental fat-soluble vitamins in a water miscible form should be administered. Possibly supplemental folic acid should also be administered prophylactically to patients on long-term therapy. There is no clear-cut answer to this question. The administration of fat-soluble vitamins to patients (at least in large doses) may carry some risk in itself and must be monitored carefully (by serum calcium levels, etc.). Drug interaction with cholestyramine is a possibility which must be kept in mind especially when administered with a potentially toxic drug such as digitalis (in particular, digitoxin).

The usual adult dose of cholestyramine is 12–24 gm daily, although larger doses have been used. The traditional method of administering the resin is in three or four divided doses as a suspension in water or fruit juice just before meals. We have found it useful to administer the resin once or twice a day without reference to meals. This offers the convenience of a simplified mode of administration. In addition, this may offer the advantage of minimizing drug interaction by separating the time of administration of the resin from that of other drugs taken concomitantly.

Side effects are discussed at length. The potential side effects are multiple; however, the most common problem is that of constipation.

ACKNOWLEDGMENT

I wish to acknowledge the invaluable assistance rendered by Mrs. Margarita Martinez and Miss Nan Nagle in the preparation of this manuscript.

REFERENCES

1. Carey, J. B., Jr. (1961). *Gastroenterology* **40**, 669–670.
2. Carey, M. D., Jr. (1961). *J. Clin. Invest.* **40**, 1028.
3. Carey, J. B., Jr. (1960). *J. Lab. Clin. Med.* **56**, 797–798.
4. Carey, J. B., Jr., and Williams G., (1961). *J. Amer. Med. Ass.* **176**, 432–435.
5. Hashim, S. A., and Van Itallie, T. B. (1960). *J. Invest. Dermatol.* **35**, 253–254.
6. Editorial (1965). *Pharm. Dig.* **28**, 542.
7. Gallo, D. G., Bailey, K. R., Sheffner, A. L., Sarett, H. P., and Cox, W. M., Jr. (1966). *Proc. Soc. Exp. Biol. Med.* **122**, 328–334.

8. Hyun, S. A., Vahouny, G. V., and Treadwell, C. R. (1963). *Proc. Soc. Exp. Biol. Med.* **112**, 496–501.
9. Howard, R. P., Brusco, O. J., and Furman, R. H. (1966). *J. Lab. Clin. Med.* **68**, 12–20.
10. Connor, W. E. (1968). *Med. Clin. N. Amer.* **52**, 1249.
11. Hashim, S. A., and Van Itallie, T. B. (1965). *J. Amer. Med. Ass.* **192**, 289–293.
12. Casdorph, H. R. (1967). *Calif. Med.* **106**, 293–295.
13. Casdorph, H. R. (1967). *Circulation* **34–35**, II–6.
14. Casdorph, H. R. (1967). *Vasc. Dis.* **4**, 305–308.
15. Casdorph, H. R. *Prog. Biochem. Pharmacol.* **II.**
16. Casdorph, H. R. (1971). "Treatment of the Hyperlipidemic States," pp. 243–267. Thomas, Springfield, Illinois.
17. Casdorph, H. R. (1970). *Angiology* **21** (No. 10), 654–660.
18. Casdorph, H. R. (1971). *Ann. Intern. Med.* **74**, 818.
19. Casdorph, H. R. (1970). *Ann. Intern. Med.* **72**, 759.
20. Grundy, S. M., Ahrens, E. H., Jr., and Salen, G. (1971). *J. Lab. Clin. Med.* **78**, 94.
21. Glueck, C. J., Ford, S., Jr., Scheel, D., and Steiner, P. (1972). *J. Amer. Med. Ass.* **222** (No. 6).
22. Nazir, D. J., Horlick, L., Kudchodkar, B. J., and Sodhi, H. S. (1972). *Circulation* **46**.
23. Levy, R. I. *et al.* (1973). *Ann. Intern. Med.* **79**, 51–58.
24. Parkinson, T. M., Gundersen, K., and Nelson, N. A. (1969). *Circulation* **40**, III–19.
24a. Anonymous (1975). *Med. Lett.* **17**, (428), 49.
25. Parkinson, T. M., Gundersen, K., and Nelson, N. A. (1970). *Atherosclerosis* **11**, 531–537.
26. Marmo, E., Caputi, A. P., Cateldi, S., and Amelio, A. (1970). *G. Atherioscler.* **8**, 229–242.
27. Strisower, E. H. (1970). *Circulation* **42**, III–24.
28. Bazzano, G., Gray, M., and Bazzano, G. S. (1970). *Clin. Res.* **18**, 592.
29. Gundersen, K. (1972). *In* "Pharmacological Control of Lipid Metabolism" (W. L. Holmes, R. Paoletti, and D. Kritchevsky, eds.), pp. 298. Plenum Press, New York.
30. Glueck, C. J., Steiner, P. M., Scheel, D., Ford, S. (1971). *Circulation* **44**, II–59.
31. Rubulis, A., Lim, E. C., and Faloon, W. W. (1972). *Fed. Proc. Fed. Amer. Soc. Exp. Biol.* **31**, 727 (Abstr.).
32. Probstfield, J. L., Lee, G., Campion, B., and Hunninghake, D. B. (1972). *Clin. Res.* **20**, 412.
33. Bazzano, G., and Bazzano, G. S. (1972). *Clin. Res.* **20**, 24.
34. Bazzano, G., and Bazzano, G. S. (1972). *J. Amer. Med. Ass.* **220**, 828–830.
35. Ryan, J. R., and Jain, A. (1972). *J. Clin. Pharmacol.* **12**, 268–273.
36. Cooper, E. E. (1972). *Circulation* **46**, II–259.
37. Parsons, W. B. (1972). *Circulation* **46**, II–272.
38. Phillips, W. A., and Elfring, G. L. (1972). *Circulation* **46**, 273.
39. Phillips, W. A., and Schultz, J. R. (1972). *Circulation* **46**, II–273.
40. Clifton-Bligh, P., Miller, N. E., and Nestel, P. J. (1972). *Circulation* **46**, II–249.
41. Glueck, C. J., Steiner, P. M., Scheel, D., and Ford, S. (1972). *J. Amer. Med. Ass.* **222**, 676–681.

42. Nye, E. R., Jackson, D., and Hunter, J. D. (1972). *N. Z. Med. J.* **76,** 12–16.
43. Miller, N. E., Clifton-Bligh, P., Nestel, P. J., and Whyte, H. M. (1973). *Med. J. Aust.* **1,** 1223–1227.
44. Kritchevsky, D., Kim, H. K., and Tepper, S. A. (1973). *Proc. Soc. Exp. Biol. Med.* **142,** 185–188.
45. Parkinson, T. M., Schneider, J. C., Jr., and Phillips, W. A. (1973). *Atherosclerosis* **17,** 167–179.
46. Sachs, B. A., and Wolfman, L. (1973). *N.Y. State J. Med.* **73,** 1068–1070.
47. Ryan, J. R., Jain, A., and McMahon, F. G. (1973). *Clin. Pharmacol. Ther.* **14,** 146–147.
48. Goodman, D. S., Noble, R. P., and Dell, R. B. (1973). *J. Clin. Invest.* **52,** 2646–2655.
49. Harvengt, C., and Desager, J. P. (1973). *Europ. J. Clin. Pharmacol.* **6,** 19–21.
50. Webster, H. D., and Bollert, J. A. (1974). *Toxicol. Appl. Pharmacol.* **28,** 57.
51. Kauffman, R. E., and Azarnoff, D. L. (1973). *Clin. Pharmacol. Ther.* **14,** 886–890.
52. Parkinson, T. M. (1967). *J. Lipid Res.* **8,** 24–29.
53. Chung, E., and Griminger, P. (1973). *Fed. Proc. Fed. Amer. Soc. Exp. Biol.* **32,** 238 (Abstr.).
54. Borgstrom, B. (1970). *Scand. J. Gastroenterol.* **5,** 549–553.
55. Chandrasckhar, N., and Parkinson, T. M. (1967). *Circulation* **34,** II–6.
56. Parkinson, T. M. (1967). *Nature* (*London*) **215,** 415–416.
57. Dryden, L. P., Bitman, J., Wrenn, T. R., and Weyant, J. (1973). *J. Nutr.* **103,** 36–42.
58. Cecil, H. C., Harris, S. J., Bitman, J., and Dryden, L. P. (1973). *J. Nutr.* **103,** 43–48.
59. Howard, A. N., and Hyams, D. E. (1971). *Brit. Med. J.* **3,** 25–27.
60. Howard, A. N., Hyams, D. E., and Evans, R. C. (1972). *In* "Pharmacological Control of Lipid Metabolism" (W. L. Holmes R. Paoletti, and D. Kritchevsky, eds.), pp. 179–187. Plenum, New York.
61. Howard, A. N., Hyams, D. E., and Evans, R. C. (1973). *Angiology* **24,** 22–28.
62. Howard, A. N., and Evans, R. C. (1974). *Atherosclerosis,* **20,** 105.
63. Gundersen, K., personal communication.
64. Howard, A. N., personal communication.
65. Grundy, S. M., personal communication.
66. Phillips, W. E. J., and Brien, R. L. (1970). *J. Nutr.* **100,** 289–292.
67. Keys, A., Grande, F., and Anderson, J. T. (1961). *Proc. Soc. Exp. Biol. Med.* **106,** 555.
68. Fahrenbach, M. J., Riccardi, B. A., and Grant, W. C. (1966). *Proc. Soc. Exp. Biol. Med.* **123,** 321.
69. Ershoff, B. H., and Wells, A. F. (1962). *Proc. Soc. Exp. Biol. Med.* **110,** 580.
70. Riccardi, B. A., and Fahrenbach, M. J. (1967). *Proc. Soc. Exp. Biol. Med.* **124,** 749.
71. Leveille, G. A., and Sauberlich, H. E. (1966). *J. Nutr.* **88,** 209–214.
72. Pick, R., Savitri, J., and Katz, L. N. (1965). *Circulation* **32,** II–26.
73. Heaton, K. W., Heaton, S. T., and Barry, R. E. (1971). *Scand. J. Gastroenterol.* **6/3,** 281–286.
74. Eastwood, M. A., and Hamilton, D. (1968). *Biochim. Biophys. Acta* **152,** 165–173.
75. Editorial (1970). *Lancet,* **ii,** 1023–1024.

76. Eastwood, M. A., and Girdwood, R. H. (1968). *Lancet,* **ii,** 1170–1172.
77. Cayen, M. N. (1970). *Amer. J. Clin. Nutr.* **23/9,** 1234–1240.
78. Stanley, M., Paul, D., Gacke, D., and Murphy, J. (1972). *Gastroenterology* **62/4,** 816.
79. Gross, L., and Figueredo, R. (1973). *J. Amer. Geriat. Soc.* **21/12,** 552–556.
80. Editorial (1965). *Pharm. Dig.* **28,** 542.
81. Kunin, R. "Ion Exchange Resins," 2nd ed. Wiley, New York.
82. Bergman, F., Heddman, P. A., and van der Linden, W. (1966). *Acta Endocrinol.* (Copenhagen) **53,** 256–263.
83. Editorial (1967). *Modern Drugs* pp. 235–236.
84. Bergstrom, S. (1961). *Fed. Proc. Fed. Amer. Soc. Exp. Biol.* **20,** 121–126.
85. Van Itallie, T. B., and Hashim, S. A. (1963). *Med. Clin. N. Amer.* **47,** 629–648.
86. Siperstein, M. D., Nichols, C. W., and Chaikoff, I. L. (1953). *Science* **117,** 386.
87. Reinke, R. T., and Wilson, J. D. (1966). *Clin. Res.* **14,** 49.
88. Goodman, D. S., and Noble, R. P. (1968). *J. Clin. Invest.* **47,** 231.
89. Casdorph, H. R., Juergens, J. L., Orvis, A. L., and Owen, C. A., Jr. (1963). *Proc. Soc. Exp. Biol. Med.* **112,** 191.
90. Goodman, D. S., and Noble, R. P. (1966). *Geriatrics* **34,** 112–113.
91. Robinson, H. J., Kelley, K. L., and Lehman, E. G. (1964). *Proc. Soc. Exp. Biol. Med.* **115,** 112–115.
92. Kelley, K. L., Lehman, E. G., and Robinson, H. J. (1963). *Fed. Proc. Fed. Amer. Soc. Exp. Biol.* **22,** 434.
93. Schaffner, F. (1964). *Gastroenterology* **46,** 67–70.
94. Basu, S. G. (1965). Master's Thesis, The University of Texas, Austin, Texas.
95. Bressler, R., Nowlin, J., and Bogdonoff, M. D. (1966). *Southern Med. J.* **59,** 1097–1103.
96. Longenecker, J. B., and Basu, S. G. (1965). *Fed. Proc. Fed. Amer. Soc. Exp. Biol.* **24,** 375.
97. Whiteside, C. H., Harkins, R. W., Fluckiger, H. B., and Sarett, H. P. (1965). *Amer. J. Clin. Nutr.* **16,** 309–314.
98. "Physician's Desk Reference," 28th ed., pp. 986–987. Medical Economics Company, Oradell, New Jersey, 1974. Questran (Cholestyramine).
99. Wessler, S., and Acioli, L. A. (1968). *J. Amer. Med. Ass.* **206,** 1285.
100. Danhof, I. E. (1966). *Amer. J. Clin. Nutr.* **18,** 343–349.
101. Hashim, S. A., Bergen, S. S., Jr., and Van Itallie, T. B. (1961). *Proc. Soc. Exp. Biol. Med.* **106,** 173–175.
102. Hashim, S. A., Bergen, S. S., and Van Itallie, T. B. (1960). *Clin. Res.* **8,** 201.
103. Zurier, R. B., Hashim, S. A., and Van Itallie, T. B. (1965). *Gastroenterology* **49,** 490–495.
104. Harkins, R. W., and Hagerman, L. M. (1965). *Fed. Proc. Fed. Amer. Soc. Exp. Biol.* **24,** 375.
105. Harkins, R. W., Hagerman, L. M., and Sarett, H. P. (1965). *J. Nutr.* **87,** 85–92.
106. Harkins, R. W., Whiteside, C. H., Fluckiger, H. B., and Sarett, H. P. (1965). *Proc. Soc. Exp. Biol. Med.* **118,** 399–402.
107. Datta, D. V., and Sherlock, S. (1963). *Brit. Med. J.* **1,** 216–219.
108. West, R. J., and Lloyd, J. K.: (1973). *Arch. Dis. Childhood* **48,** 370–374.
109. "The Bio-Science Handbook," 10th ed., pp. 140. Bio Science Lab., Van Nuys, California, 1973.

256   H. RICHARD CASDORPH

110. Caldwell, J. H., and Greenberger, N. J. (1970). *J. Clin. Invest.* **49**, 16A.
111. Gallo, D. G., and Sheffner, A. L. (1964). *Fed. Proc. Fed. Amer. Soc. Exp. Biol.* **23**, 323.
112. Gallo, D. G., Bailey, K. R., and Sheffner, A. L. (1965). *Proc. Soc. Exp. Biol. Med.* **120**, 60–65.
113. Northcutt, R. C., Stiel, J. N., Hollifield, J. W., and Stant, E. G. (1969). *J. Amer. Med. Ass.* **208**, 1857–1861.
114. Carey, J. B., Jr. (1961). *Gastroenterology* **41**, 285–287.
115. Huff, J. W., Gilfillan, J. L., and Hunt, V. M. (1963). *Proc. Soc. Exp. Biol. Med.* **114**, 352–355.
116. Beher, W. T., Beher, M. E., and Rao, B. (1966). *Proc. Soc. Exp. Biol. Med.* **122**, 881–884.
117. Billiau, A., and Van den Bosch, J. (1964). *Arch. Int. Pharm. Ther.* **150**, 46–51.
118. Levy, R. I. (1972). *Ann. Intern. Med.* **77**, 267–294.
119. Fallon, H. J., and Woods, J. W. (1968). *J. Amer. Med. Ass.* **204**, 1161.
120. Fuson, R. L., Whalen, R. E., Hackel, D. B., Hudson, B. H., and Sabiston, D. C. (1967). *Surg. Forum* **18**, 354–355.
121. Bergen, S. S., Jr., Van Itallie, T. B., Tennent, D. M., and Sebrell, W. H. (1959). *Circulation* **20**, 981.
122. Bergen, S. S., Jr., Van Itallie, T. B., Tennent, D. M., and Sebrell, W. H. (1959). *Proc. Soc. Exp. Biol. Med.* **102**, 676–678.
123. Berkowitz, D. (1963). *Amer. J. Cardiol.* **12**, 834–840.
124. Horan, J. M., DiLuzio, N. R., and Etteldorf, J. N. (1964). *J. Pediat.* **64**, 201–209.
125. Steinberg, D. (1962). *Trans. N.Y. Acad. Sci.* **24**, 704–723.
126. Weizel, A. *et al.* (1969). *Proc. Soc. Exp. Biol. Med.* **130**, 149–150.
127. Jones, R. J., and Dobrilouic, L. (1970). *J. Lab. Clin. Med.* **75**, 953–967.
128. Gallo, D. G., and Sheffner, A. L. (1965). *Fed. Proc. Fed. Amer. Soc. Exp. Biol.* **24**, 325.
129. Gallo, D. G., and Sheffner, A. L. (1965). *Proc. Soc. Exp. Biol. Med.* **120**, 91–93.
130. Schaffner, F., Klion, F. M., and Latuff, A. J. (1965). *Gastroenterology* **48**, 293–298.
131. Wells, R. F., Knepshield, J. H., and Davis, C. (1968). *Amer. J. Dig. Dis.* **13**, 86–94.
132. Langer, T., Levy, R. I., and Fredrickson, D. S. (1969). *Circulation* **40**, III–14.
133. Levy, R. I., Morganroth, J., and Rifkind, B. M. (1974). *New Eng. J. Med.* **290**, 1295–1301.
134. Saral, R., and Spratt, J. L. (1967). *Arch. Int. Pharmacodyn.* **167**, 10–18.
135. Caldwell, J. H., Bush, C. A., and Greenberger, N. J. (1971). *J. Clin. Invest.* **50**, 2638–2644.
136. Solomon, H. M., and Abrams, W. B. (1972). *Amer. Heart J.* **83**, 277–280.
137. Casdorph, H. R. (1975). *Angiology* **26**, 671–682.
138. Gross, L., and Brotman, M. (1970). *Ann. Intern. Med.* **72**, 95–96.
139. Briscoe, A. M., and Ragan, C. (1963). *Amer. J. Clin. Nutr.* **13**, 277–284.

Chapter 8

TREATMENT OF
HYPERLIPOPROTEINEMIA
IN CHILDREN

C. J. GLUECK, R. W. FALLAT, AND R. C. TSANG

*General Clinical Research Center, Lipoprotein Research Laboratory,
and the Fels Division of Pediatric Research and Newborn Division,
Department of Pediatrics, University of Cincinnati Medical Center,
Cincinnati, Ohio*

I. Atherosclerosis: Possible Genesis
in Childhood

Ischemic coronary and cerebral vascular diseases secondary to atherosclerosis are the major causes of morbidity and mortality in the United

States and in the industrial nations of the world (Hilleboe, 1967). A series of authors have suggested that the genesis of the atherosclerotic lesion is in childhood (Strong *et al.*, 1958; Strong and McGill, 1969; Kannel and Dawber, 1972; Glueck *et al.*, 1974c, 1975d). The atherosclerosis apparently progresses through a series of linked steps:

1. Development of arterial fatty streaks in infancy and childhood
2. Formation of fibrous plaque in the second to third decade of life
3. Development of complicated fibrotic and calcific mature atherosclerotic lesions in the fourth and later decades
4. Subsequent episodes of clinical atherosclerotic cardiovascular disease

Some fatty streaks are probably resorbed whereas others develop into fibrotic plaques (Strong *et al.*, 1958; Strong and McGill, 1969). In autopsy studies of Korean and Vietnam War casualties, Enos *et al.* (1955) and McNamara *et al.* (1971) found that approximately 50–70% of the men had extensive areas of aortic and coronary fibrous plaques in the second, third, and fourth decades of life. If complicated occlusive arterial lesions evolve from fibrous plaques, the data of Enos and McMillan *et al.* suggests that nearly half of all American males are already predisposed to atherosclerotic cardiovascular disease (ASCVD) by the end of the second or third decade. On the basis of these types of studies of arterial pathology and the known association of dietary composition, hyperlipidemia, and ASCVD, the atherosclerosis study group (1970) suggested overall national reduction of dietary cholesterol and saturated fatty acid intake for all age groups.

Within this frame of reference, familial and acquired elevations of plasma cholesterol and triglyceride in the pediatric age group may well represent risk factors pertinent to development of atherosclerosis in adult life (Kannel and Dawber, 1972; Glueck *et al.*, 1974a; Tamir *et al.*, 1972; Kwiterovich *et al.*, 1974; McGandy, 1971).

Familial hypercholesterolemia, hypertriglyceridemia, and combined hyperlipidemia are highly associated with the development of premature ischemic heart disease (Kwiterovich *et al.*, 1974; Stone *et al.*, 1973; Slack and Nevin, 1968; Slack, 1969; Goldstein *et al.*, 1973a). Carlson and Bottiger (1972) have shown that in addition to cholesterol, increases in plasma triglycerides are clearly associated with the development of premature heart disease. Early diagnosis of hyperlipoproteinemias in children coupled with prudent approaches to diet and/or drug therapy might provide a useful long-term approach to primary prevention of atherosclerosis. This chapter focuses upon recent studies of hypercholesterolemia and hypertriglyceridemia in children including the following areas: (1) lipids and lipoproteins in evaluation of the patient; (2)

patient evaluation, physical findings, and clinical events; (3) familial versus acquired hyperlipoproteinemia; (4) dietary recommendations for acquired hypercholesterolemia; and (5) diet and drug therapy for familial hyperlipoproteinemias.

The association of familial and acquired elevations of cholesterol and triglyceride with ASCVD, the relatively high prevalence of the familial disorders, and the common occurrence of acquired hypercholesterolemia lend considerable importance to studies in these areas. If atherosclerosis prevention is to be most effective, there is considerable urgency for establishment of prophylactic measures early in childhood.

II. Etiology and Classification of Pediatric Hyperlipoproteinemias

A. Lipids and Lipoproteins in Evaluation of the Patient

A useful frame of reference for initial sampling of plasma lipids and lipoproteins is provided by the Fredrickson and Levy (1972) lipoprotein phenotyping system. Five major lipoprotein phenotypes are currently recognized. The predominant hyperlipoproteinemias in the pediatric age group are familial and acquired hypercholesterolemia (type II hyperlipoproteinemia) and familial and acquired hypertriglyceridemia (type IV lipoprotein phenotype). Type II hyperlipoproteinemia is subdivided into types IIA (hypercholesterolemia without associated hypertriglyceridemia) and IIB (hypercholesterolemia with associated hypertriglyceridemia) (Fredrickson and Levy, 1972; Lloyd, 1972; Levy and Rifkind, 1973; Friedewald et al., 1972). Type IV hyperlipoproteinemia can be diagnosed in 20–25% of children born to parents with familial hypertriglyceridemia (Glueck et al., 1973c).

Type I hyperlipoproteinemia (familial hyperchylomicronemia), an extremely rare disorder in both adults and children, presents in childhood and is characterized by fasting hyperchylomicronemia, eruptive xanthomas, hepatosplenomegaly, and pancreatitis. Type III hyperlipoproteinemia rarely if ever occurs before age 20 (Fuhrmann et al., 1971; Fredrickson and Levy, 1972).

Type V hyperlipoproteinemia, a more common disorder in adults, is only rarely observed in children (Fredrickson and Levy, 1972; Lloyd, 1972). Types I and V hyperlipoproteinemias have been recently reviewed in depth by Fredrickson and Levy (1972), Levy and Rifkind (1973), and (Lloyd and Wolff, 1969a,b; Lloyd, 1972). Because of their relative rarity in children, types I and IV will not be specifically covered in this chapter.

Initial lipid determinations in children should include venous blood for cholesterol and triglyceride determination with the child having been on a 12- to 14-hour fast (Tables I and II). Quantitative measurements of cholesterol and triglyceride alone provide a reasonably accurate determination of lipoprotein phenotype in most subjects (Fredrickson and Levy, 1972). Lipoprotein electrophoresis is for the most part qualitative and may not be required for clinical studies. The most accurate diagnosis of familial hypercholesterolemia and hypertriglyceridemia in children is provided through quantitation of plasma low-density lipoprotein cholesterol and very low-density lipoprotein cholesterol following the National Heart and Lung Institute system (Fredrickson and Levy, 1972; Glueck *et al.*, 1973a; Levy and Rifkind, 1973; Kwiterovich *et al.*, 1974). Currently the age-adjusted "suggested" normal limits for cholesterol and triglycer-

TABLE I

HYPERCHOLESTEROLEMIA, PRIMARY AND SECONDARY,
TYPES IIA AND IIB LIPOPROTEIN PHENOTYPE

1. Obtain at least two initial fasting plasma samples on habitual diet, weight stable. In infants and young children dietary history for 2- to 4-week period prior to sampling important
2. Lipids and lipoproteins
    Elevated total cholesterol, LDL, normal triglycerides (type IIA)
    Elevated total cholesterol, LDL, elevated triglycerides, VLDL cholesterol (type IIB), absence of floating LDL
3. By history, physical examination, laboratory, and dietary information assess for secondary hypercholesterolemia
4. Causes of secondary hypercholesterolemia
    a. Habitual diet high in cholesterol and saturated fatty acids
    b. Hypothyroidism
    c. Nephrotic syndrome
    d. Biliary obstruction, hepatic disease
    e. Ketogenic diets
    f. Acute porphyria
    g. Androgenic steroids (high dose)
    h. Corticosteroid excess
5. After evaluation for secondary hypercholesterolemia, attempt to confirm primary hypercholesterolemia
    a. Absence of common disorders which cause secondary hypercholesterolemia
    b. Absence of diet-induced hypercholesterolemia. On low cholesterol < 300 mg/day, polyunsaturate-rich, P/S 2:1 diet, nearly all children with dietary hypercholesterolemia will normalize cholesterol levels
    c. Confirmation by family screening
    d. Presence of tendinous xanthomas (present in about 7 to 10% of children with familial hypercholesterolemia)

TABLE II

HYPERTRIGLYCERIDEMIA, PRIMARY AND SECONDARY, TYPE IV LIPOPROTEIN PHENOTYPE

1. Obtain at least two fasting plasma samples on habitual diet with weight stable
2. Lipids and Lipoproteins
   Elevated triglycerides, VLDL cholesterol. Normal to high normal cholesterol and normal LDL cholesterol
3. By history, physical examination, laboratory, and dietary information evaluate for secondary hypertriglyceridemia
4. Causes of secondary hypertriglyceridemia in children
   a. Poorly controlled diabetes mellitus
   b. Hypothyroidism
   c. Nephrotic syndrome
   d. Glycogen storage disease
   e. Dysproteinemias, lupus erythematosus
   f. Pancreatitis
   g. Uremia
   h. Storage diseases (Gaucher, Niemann-Pick, lecithin-cholesterol-acyltransferase deficiency)
   i. Estrogen-progestin oral contraceptives
   j. Pregnancy
   k. Corticosteroid excess
   l. Excess alcohol intake
5. Primary and familial hypertriglyceridemia
   a. Absence of disorders which cause secondary hypertriglyceridemia
   b. Persistence of elevated triglyceride levels in face of good control of blood sugar in juvenile diabetes
   c. Confirmation by family screening

ides of Fredrickson and Levy (1972) are used in pediatric studies. Normal limits for cord blood cholesterol, LDL, and triglycerides have been suggested by Tsang et al. (1974), Kwiterovich et al. (1973), Goldstein et al. (1973b), Greten et al. (1973), and Darmady et al. (1972). Upper normal limits based on much larger pediatric population bases from ongoing pediatric lipid prevalence studies of the Lipid Research Centers will soon be available.

A minimum of two fasting samples should probably be obtained at weekly intervals as baseline data before a program of differentiation between primary and secondary hyperlipidemia, family sampling, and therapy is contemplated (Tables I and II). Since the diagnosis of hyperlipidemia in children has lifelong implications for diet or drug therapy (Lloyd and Wolff, 1969a, b; Lloyd, 1972), the diagnosis should be based on several index samples. After repetitive confirmation of elevations of cholesterol or triglyceride, or both, the physician should rule out secondary causes of hypercholesterolemia and/or hypertriglyceridemia (Tables I and II). In addition, a rather complete diet history with assessment of the amount of cholesterol and saturated fats eaten should also be taken (Tables I and II). When the diagnosis of primary hyperlipoproteinemia is made, screening of all available family members should be done in an attempt to identify affected siblings, parents and grandparents, and to confirm the familial nature of the hyperlipidemia in the kindred (Tables I and II).

B. Physical Findings, Symptoms, and Clinical Events

In a majority of children under age 21 with primary hypercholesterolemia or hypertriglyceridemia, there are no abnormal physical findings and limited number of symptoms and clinical events (Kwiterovich et al., 1974; Levy and Rifkind, 1973; Lloyd and Wolff, 1969a,b; Lloyd, 1972; Harlan et al., 1966). Children with familial hypertriglyceridemia and the type IV lipoprotein phenotype have no characteristic symptoms or specific physical findings (Glueck et al., 1973b,c), but may be moderately obese (Glueck et al., 1973c; Segall et al., 1970a). Approximately 7–10% of children heterozygous for familial hypercholesterolemia have achilles tendon xanthomas (Glueck et al., 1973a; Kwiterovich et al., 1974). Recurrent acute achilles tendinitis and tenosynovitis may antedate the presence of palpable tendinous xanthomas in children heterozygous for type II (Glueck et al., 1968; Shapiro et al., 1974). Arcus cornea juvenalis may rarely be present. Children heterozygous for type II do not have angina pectoris or unexplained cardiac murmurs (Glueck et al., 1973a; Kwiterovich et al., 1974).

In the rare children homozygous for familial hypercholesterolemia, tendon and tuberous xanthomas, as well as aortic stenosis secondary to aortic-ring atheromas are often present (Khachadurian, 1968; Moutafis et al., 1971). Myocardial ischemia and infarctions are very common in children homozygous for familial type II, most of whom do not survive into the third decade of life.

C. Familial versus Acquired Hyperlipoproteinemia

National diet habits may shift the distribution of cholesterol sharply to the right and increase the skew toward higher levels. In several pediatric population studies, 8–25% of children in industrial societies have had moderate to substantial elevations of plasma cholesterol, for the most part, related to dietary habits. Schilling et al. (1964) reported a marked age-related increase in plasma cholesterol in the late second decade of life in American males, so that by ages 20, 30, and 40, virtually all American males were relatively hypercholesterolemic by world-wide norms. Golub-jatnikov et al. (1972) measured serum cholesterols in rural Mexican and Wisconsion school children and reported that the mean cholesterol of the Wisconsin children was approximately twice that of rural Mexicans. They concluded that diet modification should be started in the preschool and school-age period to achieve maximum success. Clarke et al. (1970) reported that 13% of adolescent Vermont children had cholesterol levels above 200 mg/100 ml. Starr (1971) found that 7% of school children, ages 6–14, had cholesterol levels above 220 mg/100 ml. McGandy (1971) studied 1236 boys in eastern United States school systems and found plasma cholesterol levels to be in the 180–200 mg/100 ml range. Children in the upper and lower quartiles of the cholesterol frequency distribution at the first examination remained in these quartiles of the distribution at follow-up sampling. McGandy concluded that dietary and environmental forces strongly associated with arteriosclerotic cardiovascular disease in middle life began to operate during adolescence. A high prevalence of pediatric hypercholesterolemia has not been limited to American studies. Godfrey et al. (1972) reported that 32 out of 1292 Australian school children had plasma cholesterol levels over 238 mg/100 ml. The frequency of acquired hypertriglyceridemia in free-living populations is not yet well known. Pediatric population studies of the Lipid Research Clinics should furnish data on prevalence of acquired and familial hypertriglyceridemia within the next several years.

If 6–15% of children in westernized societies have elevated plasma cholesterol, what proportion of these have familial elevations of cholesterol? Studies by Tsang et al. (1974) have estimated that the prevalence of "monogenic" familial hypercholesterolemia in unselected neonates is approximately 1/250, while studies by Goldstein et al. (1973a,b) have suggested a somewhat lower prevalence of approximately 1 in 600. Assuming that a conservative estimate of the prevalence of familial hypercholesterolemia would be approximately 1 in 300 to 1 in 500 unselected children or adults, the prevalence of hypercholesterolemia acquired secondary to diet is much more common than that of familial hypercholesterolemia.

Familial hypercholesterolemia, familial combined hyperlipidemia, and familial hypertriglyceridemia all appear to be transmitted as autosomal dominant traits (Goldstein et al., 1973a; Fredrickson and Levy, 1972; Glueck et al., 1973b,c; Kwiterovich et al., 1973, 1974; Tsang et al., 1974). However, the distinction between familial and acquired hypercholesterolemia is made difficult by the current absence of widely utilizable biochemical tests which differentiate specific forms of familial hypercholesterolemia and differentiate these from acquired hypercholesterolemia.

Currently, measurements used in distinguishing familial from acquired hyperlipidemia include measurements of lipid and lipoprotein levels, assessment for other disease states that might cause secondary hyperlipidemia, and extensive kindred sampling (Tables I and II). More specific techniques may hopefully be developed along the lines of the recent elegant studies by Goldstein and Brown (1973) and Brown *et al.* (1974) who demonstrated that cultured human fibroblast cells from type II homozygotes failed to suppress cholesterol synthesis normally when exposed to human low-density lipoproteins. HMG-CoA reductase activity in fibroblasts from homozygotes was increased 60 times that of normal cells, and synthesis in fibroblast cells from heterozygotes fell somewhere in the range below that of homozygous cells and above that of normals.

III. Acquired Hypercholesterolemia: Dietary Recommendations for Unselected Pediatric Populations

The high prevalence of hypercholesterolemia in children from adult societies with unacceptably high ASCVD event rates (Kannel and Dawber, 1972) has led to dietary recommendations aimed at the amelioration of atherosclerosis. The Atherosclerosis Study Group (1970) made a series of recommendations for the general public from infancy through adulthood based upon the statistical association of dietary composition and development of atherosclerosis as follows: (1) caloric intake be adjusted to achieve and maintain optimal weight; (2) dietary cholesterol be restricted to <300 mg/day; and (3) dietary saturate fat intake be substantially reduced.

Dietary fat should provide <35% of total calories. Fat calories should be equally distributed among saturated, monounsaturated, and polyunsaturated sources. These suggested changes would diminish cholesterol intake by 50%, decrease saturated fatty acid intake from 15 to 18% of calories to 10% or less, and would increase polyunsaturated fats from 3 to 5% to 10% of calories. From a practical point of view, this would diminish the intakes of eggs, butter, cheese, and ice cream and moderately reduce the intake of meats with high levels of saturated fat (Schubert, 1973). The Atherosclerosis Study Group further stated that "with these dietary principles, requirements for optimal nutrition can be met for all sectors of the population including infants, children, adolescents, pregnant and lactating women and older persons." It was emphasized that lifelong diet from infancy to adulthood be altered in an attempt to prevent development of elevated blood cholesterol levels, and atherosclerosis.

Two pediatric groups have subsequently criticized these recommendations. The committee on nutrition of the American Academy of Pediatrics (1972) concluded that "such dietary intervention is at present experimental and recommends against dietary changes for all children." The committee felt that the study group recommendations would be "more persuasive" if dietary cholesterol restriction in unselected populations could be shown to reduce the frequency of coronary heart disease. The committee on nutrition concluded that dietary intervention should be tested in persons with a high risk of inherited coronary heart disease, specifically those with familial type II hyperlipoproteinemia. They felt that modified cholesterol-poor, polyunsaturate-rich diet for persons with familial hypercholesterolemia would provide data relevant to safety, acceptability, and effectiveness of the diet for general populations. A second group of pediatric researchers (Mitchell et al., 1972) reached a similar conclusion. "There is no scientific justification at this time for recommending to the population at large that diets of all children be radically altered in the hope of preventing premature heart disease."

There is general agreement, however, that children whose hypercholesterolemia is secondary to habitual cholesterol-rich diet benefit from prudent restriction of cholesterol and saturated fatty acid intake (McGandy, 1971; Golubjatnikov et al., 1972; Stare and McWilliams, 1973). Stare and McWilliams (1973) effected substantial plasma cholesterol lowering in 500 adolescent males by reduction of fat from 39 to 33% of calories and increasing the P/S ratio from 0.2 to 1.0. Very little disruption of food habits was necessary for this fat and cholesterol modified diet.

In the small number of children with well-defined familial elevations of cholesterol, triglyceride, or both, early diagnosis and prudent early intervention appear to be appropriate in an attempt to reduce cholesterol and triglyceride levels to normal.

IV. Diet and Drug Therapy of Familial Hyperlipoproteinemias

Pediatric hypertriglyceridemia and hypercholesterolemia at birth and later in life may represent a heterogeneous group of familial disorders including the following: (1) familial hypercholesterolemia (type II hyperlipoproteinemia); (2) familial combined hyperlipidemia; and (3) familial hypertriglyceridemia (type IV hyperlipoproteinemia).

A. Familial Hypercholesterolemia

Familial hypercholesterolemia (essential hypercholesterolemia, hypercholesterolemic xanthomatosis, familial type II hyperlipoproteinemia) is an inherited (autosomal dominant) disorder of lipoprotein metabolism (Fredrickson and Levy, 1972; Goldstein et al., 1973a,b; Kwiterovich et al., 1973, 1974; Schrott et al., 1972; Slack and Nevin, 1968). Heterozygotes are characterized by major increments in cholesterol and LDL cholesterol. Homozygotes have extraordinary elevations of cholesterol (800–1000 mg/100 ml) and usually have ASCVD before age 20. The homozygous state is distinctively characterized by the lack of feedback responsiveness of cultured human fibroblast to LDL and a basal 40- to 60-fold increment in synthesis of HMG-CoA reductase molecules (Brown et al., 1974; Goldstein and Brown, 1973). Familial hypercholesterolemia is probably the most commonly identified form of familial hyperlipoproteinemia in children (Kwiterovich et al., 1974). Approximately 50% of children born to matings of familial hypercholesterolemia times normal will have hypercholesterolemia (Kwiterovitch et al., 1973, 1974; Schrott et al., 1972).

Ascertainment of familial hypercholesterolemia in infants as compared to older children appears to have substantial promise but is also an area which requires considerable clarification (Tsang et al., 1974; Fredrickson and Breslow, 1973). Kwiterovich et al. (1973) have shown that measurement of LDL cholesterol in cord blood permits diagnosis of neonatal familial hypercholesterolemia when one parent is known to have the disorder. Elevated levels of LDL cholesterol persisted during the first 2 years of life and the children were easily identified as having hypercholesterolemia at follow-up. Tsang et al. (1974) have concluded that measurement of cord blood cholesterol and LDL cholesterol in unselected populations coupled with complete kindred studies may be useful in the diagnosis of familial hypercholesterolemia in as many as 1 in 200 live births. At follow-up at age 1, most children with neonatal familial hypercholesterolemia on high cholesterol intake had distinctive elevations of cholesterol. However, a dissenting study has been published by Darmady et al. (1972) who concluded that cord blood cholesterol determinations had little utility in the diagnosis of familial hypercholesterolemia. In a longitudinal study of cholesterol in 302 infants, Darmady et al. reported that only 5 of 30 infants with a cord blood cholesterol above 100 mg/100 ml had plasma cholesterol above 240 mg/100 ml at age 1.

1. THERAPY OF PEDIATRIC FAMILIAL HYPERCHOLESTEROLEMIA

Only a limited number of studies are currently on hand which detail the effects of dietary intake on plasma cholesterol during the first 2 years of life in infants identified as having familial hypercholesterolemia (Glueck et al., 1972b; Glueck and Tsang, 1972; Tsang et al., 1974). As a general rule, hypercholesterolemic neonates with "monogenic" familial hypercholesterolemia maintain elevated cholesterol and LDL cholesterol levels on whole milk and regular American baby diets. In contrast, infants with neonatal familial type II usually achieve normal cholesterol levels at ages 1 and 2 on diets low in cholesterol and low in saturated fats. For children ages 2–5, therapeutic low cholesterol, low saturated fat diets have been designed (Larsen *et al.*, 1974) which provide adequate protein (Food and Nutrition Board, 1973; Stearns *et al.*, 1958) for rapid total body growth at this age.

In older children heterozygous for familial hypercholesterolemia, diets have been utilized which are generally low in cholesterol, usually 150–300 mg of cholesterol per day, and rich in polyunsaturates, with a P/S of 1.5 to 1 or greater (Table III) (Fredrickson *et al.*, 1970; Lloyd and Jukes, 1961). Lloyd and Wolff (1969b) reduced cholesterol levels in two children with type II to just above normal using a low cholesterol, low

TABLE III

SUGGESTED TREATMENT REGIMEN FOR CHILDREN WITH HYPERCHOLESTEROLEMIA, TYPES IIA AND IIB PHENOTYPES

1. Hypercholesterolemia secondary to habitual diet high in cholesterol and saturated fatty acids
   a. Reduce cholesterol intake to < 300 mg/day, and diminish saturated fat intake while increasing polyunsaturate intake to elevate P/S ratio to 1.5/1.
2. Hypercholesterolemia secondary to other primary disease states (hypothyroidism, etc., cf. Table I)
   a. Where possible treat primary disease state
   b. Where possible discontinue steroid medications
3. Primary familial hypercholesterolemia, children heterozygous for types IIA and IIB phenotypes

Age	Suggested diet	Suggested drug treatment (if any)
<1	None	None
1–5	<200 mg cholesterol, P/S 2:1	None
6–12	<200 mg cholesterol, P/S 2:1	None if total plasma cholesterol consistently <240 mg/100ml, LDL cholesterol <170mg/100 ml while on diet. If total cholesterol >240, LDL >170, add cholestyramine 12 gm active resin per day. If control of total and LDL not optimal may go to 16 gm/day
13–20	<200 mg cholesterol, P/S 2:1	None if total plasma cholesterol consistently <240 mg/100 ml, LDL cholesterol <170 mg/100 ml while on diet. If total cholesterol >240, LDL >170, add cholestyramine 12 gm active resin per day. If control of total and LDL cholesterol not optimal may go to 20 gm/day

saturate diet alone. Segall *et al.* (1970b, 1971) reported up to a 24% reduction in plasma cholesterol in children taking a corn oil-rich diet. Decrements of 12% in cholesterol were reported by our laboratory using a low cholesterol polyunsaturate-rich diet in outpatient children with type II (Glueck and Tsang, 1972; 1973a). Kwiterovich *et al.* (1970) have reported a 14% fall in cholesterol using a similar diet. Diet alone is sufficient to reduce cholesterol and LDL cholesterol to normal levels in about one-third of children heterozygous for familial hypercholesterolemia (Glueck *et al.*, 1973a). Children homozygous for familial hypercholesterolemia have insubstantial changes in plasma cholesterol on diet (Khachadurian, 1968; Moutafis *et al.*, 1971).

The relative roles of cholesterol restriction, saturated fat restriction, and polyunsaturated fat enrichment remain poorly defined in such diets. Reiser (1973) has seriously questioned the necessity for enrichment for polyunsaturates and reduction in saturates in cholesterol-restricted diets.

In children heterozygous for familial hypercholesterolemia who consistently maintain elevated levels of cholesterol and LDL despite the best diet response, a series of combined diet and drug approaches have been suggested (Table III). Several investigators have combined diet with cholestyramine therapy (Glueck *et al.*, 1973a; Levy *et al.*, 1973; Lloyd, 1972; Kwiterovich *et al.*, 1970). Horan *et al.* (1964) gave cholestyramine (15 gm/day) to two children with familial hypercholesterolemia and reported 12–40% reductions in cholesterol, but poor drug adherence. Kwiterovich *et al.* (1970) have reported excellent short-term response to cholestyramine in children with hypercholesterolemia. West and Lloyd (1973) gave up to 24 gm of cholestyramine per day and reported an average reduction in plasma cholesterol of 36%. Drug effect in this study was observed with and without a cholesterol-restricted diet. West and Lloyd (1973) reported a concurrent fall in serum folic acid levels and a decrease in red cell folate in some children. Glueck *et al.* (1973a) studied 20 type II children, ages 7–21, shown to be nonresponsive to diet. Cholestyramine, 12 gm/day, was added to diet. Ten of these 20 children had good drug adherence. Cholesterol and LDL cholesterol fell to normal or near normal levels in six of these ten children on cholestyramine and diet for 6 and 12 months, but did not fall to normal levels in the children with poor or fair drug adherence. Transient but recurrent symptoms of nausea or abdominal fullness were noted in about 20% of the children, but in no case was any symptom severe enough to necessitate discontinuance of the drug. Neither low cholesterol, polyunsaturate-rich diet alone, or diet plus cholestyramine lowered plasma levels of vitamins A and E below normal levels (Glueck *et al.*, 1974b). Growth and weight gain progressed normally in all individuals during the 18-month period while on diet alone or diet plus cholestyramine. Colestipol resin, with cholesterol lowering effect similar to cholestyramine (Glueck et al., 1972a) has also been used in children with excellent results, including normalization of total and LDL cholesterol on 60% of the subjects (Glueck et al., 1975).

Despite the apparent lack of short-term side effects of cholestyramine, caution must be exercised in the chronic use of any cholesterol-lowering drug in children. We routinely followed indexes of growth and maturation, calcium, phosphorus, electrolytes, hematologic and coagulation parameters, and liver, thyroid, and renal function tests (Glueck et al., 1973a). Long-term ingestion of any lipid-lowering drug may have deleterious psychological effects in children. Consultation with psychologists or child psychiatrists may be useful on long-term regimens.

Two other approaches to children heterozygous for familial hypercholesterolemia have been reported. Segall et al. (1970b, 1971) added clofibrate to diet, giving an overall reduction of 33% in plasma cholesterol. Buchwald et al. (1970) performed ileal bypass surgery in six children heterozygous for hypercholesterolemia. The mean serum cholesterol was 303 preoperatively and 202 postoperatively giving a 33% average reduction. Buchwald et al. reported no postoperative diarrhea, no requirement for drug management, no serious complications of surgery, and no specific vitamin maintenance therapy besides 1000 μg of vitamin B_{12}, which had to be given every 2 months. Other pediatric surgical series have not yet been reported to confirm the Buchwald data.

In the rare children homozygous for hypercholesterolemia, multiple drug therapy is necessary in addition to diet in an attempt to control cholesterol levels (Khachadurian, 1968; Moutafis et al., 1971; Lloyd, 1972; Fredrickson and Levy, 1972). In homozygotes, combinations of cholestyramine and nicotinic acid appear to have the best effect, but usually cholesterol and LDL cholesterol cannot be lowered to even relatively normal levels. Clofibrate and cholestyramine have also been utilized. In homozygotes unresponsive to multiple drug therapy, partial ileal bypass has been used with varying degrees of effectiveness (Buchwald et al., 1970). Portal–caval shunt surgery has recently, under experimental conditions, provided a dramatic reduction of cholesterol in patients homozygous for familial hypercholesterolemia (Stein et al., 1975).

2. Possible Detrimental Effects Related to Diet and Drug Therapy

Some investigators have concluded that low cholesterol, polyunsaturate-rich diets or drugs, or both might have long-range detrimental effects. Reiser and Sidelman (1972) reported that high dietary cholesterol intake during the suckling period in pigs and rats might establish a precedent for maintaining a low serum cholesterol level in adulthood, regardless of high cholesterol intakes. When this hypothesis was studied in humans, Glueck et al. (1972b) concluded that very early low cholesterol intake did not appear to impair subsequent ability to maintain normal cholesterol homeostasis when cholesterol intake increased in later infancy. Fomon (1971) has wondered if the rate of myelination in the brain and nervous system might be reduced by an intake of low cholesterol diets and diets low in long-chain fatty acids. Fomon (1971) has stated that "it is possible that diets based on vegetable oils may interfere with myelination of the brain because they do not provide cholesterol as such, and because polyunsaturated fatty acids present in such diets alter the quantity and type of fats that circulate in the body." Fumagalli et al. (1969) have reported that drugs inhibiting the biosynthesis of cholesterol also retarded rate of brain myelination. Both of these hypotheses are being closely followed in groups of children with familial hypercholesterolemia. On diet and cholestyramine, or both, over a 2- to 3-year follow-up period, no symptomatic neurological side effects have been observed, but a watchful stance must be taken (Glueck et al., 1973a, 1974b).

A third area of concern is related to the possibility that a diet low in cholesterol, and high in polyunsaturated fatty acids may lead to the formation of gallstones (Hofmann et al., 1973; Sturdevant et al., 1973). Sturdevant et al. (1973) reported from autopsy material that 34% of men consuming a vegetable oil-rich experimental diet developed gallstones. In contrast, 14% of the control subjects who received a conventional diet containing 40% animal fat calories had gallstones. Sturdevant et al. (1973) suggested that some property of the experimental diet, high P/S ratio or high plant sterol to cholesterol ratio, or both, might promote gallstone formation.

A fourth area of concern is related to the use of skim milk in children. Skim milk in preference to whole milk has frequently been fed to infants beginning at 4 to 6 months of age as an adjunct to low cholesterol diet with an aim to prevent atherosclerosis. Fomon (1967) warns that when skim milk serves as a major source of calories during infancy, total calories and intake of fat were likely to be undesirably low, and excess solute content might predispose to dehydration. Fomon (1967) concluded that skim milk was not satisfactory as a food for infants, although it was a reasonable food for children beyond infancy.

A fifth area of concern has been that diets rich in polyunsaturates and poor in saturates with or without added cholestyramine might eventually induce vitamin E deficiency (American Academy of Pediatrics, 1963). Glueck et al. (1974b) reported that long-term low cholesterol, high P/S diets in children with and without added cholestyramine did not appreciably lower plasma vitamin A and E levels.

Since therapy with diet and drug is likely to be lifelong in children with pediatric familial hypercholesterolemia, close attention to these types of concerns is clearly warranted.

B. Familial Combined Hyperlipidemia

Goldstein *et al.* (1973a) have focused on a disorder, familial combined hyperlipidemia, which they had suggested is genetically distinct from familial hypercholesterolemia. They found affected relatives with hypercholesterolemia, hypertriglyceridemia, or mixed elevations of both cholesterol and triglyceride, and proposed that familial combined hyperlipidemia appeared to be transmitted as a simple Mendelian autosomal dominant trait. Not enough experience with children with the type IIA or IIB phenotype from these kindreds (Glueck *et al.*, 1973b) has been accumulated to determine if their response to diet or diet plus cholestyramine differs from that of children from kindreds with familial hypercholesterolemia.

C. Familial Hypertriglyceridemia

Although familial hypertriglyceridemia (Glueck *et al.*, 1973c) is apparently less often expressed in childhood than familial hypercholesterolemia, it is important to diagnose because of exceptional sensitivity to diet. Studies have recently been completed in 130 children of 36 index cases with familial hypertriglyceridemia. Approximately 20–25% of the children at various age groups had elevated triglycerides. Weight reduction and the NIH type IV (Fredrickson *et al.*, 1970) diet were effective in reducing triglyceride levels to normal (Table IV). Similar effectiveness of diet has been reported by Segall *et al.* (1970a).

TABLE IV

Suggested Treatment Regimen for Children with Hypertriglyceridemia, Type IV Phenotype

1. Hypertriglyceridemia secondary to other primary disease states (diabetes mellitus out of control, etc., cf. Table III)
    a. Where possible treat primary disease state
    b. Where possible discontinue steroid or estrogen medications
2. Primary familial hypertriglyceridemia
    a. Reduction toward "ideal" body weight
    b. Maintenance (if hypertriglyceridemia persists after weight reduction) on carbohydrate-restricted NIH type IV diet

V. Conclusions

Elevations of cholesterol and/or triglyceride are relatively common in childhood and represent a mix heavily weighted to acquired as compared to familial hyperlipoproteinemias. Initial diagnostic approaches should be directed at differentiation between elevated lipid levels secondary to diet habits or other primary diseases, and familial hyperlipoproteinemias. Complete family screening in kindreds where children appear to have primary hyperlipidemias is important since common familial pediatric lipid disorders, familial hypercholesterolemia, familial combined hyperlipidemia, and familial hypertriglyceridemia appear to be inherited as autosomal dominant traits. Repeated evaluation at selected intervals up to age 30 is necessary in normolipemic children from kindreds with familial hypertriglyceridemia and familial combined hyperlipidemia because of the phenomenon of delayed phenotypical expression. In children with hypercholesterolemia, secondary to habitual diet rich in cholesterol and polyunsaturates, prudent cholesterol and saturated fat reduction will almost uniformly reduce cholesterol to normal levels. Infants and children under age 6 with familial hypercholesterolemia are quite sensitive to a low cholesterol diet rich in polyunsaturates and poor in saturates, and most will retain relatively normal levels of cholesterol and LDL cholesterol on this regimen. About one-third of older children heterozygous for familial hypercholesterolemia will normalize on diet alone, while cholestyramine resin is the therapy of choice for the remainder (of whom 70% should normalize on diet plus added cholestyramine resin). Children with familial hypertriglyceridemia are very responsive to weight reduction and the carbohydrate restricted NIH type IV diet, with essentially all children normalizing their triglycerides on such a regimen.

In relatively short-term studies, undesirable side effects of low cholesterol polyunsaturate-rich diets and cholestyramine resin or both have not been significant. A watchful stance must be maintained over the long term for both side effects and any effects on normal growth and maturation as well as any psychological effects of chronic diet and drug therapy.

Diet and drug therapy of pediatric hyperlipoproteinemias is itself currently in its infancy. Children with severe familial elevations of cholesterol and triglyceride should be identified as early as possible and probably should be followed on some combination of diet and drug therapy or both with close attention to any long-term detrimental side effects. Data is not yet available to answer the most important question in these children who would otherwise be heir to an unacceptable cardiac risk as adults, that is, will long-term reduction of cholesterol and triglyceride prevent the development of atherosclerosis?

ACKNOWLEDGMENTS

Supported in part by the General Clinical Research Center Grant RR-00068-12. A portion of this work was done during Dr. Glueck's tenure as an Established Investigator of the American Heart Association, 1971–1976.

REFERENCES

American Academy of Pediatrics, Report of the Committee on Nutrition. (1963). *Pediatrics* **31**, 324.
American Academy of Pediatrics, Committee on Nutrition. (1972). *Pediatrics* **49**, 305.
Atherosclerosis Study Group and Epidemiology Study Group of the Intersociety Commission for Heart Disease Resources. (1970). *Circulation* **42**, A-55.
Brown, M. S., Dana, S. E., and Goldstein, J. L. (1974). *J. Biol. Chem.* **249**, 789.
Buchwald, H., Moore, R. B., Frantz, I. D., Jr., and Varco, R. L. (1970). *Surgery* **68**, 1101.
Carlson, L. A., and Bottiger, L. E. (1972). *Lancet* i, 865.
Clarke, R. P., Merrow, S. B., Morse, E. H., and Keyser, D. E. (1970). *Amer. J. Clin. Nutr.* **23**, 754.
Darmady, J. M., Fosbrooke, A. S., and Lloyd, J. K. (1972). *Brit. Med. J.* **2**, 685.
Enos, W. F., Beyer, J. C., and Holmes, R. H. (1955). *J. Amer. Med. Ass.* **158**, 912.
Fomon, S. J. (1967). "Infant Nutrition." Saunders, Philadelphia, Pennsylvania.
Fomon, S. J. (1971). *Bull. N.Y. Acad. Med.* **47**, 569.
Food and Nutrition Board. (1973). Recommended Dietary Allowances (Revised) Publ. No. 2216, Nat. Acad. Sci., Nat. Res. Council, Washington, D.C.
Fredrickson, D. S., and Breslow, J. L. (1973). *Annu. Rev. Med.* **24**, 315.
Fredrickson, D. S., and Levy, R. I. (1972). *In* "The Metabolic Basis of Inherited Disease" (J. B. Stanbury, J. B. Wyngaarden, and D. S. Fredrickson, eds.), 3rd ed., p. 545. McGraw-Hill, New York.
Fredrickson, D. S., Levy, R. I., Jones, E., Bonnell, M., and Ernst, N. (1970). "The Dietary Management of Hyperlipoproteinemia: A Handbook for Physicians," p. 83. U.S. Dept. of Health, Education and Welfare, Public Health Service, Washington, D.C.
Friedewald, W. T., Levy, R. I., and Fredrickson, D. S. (1972). *Clin. Chem.* **18**, 499.
Fuhrmann, W., Scheonborn, W., Huth, H., and Reimers, H. J. (1971). Familial hyperlipoproteinemia, type III. *Proc. 13th Int. Congr. Pediat.* **7**, 199. Wiener Medizinschem Akademie.
Fumagalli, R., Smith, M. E., Urna, G., and Paoletti, R. (1969). *J. Neurochem.* **16**, 1329.
Glueck, C. J., and Tsang, R. C. (1972). *Amer. J. Clin. Nutr.* **25**, 224.
Glueck, C. J., Levy, R. I., and Fredrickson, D. S. (1968). *J. Amer. Med. Ass.* **206**, 2895.
Glueck, C. J., Ford, S., Scheel, D., and Steiner, P. (1972a). *J. Amer. Med. Ass.* **222**, 676.
Glueck, C. J., Tsang, R. C., Balistreri, W., and Fallat, R. (1972b). *Metabolism* **21**, 1181.
Glueck, C. J., Fallat, R., and Tsang, R. (1973a). *Pediatrics* **52**, 669.
Glueck, C. J., Fallat, R., Buncher, C. R., Tsang, R., and Steiner, P. (1973b). *Metabolism* **22**, 1403.

Glueck, C. J., Tsang, R., Fallat, R., Buncher, C. R., Evans, G., and Steiner, P. (1973c). *Metabolism* 22, 1287.

Glueck, C. J., Fallat, R., Tsang, R., and Buncher, C. R. (1974a). *Amer. J. Dis. Child.* 127, 70.

Glueck, C. J., Tsang, R. C., Fallat, R. W., and Scheel, D. (1974b). *Pediatrics* 54, 51.

Glueck, C. J., Fallat, R. W., and Tsang, R. C. (1974c). *Amer. J. Dis. Child.* 128, 569.

Glueck, C. J., Fallat, R. W., and Moulton, R. (1974d). *In* "Cardiac Diagnosis and Treatment" (N. Fowler, ed.). Harper Hoeber, New York.

Glueck, C. J., Fallat, R. W., Mellies, M., and Tsang, R. C. (1975). *Pediatrics* in press.

Godfrey, R. C., Stemhouse, N. S., Cullen, K. J., and Blackman, V. (1972). *Aust. Paediat. J.* 8, 72.

Goldstein, J. L., and Brown, M. S. (1973). *Proc. Nat. Acad. Sci. U.S.* 70, 2804.

Goldstein, J. L., Hazzard, W. R., Schrott, H. G., Bierman, E. L., and Motulsky, A. G. (1973a). *J. Clin. Invest.* 52, 1533.

Goldstein, J. L., Albers, J. J., Hazzard, W. R., Schrott, H. R., Bierman, E. L., and Motulsky, A. S. (1973b). *J. Clin. Invest.* 52, p. 35A (Abstr. 127).

Golubjatnikov, R., Paskey, T., and Inhorn, S. L. (1972). *Amer. J. Epidemiol.* 96, 36.

Greten, H., Wengeler, H., and Wagner, H. (1973). *Nutr. Metabol.* 15, 128.

Harlan, W. R., Jr., Graham, J. B., and Estes, E. H. (1966). *Medicine* 45, 77.

Hilleboe, H. E. (1967). *In* "Cowdry's Arteriosclerosis" (H. T. Blumenthal, ed.), p. 623. Thomas, Springfield, Illinois.

Hofmann, A. F., Northfield, T. C., and Thistle, J. L. (1973). *New Engl. J. Med.* 288, 46.

Horan, J. M., Di Luzio, N. R., and Etteldorf, J. N. (1964). *J. Pediat.* 64, 201.

Kannel, W. B., and Dawber, T. R. (1972). *J. Pediat.* 80, 544.

Khachadurian, A. K. (1968). *J. Atheroscler. Res.* 8, 177.

Kwiterovich, P. O., Levy, R. I., and Fredrickson, D. S. (1970). *Circulation* 42, III-11.

Kwiterovich, P. O., Levy, R. I., and Fredrickson, D. S. (1973). *Lancet* i, 118.

Kwiterovich, P. O., Jr., Fredrickson, D. S., and Levy, R. I. (1974). *J. Clin. Invest.* 53, 1237.

Larsen, R., Glueck, C. J., and Tsang, R. C. (1974). *Amer. J. Dis. Child.* 128, 67.

Levy, R. I., and Rifkind, B. M. (1973). *Amer. J. Cardiol.* 31, 547.

Levy, R. I., Fredrickson, D. S., Stone, N. J., Bilheimer, D. W., Brown, W. V., Glueck, C. J., Gotto, A. M., Herbert, P. N., Kwiterovich, P. O., Langer, T., LaRosa, J., Lux, S. E., Rider, A. K., Shulman, R. S., and Sloan, H. R. (1973). *Ann. Intern. Med.* 79, 51.

Lloyd, J. K. (1972). *Aust. Paediat. J.* 8, 264.

Lloyd, J. K., and Jukes, H. R. (1961). *Lancet* i, 312.

Lloyd, J. K., and Wolff, O. H. (1969a). *In* "Disturbances in Serum Lipoproteins in Endocrine and Genetic Diseases of Childhood" (L. I. Gardner, ed.), p. 937. Sanders, Philadelphia, Pennsylvania.

Lloyd, J. K., and Wolff, O. H. (1969b). *J. Atheroscler. Res.* 10, 135.

McGandy, R. B. (1971). *Bull. N.Y. Acad. Med.* 47, 590.

McNamara, J. J., Molot, M. A., Stremple, J. F., and Cutting, R. T. (1971). *J. Amer. Med. Ass.* 216, 1185.

Mitchell, S., Blount, S. G., Jr., Blumenthal, S., Jesse, M. J., and Weidman, W. H. (1972). *Pediatrics* 49, 165.

Moutafis, C. D., Myant, N. B., Mancini, M., and Oriente, P. (1971). *Atherosclerosis* 14, 247.

Reiser, R. (1973). *Amer. J. Clin. Nutr.* 26, 524.

Reiser, R., and Sidelman, Z. (1972). *J. Nutr.* **102**, 1009.

Schilling, F. J., Christakis, G. J., Bennett, N. J., and Coyle, J. F. (1964). *Amer. J. Pub. Health* **54**, 461.

Schrott, H. G., Goldstein, J. L., Hazzard, W. R., McGoodwin, M. C., and Motulsky, A. G. (1972). *Ann. Intern. Med.* **76**, 711.

Schubert, W. K. (1973). *Amer. J. Cardiol.* **31**, 581.

Segall, M. M., Fosbrooke, A. S., Lloyd, J. K., and Wolff, O. H. (1970a). *Arch. Dis. Childhood* **45**, 73.

Segall, M. M., Fosbrooke, A. S., Lloyd, J. K., and Wolff, O. H. (1970b). *Lancet* i, 641.

Segall, M. M., Fosbrooke, A. S., Lloyd, J. K., and Wolff, O. H. (1971). *Amer. Heart J.,* p. 707.

Shapiro, J. R., Fallat, R. W., Tsang, R., and Glueck, C. J. (1974). *Amer. J. Dis. Child.* **128**, 486.

Slack, J. (1969). *Lancet* ii, 1380.

Slack, J., and Nevin, N. C. (1968). *J. Med. Genet.* **5**, 4.

Stare, F. J., and McWilliams, M. (1973). "Living Nutrition." Wiley, New York.

Starr, P. (1971). *Amer. J. Clin. Pathol.* **56**, 515.

Stearns, G., Newman, K. J., and McKinley, J. B. (1958). *Ann. N.Y. Acad. Sci.* **69**, 857.

Stein, E. A., Mieny, C., Spitz, L., Saanon, I., Pettifon, J., Heimann, K. W., Bersohn, I., and Dinner, M. (1975). *Lancet* i, 832.

Stone, N. J., Levy, R. I., Fredrickson, D. S., and Vemter, J. (1973). *Circulation* **48** (Suppl. IV), II-14.

Strong, J. P., and McGill, H. C., Jr. (1969). *J. Atheroscler. Res.* **9**, 251.

Strong, J. P., McGill, H. C., Jr., Tejada, C., and Holman, R. L. (1958). *Amer. J. Pathol.* **34**, 731.

Sturdevant, R. A. L., Pearce, M. L., and Dayton, S. (1973). *New Engl. J. Med.* **288**, 24.

Tamir, I., Bojamower, Y., Levtow, O., Heldenberg, D., Dickerman, Z., and Werbin, B. (1972). *Arch. Dis. Childhood* **47**, 808.

Tsang, R., and Glueck, C. J. (1974). *Amer. J. Dis. Child.* **127**, 78.

Tsang, R. C., Fallat, R. W., and Glueck, C. J. (1974). *Pediatrics* **53**, 458.

West, R. J., and Lloyd, J. K. (1973). *Arch. Dis. Childhood* **48**, 370.

Chapter 9

FEMORAL ANGIOGRAPHY TO EVALUATE HYPERLIPIDEMIA THERAPY

DAVID H. BLANKENHORN

Department of Medicine, Cardiology Section,
University of Southern California School of Medicine,
Los Angeles, California

I. The Need for Additional Indicators to Guide Therapy in Hyperlipidemia

The major reason for treating the common forms of hyperlipidemia is to prevent heart attacks and stroke. Evaluation of the efficacy of the prescribed therapy in preventing vascular damage is based on detection of end points measuring myocardial or cerebral ischemia. Although these end points furnish unequivocal evidence for assessment of therapy, they provide information only after tissue damage has occurred. To realize full advantage of the spectrum of therapeutic agents available, substitution of a more satisfactory regimen for one which is not succeeding should be as early as possible, before damage to the heart or central nervous system has occurred.

The immediate end points now used to evaluate therapy in individual patients are blood lipid level and signs of drug toxicity. Because side effects usually present an immediate risk, and benefits of lowering serum lipid are realized in the future, there is a natural tendency to discontinue hyperlipidemic therapy at the first sign of toxicity. This approach reduces short-term problems in patient management, but frequently leaves the patient still at long-term risk of vascular damage.

It has been contended that high cholesterol level has been established as a risk factor, so that lowering the level is a separate and independent goal. A central part of the evidence supporting cholesterol level as a risk factor is experimental atheroma regression/progression study of animals. This clearly documents the overall relationship between cholesterol level and atheromatosis, but introduces conflicting evidence of the effectiveness of this factor used singly to guide therapy. For example, Kottke et al. (1974) found that "spontaneous genetic atherosclerosis can be made to regress" with ileal bypass. However, the cholesterol level does not show significant differences with or without surgery. In pigeons 3 months postoperative the level was 178 mg/100 ml, and at 6 months, 215. Control pigeon levels were 186 and 233 mg/100 ml, respectively. Armstrong (1974) tested the effect of supplementary dietary cholesterol in rhesus monkeys and found it was possible to induce arterial damage with very small supplements which did not change blood cholesterol levels.

Techniques are needed that provide methods of measurement of arterial walls. Angiography is used for studying blood vessels, but has not been used to evaluate hyperlipidemia therapy. Angiograms can be used for this purpose, but the technique departs from usual clinical practice, because evaluation combines results of human interpretation with computer–densitometer measurements.

II. Elements of Angiographic Interpretation

It is commonly assumed that all information potentially available from angiograms can be discerned by skilled readers because they are particularly adept at feature recognition and pattern classification. However, they are limited in other aspects of angiographic interpretation, in particular, precise definition of vessel edge location and lesion size.

The conventional use of femoral angiography is for planning surgery to relieve lower limb ischemia. Feature recognition and pattern classification appear to be the most important aspects of what the surgeon must know the following.

1. Is the vessel completely occluded? If so, how long is the occluded area?

2. Are collateral vessels present? If so, what are their location and probable capacity for blood flow?

3. Are vessels distal to the proposed surgical site capable of accepting more blood? Is there adequate "run off"?

Analysis of the process by which readers answer these questions suggests that feature recognition is more important than measurement. For example, evaluation of "run off" consists of comparing the filling pattern of lower leg vessels with an idealized pattern known on the basis of prior experience to lead to favorable postoperative results.

To evaluate hyperlipidemia therapy this quantitative information is needed before therapy and at later intervals:

1. Are atheromas present?

2. What is the stage of development of any atheromas present?

3. What is the size and number of any atheromas present?

The first two questions can be answered by feature recognition techniques and human readers do this quite well (Dejdar et al., 1967). However, visual estimation cannot quantitate atheroma size and number with precision. Human limitation in counting and estimating lesion size is difficult even when blood vessels can be examined directly at autopsy (Cranston et al., 1964). Physiological and psychological factors interact in a complex way rendering visual assessment "extremely variable, inaccurate, and poorly reproducible" (Cranston et al., 1964). Training and experience with lesion grading does not significantly improve the performance of human observers. Robbins et al. (1966) have shown that human variability and lack of reproducibility in repeated evaluations also detract from quantitative interpretation of angiograms. Visual X-ray reading is subject to the same size and number counting problem as gross examination of autopsy specimens, plus a further difficulty imposed by a need to differentiate subtle differences in shades of gray.

To measure and count vascular lesions on an angiogram it is necessary to locate vessel edges precisely. As illustrated in Fig. 1, lesions are located by scanning up and down the vessel shadow looking for indentations. In the "mind's eye" an expected vessel edge is projected across indentations followed by counting their number and estimating their size. Precise assessments require precise location of the nominal vessel edge and the edge of indentations. This requires an ability to differentiate shades of gray. Figure 2 illustrates a femoral angiogram with vessel edges sharply outlined. A small plaque may be present in the midportion of the vessel

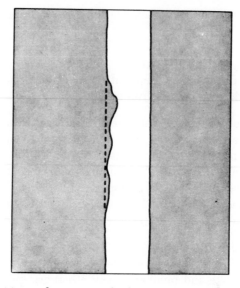

Fig. 1. Recognition and counting of atheromas on an angiogram. The "mind's eye" projects a vessel edge through indentations in the vessel shadow.

just above one group of side branches. To project a nominal vessel edge across the possible plaque a closer look is required, i.e., magnification. Magnification (Figs. 3 and 4) reveals that the vessel edge is not "sharp," but is formed by transition in shades of gray.

All angiograms show this effect, although it can be reduced by selection of exposure and development conditions to produce high-contrast angiograms. Currently available diagnostic X-ray sources are not monochromatic and produce rays with varying energy content. Lower energy radiation is absorbed by an amount of contrast medium which allows higher energy radiation to penetrate. This produces a zone where differential X-ray penetration occurs at the edge of a column of contrast medium. For this reason, and others beyond the scope of this discussion, the edge of any vessel shadow is always delineated by shades of gray. Computer-controlled densitometers are of great assistance in locating vessel edges because differentiation of shades of gray can be performed rapidly and with precision.

Lesion counting and sizing on radiograms presents an additional problem. There is need to produce spatial reconstruction of vessel indentations if all lesions are to be counted and measured. This is dictated by the geometry of lesions in relation to the X-ray source and film plane. As shown in Fig. 5, atheromas perpendicular to the axis of the film will be registered as edge indentations, lesions parallel to the film will not.

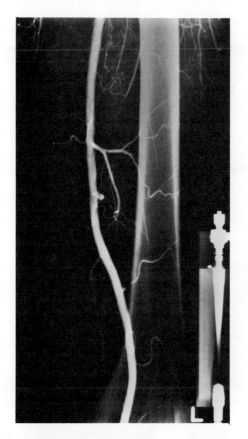

FIG. 2. Femoral angiogram reduced 3:1. The vessel edges appear sharp. The object at lower right is a test object placed beside the leg to evaluate radiographic exposures.

Human interpreters have difficulty estimating the size of spatial reconstructions in the "mind's eye" even in stereoptic views. Computer-controlled densitometers can be used to perform these reconstructions from multiple films.

A different approach to the geometric problem in lesion counting and sizing is possible when film densities can be measured accurately. The integrated density of vessel shadows is an accurate measure of lumen size when film density is proportional to X-ray absorption by contrast medium and contrast medium concentration is known. It is unfortunate that clinical angiographic routines used currently do not produce films meeting these requirements. The major problem is that contrast medium concentration in vessels is not known because blood flow at the time of

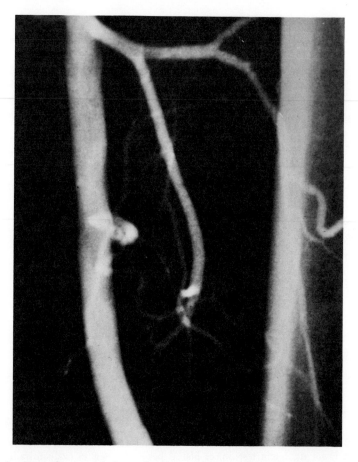

FIG. 3. Magnification of the midportion of Fig. 2 where a possible plaque may be present above a group of side branches. The vessel edge no longer appears sharply defined.

injection is not determined. If blood flow measurements were made during injection of a known volume of contrast medium, the information required for automated lesion counting and size determination could be available on a single plane X ray.

III. Computer-Densitometer Measurements of Angiograms

Computer-densitometer film measurements begin with multiple determinations of film density by a cathode ray tube flying spot scanner or an

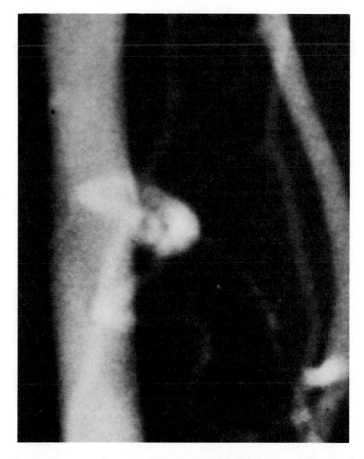

FIG. 4. Further magnification of Fig. 3. Additional blurring of the vessel edge is apparent.

image dissector densitometer. A current procedure is to scan a 5-cm length of vessel at 25-μm intervals along scan lines perpendicular to its axis. Densities on each scan line are digitized and smoothed by digital computer. To reduce computation, extraneous data is excluded by masking the film on either side of the vessel. Although only short segments of vessel are scanned at one time, segments can be linked. The femoral artery is scanned just distal to Hunter's canal at a point where the vessel passes behind the femur. We scan cephalad and link two 5-cm areas to provide lesion estimates in Hunter's canal where lesions are known to be most common (Lindbom, 1950).

Vessel edge points are located by examining the rate of change of

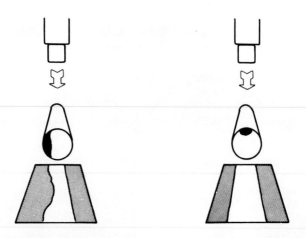

FIG. 5. The effect of geometry of plaque formation upon edge irregularity of angiographic shadows. When plaques are parallel to the X-ray film plane (as shown on the right) they may not cause edge irregularity.

density on each scan line. A cubic equation fit by least squares to 15 adjacent points is moved across each scan line and the rate of change of the second derivative used to locate vessel edge points. When vessel edge locations are identified in each scan line, the vessel image is recreated for inspection and if it is satisfactory, additional computations are made. The final result is a photograph of the vessel showing edge points and the results of additional computation.

The additional computations with which we have had most experience are "edge roughness" and "lumen variation." Edge roughness measures fine irregularity of the vessel wall and is insensitive to large plaques. It is determined by comparing the effect of various degrees of smoothing of edge data points. Figure 6 shows edge roughness determined by measuring the area included between two edge lines with 10 and 200 point smoothing. The area between the two edge measures on each side of the vessel is averaged and printed out for each centimeter.

The algorithm for lumen variation is sensitive to large plaques. Edge points are used to locate the center of the vessel on each scan line. Each scan line is next adjusted laterally until all center points are in vertical alignment. Next, two parallel lines are moved outward from the vessel center until 95% of all edge points are included. The center of each scan line is then returned to its original position and the average area between lateral lines and vessel edge determined and printed out for each centimeter of vessel.

A third algorithm which has shown promise, but has not been applied routinely to clinical films, is "integrated cross-sectional density." A line-by-

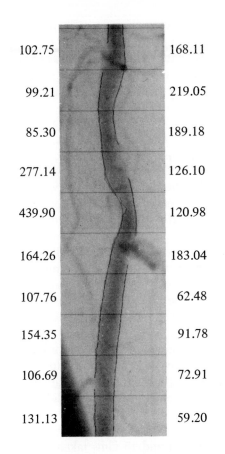

102.75	168.11
99.21	219.05
85.30	189.18
277.14	126.10
439.90	120.98
164.26	183.04
107.76	62.48
154.35	91.78
106.69	72.91
131.13	59.20

FIG. 6. A femoral angiogram with vascular shadow edge roughness determined by densitometer–computer image processing. The vessel is divided into centimeters and edge roughness is computed separately for each side of the vessel. Edge lines are omitted where branches or other features of the angiogram interfere with the measurement. The shadow at lower left is the edge of the femur.

line average of densities between film edges is determined and from this a mean background density is subtracted. Mean background density is estimated by averaging the density in two parallel strips immediately along side each vessel segment. Integrated cross-sectional density is an attractive means of measurement because it does not have the geometric uncertainty of edge indentation measurement. The major limitation of this algorithm for clinical use is lack of information concerning contrast density concentration in vessels being radiographed. A lesser problem is a sensitivity to changing soft tissue thickness, which makes it difficult to use for some patients. Integrated cross-sectional density has been used for radiographic postmortem atherosclerosis assessment, an instance in which contrast medium concentration can be controlled. The combination of integrated cross-sectional density with edge roughness algorithms improves the diagnostic potential of both measurements in distal aorta (Blankenhorn *et al.*, 1974).

IV. Clinical Angiographic Techniques for Computer-Densitometer Measurement

Angiograms taken by the usual clinical procedure with pressure injection through a catheter can be processed by computer-densitometer. Better results are obtained with a modification that fills the main trunk of the femoral artery from groin to knee, but does not fully delineate collateral circulation or peripheral run off. Patients are premedicated lightly and positioned on the X-ray table supine with thighs and legs straight. The thigh is centered so that the superior portion of the patella is at the lower limit of the film. The foot is stabilized by soft restraints at an angle of 45 degrees external rotation. High-contrast film is exposed using a stationary grid, a par speed screen, and a constant potential generator. Exposure factors are varied in accordance with the thickness of the thigh, but range in average-sized individuals from 74 to 80 kV, 100 mA, and 0.05–0.08 seconds. An 18-gauge Seldinger needle is inserted in the groin and into the femoral artery after lidocaine infiltration. Sixty percent meglumine diatrizoate is injected through a short connector tubing over a period of 2 to 3 seconds. The first film is exposed before contrast injection is started and 15 films taken, two per second.

A relatively slow injection of contrast medium is preferred to high-speed pressure injection because it minimizes motion artifact and allows films to be exposed through several cardiac cycles. Limiting the amount of contrast medium also has the advantage of reducing patient discomfort. This is of considerable importance when angiography is used to evaluate hyperlipidemia therapy. These circumstances greatly differ from presurgical ones, and the two classes of patients have different attitudes concerning the procedure. When patients are angiogrammed to plan surgery for lower limb ischemia they have experienced chronic leg pain and view X-ray examination as a one time experience which may lead to surgery for relief of pain. Patients with hyperlipidemia usually have not had leg pain and anticipate that the repeated X-ray examination will be required to judge their progress. Hyperlipidemic patients, quite reasonably, are willing to accept much less discomfort during the angiogram.

In 27 serial examinations the degree of discomfort experienced by patients premedicated with 50 mg of diphenhydramine hydrochloride and 30 mg of pentazocine hydrochloride intramuscularly both given 20 minutes before femoral angiography was considered mild to moderate in 24 and severe in 3. No complications have occurred from femoral angiography during a total of 160 examinations except for one episode of asystole. A 42-year-old man with type II hyperlipidemia known to have advanced coronary disease with recurrent angina pectoris developed 40 seconds of asystole during femoral arterial puncture, but made an uneventful recovery. Another patient with previous myocardial infarction developed recurrent infarction during a coronary angiographic procedure which followed immediately after femoral angiography.

V. Prevalence of Femoral Atheromas in Hyperlipidemia

Visual inspection of films is adequate to ascertain a high prevalence of femoral atherosclerosis in the common forms of hyperlipidemia. Table I presents data on 28 patients with type II and type IV hyperlipidemia, ages 20–51 years. It is noteworthy that only two of these patients complained of claudication, whereas 18 of 28 had previous myocardial infarction or angina pectoris. Observations by Kannel *et al.* (1970) on claudication in Framingham are relevant. A population sample of men, ages 30–74, showed parallel rates of development of myocardial ischemia and claudication, but the rate of development of claudication was approximately one-fourth the rate for myocardial ischemia. The principle hazard for those who developed claudication was mortality from coronary atherosclerosis.

Our studies and those from Framingham refute a common misconception about femoral atherosclerosis. Because crippling claudication and overt ischemia are most commonly found in patients with diabetes or advanced age, it is assumed that femoral involvement is a late manifestation of atherosclerosis. However, femoral atheromas are common in middle-aged individuals, but cause few leg symptoms. The major clinical significance of femoral involvement is what this indicates about other vascular beds.

VI. Do Femoral Atheromas Regress with Therapy?

A priori, the prevalence of femoral atheromas can be stated to increase with age because pronounced atherosclerosis is rare in children, but

TABLE I

PREVALENCE OF FEMORAL ATHEROSCLEROSIS IN TYPE II AND TYPE IV HYPERLIPIDEMIA

| | Type II | | | Type IV | |
Age	Vascular disease[a]	Percentage involvement[b]	Age	Vascular disease[a]	Percentage involvement[b]
20	—	0	24	—	35
28	—	11	36	—	8
30	—	7	38	MI	100
36	—	15	44	MI	67
36	MI	19	45	MI	44
38	MI	31	46	MI	100
40	MI	47	49	MI	92
41	MI	16	49	MI	38
42	MI-C	100	49	A	56
43	—	38	50	MI	27
45	MI	31			
47	MI	40			
48	MI-C	100			
49	MI	51			
49	—	56			
50	MI	100			
51	MI	100			
51	—	50			

[a] MI, myocardial infarction; C, claudication; A, positive exercise test; coronary artery disease by angiography.

[b] Percentage of total arterial edge involved by recognizable atheromas.

common in middle years. What is not known is whether atheroma progression occurs as a steady increase or episodically because of the small number of patients who have had serial examination of their femoral arteries. If lesion increase is a steady process, therapy can be judged successful when two measurements show no change. If lesion growth is episodic, therapy evaluation will be more difficult. We have performed second angiograms after an interval of approximately 12 months in 20 hyperlipidemic patients and 20 men with premature myocardial infarction enrolled in a risk reduction program. Films are being evaluated sequentially, first by a panel of four independent examiners who read them as unknowns, and next by computer processing. At the time of this writing, 16 films from men with premature myocardial infarction have been evaluated by the panel and there is a consensus that lesions are smaller in two patients. A lesion showing unequivocal improvement in the opinion of the panel is shown in Fig. 7. Fourteen patients show lesser change and there is no consensus among the panel readers as to whether lesions are increasing, decreasing, or unchanged. Computer measurement on these films are not complete.

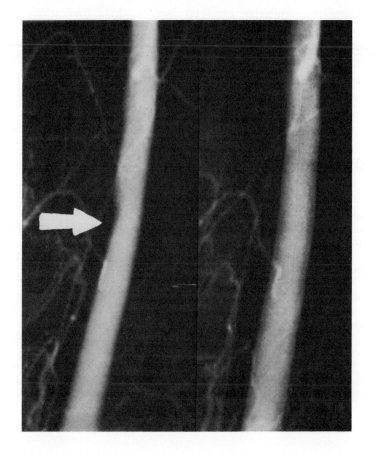

FIG. 7.   Portions of two femoral angiograms made at an interval of 1 year. The lesion indicated by the arrow is not seen 1 year later.

VII. Femoral Atherosclerosis and Coronary Atherosclerosis

Since measurements of any sort are more easily performed on peripheral arteries than coronary arteries, the prospect of using the femoral vessel as a surrogate to evaluate the course of coronary atherosclerosis is attractive, but not a new idea. Kannel *et al.* (1970) have suggested study of arterial pulse contours for this purpose, whereas Hartman *et al.* (1971) suggested peripheral plethysmographic flow measurements. Femoral angiography as described here adds additional more direct information about the state of femoral arterial walls. If the usual direction of lesion regression/progression corresponds in the two vessels, these techniques will all have value. Accumulation of serial angiographic observations of both coronary and femoral arteries will be informative. At the present time we perform femoral angiography on informed human volunteers for the sole purpose of evaluating hyperlipidemia therapy. We feel justified in advocating this use of angiography in patients without symptoms because there is little risk in the procedure. Coronary angiography has measurably higher risks and we do not utilize it to evaluate the asymptomatic hyperlipidemic patient. When clinical evidence of ischemic heart disease is also present and accepted indications for coronary angiography exist, we advocate both procedures in hyperlipidemic patients.

REFERENCES

Armstrong, M. K. (1974). *Circ. Res.* **34,** 447–454.

Blankenhorn, D. H., Brooks, S. H., Selzer, R. H., Crawford, D. W., and Chin, H. P. (1974). *Proc. Soc. Exp. Biol. Med.* **145,** 1298–1300.

Cranston, W. I., Mitchell, J. R. A., Russell, R. W. R., and Schwartz, C. J. (1964). *J. Atheroscler. Res.* **4,** 29–39.

Dejdar, V. R., Roubkova, H., Cachovan, M., Kruml, J., and Linhart, J. (1967). *Arch. Kreislaufforsch.* **54,** 309–335.

Hartman, G., Ritzel, G., and Widmer, L. K. (1971). *Int. Z. Vitaminforsch.* **41,** 104–115.

Kannel, W. B., Skinner, J. J., Schwartz, M. J., and Shurtleff, D. (1970). *Circulation* **41,** 875–883.

Kottke, B. A., Unni, K. K., Carlo, I. A., and Subbiah, M. T. R. (1974). *Trans..Amer. Ass. Physicians* **87,** 263–270.

Lindbom, A. (1950). *Acta Radiol. Suppl.* **80,** 5–80.

Robbins, S. L., Rodriguez, F. L., Wragg, A. L., and Fish, S. J. (1966). *Amer. J. Cardiol.* **18,** 153–159.

Author Index

Numbers in italics refer to the pages on which the complete references are listed. Numbers in parentheses are reference numbers and indicate that an author's work is referred to, although his name is not cited in the text.

Grunnet, N., 190, 192, 202, *215, 216*
Gsell, O., 4, *39*
Guarnieri, M., 77
Gubner, R. S., 165, *178*
Günther, D., 30, *40*
Guerra, C., 138, *157*
Guest, M. J., 128, *158*
Gulick, M., 167, *177*
Gullo, L., 200, *218*
Gundersen, K., 108, *124*, 222(24, 25, 29), 223(63), 241(24), *253, 254*
Gunning, B., 168, *179*
Gurtner, H. P., 175, *178*
Gusman, H. A., 170, *178*
Gustafson, A., 54, 77
Gutierrez, L., 111, *125*
Guynn, R. W., 187, *215*
Guzman, M. A., 136, *158*

H

Hackel, D. B., 249(120), *256*
Hadler, A. J., 171, 173, 175, *178*
Haerem, J. W., 64, 77
Härtel, G., 204, *216*
Häusler, 29, *41*
Hafiez, A. A., 67, 73, 77
Hagenfeldt, L., 63, 77
Hagerman, J. S., 132, *156*
Hagerman, L. M., 69, 77, 235(104, 105), 239(105), *255*
Haggard, M. E., 44, 68, 77, 78
Hahn, P., 23, *38*
Halder, K., 68, 75
Hallal, F. J., 16, *39*
Hallberg, D., 106, *120, 122*
Hamberg, M., 73, 77, *81*
Hamilton, D., 224(74), *254*
Hamilton, J. G., 111, *122*, 153, *156*
Hamprecht, B., 98, *122*, 147, *154*
Hampton, J. R., 64, 74, 77
Hansen, A. E., 44, 68, 75, 77, 78, *81*
Hansen, J. D. L., 67, *81*
Hansen, H., 50, 75
Hansen, I. B., 45, 78
Hanzal, R. F., 12, *39*
Harding, R. S., 111, *124*
Harding, T., 174, *178*
Hardison, W. G. M., 111, *122*, 153, *156*

Harkins, R. W., 231(97), 235(104, 105, 106), 239, *255*
Harlan, W. R., Jr., 262, *274*
Harland, W. A., 50, 78
Harlow, R. D., 70, *82*
Harman, D., 49, 78
Harmuth, E., 67, *81*
Harper, H. W., 170, *177*
Harrigan, P., 105, 106, 114, 117, *120, 123*
Harris, P. L., 50, 78
Harris, R. A., 61, 78
Harris, R. L., 187, *215*
Harris, S. J., 223(58), *254*
Harrison, M. J. G., 74, 77
Hart, A., 170, *179*
Hartman, G., 289, *290*
Hartree, E. F., 189, *216*
Hartroft, W. S., 197, 198, 202, *214, 218*
Hartwich, A., 12, *39*
Harvengt, C., 222(49), *254*
Hash, A. M., Jr., 173, *178*
Hashim, S. A., 51, 58, 62, *76, 78, 80*, 108, *122*, 146, 148, *156*, 222(5, 11), 226(85), 235(103), 239(101, 102), 241(5), 242(5), 249(11), 252, *253, 255*
Hashimoto, S., 62, 76, 84, *121*
Hassan, H., 51, 78
Hasse, H. M., 29, *39*
Hasumura, Y., 189, 190, 192, 196, 197, *215, 216, 219*
Haupt, E., 37, *39*
Haupt, V., *40*
Hauser, S., 129, *158*
Haust, H. L., 58, 78, 131, *155*
Hauton, J., 110, *125*
Havel, R. J., 60, 79, 104, 105, *122*, 129, 149, *156, 157, 159*
Havenstein, N., 56, 59, 63, 79, *80*, 138, 140, *158*, 173, 175, *178*
Hawkins, R. D., 195, *215*
Hazelwood, J. C., 173, *178*
Hazzard, W. R., 115, *120*, 137, 142, 149, *154, 155*, 258, 261, 263, 266, 271, *274, 275*
Heath, D., 35, *39*
Heaton, K. W., 149, *157*, 224, *254*
Heaton, S. T., 224(73), *254*
Heberer, G., 9, 10, 13, 30, *39*
Heddman, P. A., 225(82), 237(82), *255*
Hegsted, D. M., 201, *216*

Hurley, P. J. 103, 115, *125*
Hurwitz, A., 174, 176, *180*
Huth, H., 259, *273*
Hutterer, F., 193, *219*
Huttunen, J. K., 105, *121*
Hyams, D. E., 108, 117, *122*, 223(59, 60, 61), *254*
Hyun, S. A., 110, *122*, 222(8), 227(8), 241(8), 242(8), *253*

I

Iber, F. L., 194, 196, 200, *216*, *218*
Ihrig, T. J., 204, *215*
Imlah, N. W., 174, *179*
Infante, R., 207, *214*
Ingvaldsen, P., 57, *79*
Inhorn, S. L., 263, 265, *274*
Inkpen, C. A., 61, *78*
Insull, W. J., 50, *75*
Insull, W., Jr., 45, 62, *78*, 140, *157*
Insunza, I., 194, *220*
Ipsen, J., 164, *178*
Irion, E., 74, *78*
Irsigler, K., 66, *78*
Irvine, W. T., 5, *40*
Isaac, P. F., 175, *181*
Iseri, O. A., 193, 202, 204, 212, *216*
Ishii, H., 189, 193, 194, 196, 197, 207, 213, *216*, *218*, *219*
Ishikawa, T., 145, *157*
Israel, Y., 194, 196, 204, *214*, *216*, *220*
Issekutz, B., 162, 166, *177*, *179*
Isselbacher, K. J., 195, 206, 208, 209, 210, *214*, *217*, *218*, *219*
Itil, T. M., 172, *178*
Iturriaga, H., 194, 195, *220*
Ivy, A. C., 144, *157*
Iwamoto, A., 201, *216*

J

Jackson, D., 222(42), *254*
Jacobsen, P. A., 65, *75*
Jacobson, G., 3, *38*
Jager, F. C., 50, *78*
Jahn, H., 23, *41*
Jaillard, J., 63, *78*

Jain, A., 108, *125*, 222(35, 47), *253*, *254*
James, G., 84, *121*
James, T., 45, *78*
Jampel, S., 84, *121*
Janeway, C. A., 103, *122*
Janowitz, H. D., 200, 208, *214*, *216*
Jarrell, J. J., 149, *155*
Jenkins, D., 27, *41*
Jenkins, D. W., 205, *216*
Jensen, M. K., 68, *76*
Jenson, W. K., 165, *178*
Jespersen, J., 175, *179*
Jesse, M. J., 265, *274*
Jewett, R. E., 171, *179*
Johns, W. H., 109, *122*
Johnson, A. A., 48, *78*
Johnson, M., 65, *81*
Johnson, O., 209, *215*
Johnson, R. E., 62, *78*
Johnson, R. M., 77
Johnston, D. A., 147, *157*
Johnston, P. V., 54, *81*
Jolliffe, N., 57, *78*
Joly, J. G., 189, 193, 196, 207, 213, *216*, *219*
Jones, A. L., 207, *216*
Jones, C. M., 199, *220*
Jones, D. P., 186, 187, 196, 197, 198, 199, 200, 202, 206, 207, 210, *216*, *217*
Jones, E., 267, 271, *273*
Jones, J. V., 103, 115, *125*
Jones, R. J., 144, *156*, 249(127), *256*
Jonsson, C. E., 72, *78*
Jonsson, H. T., 46, *76*
Jose, A., 169, *179*
Jourdan, M. H., 166, *177*
Jow, E., 200, *218*
Joyce, J. W., 10, *40*
Juergens, J. L., 10, *40*, 228(89), *255*
Jukes, H. R., 267, *274*
Jun, H. W., 172, *179*
Jung, Y., 120, *121*
Jungblut, R., 30, *40*
Jungmann, R. A., 74, *81*
Justice, D., 64, *77*
Juul, A. H., 109, 110, *122*

K

Kabara, J. J., 70, *78*
Kager, L., 187, *214*

Subject Index

A